CHILDREN OF THE EARTH

CHILDREN OF THE EARTH

LITERATURE, POLITICS, AND NATIONHOOD

MARC SHELL

OXFORD UNIVERSITY PRESS

New York · Oxford 1993

Oxford University Press

Oxford New York Toronto
Delhi Bombay Calcutta Madras Karachi
Kuala Lumpur Singapore Hong Kong Tokyo
Nairobi Dar es Salaam Cape Town
Melbourne Auckland Madrid

and associated companies in
Berlin Ibadan

Published by Oxford University Press, Inc.,
200 Madison Avenue, New York, New York 10016

Library of Congress Cataloging-in-Publication Data
Shell, Marc.
Children of the earth : literature, politics,
and nationhood /Marc Shell.
p. cm. Includes bibliographical references (p.) and index.
ISBN 0-19-506864-5
1. Politics and literature. 2. Nationalism in literature.
3. Self-determination, National, in literature. 4. Kinship in literature.
5. Literature and society. I. Title.
PN51.S3637 1992
809'.93358—dc20 91-32338

1 3 5 7 9 8 6 4 2

Printed in the United States of America
on acid-free paper

For Hanna and Jacob

PREFACE

Children of the Earth bears on the apparently genial motto "All men are brothers" (or "All human beings are siblings"). For millennia this dictum has had enough staying power and influence to warrant investigation into what its collapse of species with family might actually mean in the political realm: a recurring and multifarious transformation of its ideal of the familial unity of humankind into an effective politics where "only my 'brothers' are men, all 'others' are animals." Why this turn comes about and what, if anything, we might do about it are ancient, still urgent questions.

The motto "All men are brothers" has had political significance for national unity and religious universalism. But however profound its cultural origin or ineradicable its linguistic presence, it is often prematurely dismissed as merely metaphorical, as if simply to say that "brother" does not mean "brother." Throughout *Children of the Earth,* in varying historical and philosophical contexts, I explore politically symptomatic differentiations between "literal" and "metaphorical" in the realm of kinship terminology and politics. Here the differentiation of biology from sociology—and of culture from nature—is elucidated in terms of the linguistic distinctiveness of kinship words or names and in terms of a poetics, or metaphorics, of the classification of kin and kind ("a little more than kin, and less than kind"). The figural aspects of language help us to explain the often sociologically needful prejudice that familial consanguinity or its counterpart is the primary kind of kinship or identity, and that other kinds—adoption, friendship, and national belonging—are merely secondary.

For nomads and settlers alike, political analysis begins with knowing home. I introduce the problem at hand in a context at once North American and Judaeo-Christian. Chapter 1 is intentionally an American introduction to the study of kinship and nations. In focusing on the indeterminacy or deniability of blood or race (as illustrated, for example, by Mark Twain), I observe an unallayable anxiety about who's in or out of a particular kinship group and the resulting allure of a universal kinship that renders knowledge of particular kin or nation beside the point (as illustrated, for example, by Herman Melville). In American history a guiding principle is that all men are, in one way or another, equal; the concomitant struggle has often proved to be one of brother against brother, as in the War of Independence and the War Between the States.

In European history, the principal historical examples of toleration—or putting up with what we don't like—generally involve particularist societies which hold that "some men are brothers, and some are others." Why this should be the case is the subject of chapter 2. Here I examine the celebrated coexistence in Muslim Spain of the three so-called peoples of the Book—Christians, Muslims, and Jews—and discuss the political significance of universalist and particularist definitions of what a "people" is. Under Christian rule Spain experienced a politically consequential confusion of the extraordinary, universalist view ("All men are brothers") with the ordinary, particularist view ("Some men are brothers, and some men are not brothers"), a confusion that provided the impetus for such diverse institutions as bullfights, religious inquisitions and, more specifically, race slavery. By the same token, as we shall see, the modern notion of toleration, with its characteristically wavering defense of particular siblinghoods, for centuries defined itself against the universalist experience of Spanish Christendom.

The mosaic of peoples in multilingual Spain in medieval times resembles that in bilingual Québec in the present century. In that part of North America, free of "melting pot" ideologies, divisive language differences are conflated with distinctions of nation and blood. Québec is of special interest to me. My concern with the transformation from universal brotherhood to tribal otherhood was shaped as much by growing up in Québec—with its constitutionally separate linguistic and religious groups—as by my translocation in 1965 to the United States—with its insistence on the ideal of single siblinghood. Once an English colony and now a partner in the Canadian confederation, Québec still exists between poles of unity and schism that are comparable to those in the United States and Spain. Likewise, its popular struggles arise partly from ideas of nationhood brought over from the mother countries of Québec's two official national groups, France and England.

England and, more specifically, its purportedly bastard and virgin queen, Elizabeth, is the subject of chapter 4, which examines the relationship between the deformation—even collapse—of Elizabeth's family in the 1530s and the subsequent formation of the English nation. By reviving in this chapter the ancient notion that "where all human beings are siblings, all acts of sexual intercourse are incestuous," I clarify the ideal Renaissance transformation of the fear of physical incest into a desire for spiritual incest, thereby illuminating the political role of Elizabeth as mother, wife, and sister of all Englishmen. This role, as we shall see, helped in defining politics as a liberal estate and provided an early impetus for modern liberalism and other forms of liberation. Even at age eleven, Elizabeth Tudor, bereft of a mother and abandoned by her father, theorized about familial and tribal relations. Later she actualized her thinking in such a way that it still influences "the indifferent children of the earth" (*Hamlet*) and their politics of liberal nationalism.

Hamlet, the subject of chapter 5, is a canonical play about kin, kind, and king.

But its links with the Western religious and political traditions have generally been misunderstood—as have these "traditions" themselves—thanks to a pervasive predilection to ignore the extraordinary quality of the siblinghood that *Hamlet* hypothesizes. This is a Greek stoic, imperial Roman, or Roman Christian universal siblinghood where all men are alike—or may as well be alike—as if every human being were a tragic player in some politically fateful game of blindman's buff that is both desired and feared. This chapter considers the tragic implications of any such national kin or kind and also clarifies the predisposition to pass over in silence a generally insupportable tug-of-war between celibacy and incest.

The universalist aspect of imperial Rome and of Roman Christianity in France is the subject of chapter 6. Here I focus on the life and thought of Jean Racine, a "child of adoption" to a French universalist religious order. Just as Elizabeth grew up in a family of successive stepmothers, wrote a book where one being is fourfold kin to another (as parent, sibling, child, and spouse), and then became the mother, wife, and sister of a newly constituted English nation, so the young orphan Racine, bereft of father and mother, grew up in the spiritual family of Port-Royal and wrote plays about families and empires gone awry. Racine links a loss of particularized kin to Roman religious and imperial universalism; and in *Britannicus* he delineates the Western archetype of unkind monstrous cruelty that is Nero. His work and its interpretations help us understand the significance of such widespread romantic revisions of universalism as the French national slogan "Liberty, Equality, Fraternity" and illuminate the political implications of such demographic experiments as contemporary China's regulation requiring families to have no more than one child.

Family pets are not exactly monsters, but in Western societies the institution of the family pet likewise demarcates the shifting boundaries that separate familial kin from species kind. The way we think about and treat these family members says much about how we are likely to treat other humanoid creatures—and why. Here the evidence ranges from folktales to religious rules concerning what humanlike creatures we may eat or have sexual intercourse with. Chapter 7 considers how the doctrine "All men are brothers"—insofar as it turns into the dogma "Only my brothers are human beings" and lacks effective rules for treating specifically nonhuman beings with special kindness—tends toward a loving bestiality and a Neronian cruelty that is inhumane and, hopefully, also essentially nonhuman.

For the universalist, the creature called "the family pet" is somehow human (kin is kind). In the vertiginous no-man's-land of the world, however, we cannot say for sure who is kin (hence, for the universalist, also who is humankind) except *ex machina,* as when some social doctrine or literary fiction tells us credibly that the creature we thought was nonhuman is really human, or vice versa. Abraham thought the three potentially hostile strangers he hosted in his desert tent in Hebron were human, but they turned out to be angels.

In the final chapter I recall diverse characteristics adduced for restricting or extending group membership—religion, gender, language group, skin color, planetary origin, and the like. Examining whether we must have enemies, I rehearse commonplace multifaceted justifications for behaving inhumanely toward nonkin as well as kin. As we shall see, the universalist transformation of brother into other suggests that it is better to be an outsider in a particularist kinship system where there are human kin and human aliens than to be an outsider in a universalist kinship system where there are only humankind and animals.

Children of the Earth is a literary study in politics, religion, and sociology. Questioning the usual distinction between figural and literal kinship, it revives and examines anew problems of national identity and difference that gave rise to such fields as comparative literature at the beginning of the nineteenth century. In Germany Moses Mendelssohn had already warned (in *Jerusalem*) against the intolerant tendency of a well-intentioned, purportedly secular universalism. Likewise, in France Chamfort had paraphrased the national motto of the French Revolution as "Be my brother, or I will kill you." Since variations in the way we demarcate essential familial and species boundaries amount to life-and-death differences, understanding the ideal of universal siblinghood and its tenacity is important if we are to learn whether and how human beings might coexist in a condition of enduring toleration or live together without killing one another.

Many people helped and encouraged me in the writing of *Children of the Earth*. I am especially grateful to Sacvan Bercovitch, Jay Berkovitz, Stanley Cavell, Natalie Zemon Davis, Samuel R. Delany, Stephen Greenblatt, Don Levine, James Maraniss, Murray Schwartz, Robert Schwartzwald, Susan Meld Shell, Judith Shklar, and Barry Weller. Sections of some chapters were published elsewhere: in preparatory versions in *Critical Inquiry* (1991) and *Representations* (1986); in a French-language version in *Journal canadien de recherche sémiotique* (1974); and in my *Elizabeth's Glass* (1993). For photographic material and access to rare books and manuscripts I am thankful to the American Numismatic Society, the Bibliothèque nationale, the British Library, the Bodleian Library, the Beinecke Rare Books and Manuscripts Library, the Department of Biomedical Communications at Yale University, the Houghton Library, the Institute of the History of Medicine at The Johns Hopkins University, the Library of Canada, the Louvre, the Museum of Modern Art, and the Scottish Record Office. The Library of the Jewish Theological Seminary of America supplied the jacket photograph from the Italian fascist journal *La Difesa della Raza*.

I am indebted to my students at the State University of New York at Buffalo, the University of Massachusetts at Amherst, and Harvard University. And I am grateful to my colleagues at Harvard for inviting me to lecture on toleration.

Finally, I wish to thank the John D. and Catherine T. MacArthur Foundation for its support, and the British Government for a United Kingdom Commonwealth Scholarship.

Cambridge, Mass. M.S.
September 1992

CONTENTS

CHILDREN OF THE EARTH

HAMLET: My excellent good friends! How dost thou, Guildenstern? Ah, Rosencrantz! Good lads, how do you both?

ROSENCRANTZ: As the indifferent children of the earth.

—*Hamlet*

1

THE JUDGMENT OF SOLOMON

OR

An American Introduction to the Study of Brothers and Others

So, then, an elongated Siamese ligature united us. Queequeg was my own inseparable twin brother; nor could I any way get rid of the dangerous liabilities which the hempen bond entailed.

—Melville, "The Monkey-Rope," in *Moby-Dick, or, The Whale*

KNOWING WHO'S WHO

Do you know who your parents are?

—Sophocles, *Oedipus the King*

The commonplace view is that consanguineous kinship is real, or literal, kinship. Anthropologists and sociologists usually lump together all other kinds as pseudo-kinship (or kinship by extension), which they then divide into subcategories such as figural, fictive, and ritual.[1] However, the fundamental distinction between "real" kinship and "pseudo"-kinship—or between literal and figural structure—is the topic of a still-unresolved debate about whether kinship is essentially a matter of biology or sociology. For the substance or quality that makes people akin varies from culture to culture, as the skeptical Montaigne insists,[2] and it is ambiguous even within a culture. For example, which is more fundamental, my likeness to my supposed genitor or my likeness to God, who created me in his image? Which of the following is fundamental—the genes I share with my genitor, the love between my adoptive parent and myself, the milk I sucked from my mother, the blood I commingled with my blood brother, the wafer and wine I shared at a communal feast, or the dust from which all things (including myself) are made?

The literalist view, even as it belittles the figural as merely fictive, itself involves a key fiction, namely, that we can really know who are our consanguineous kin. For any particular consanguineous link is always deniable, if not always denied. Who can deny that her children or parents might not be her children or parents? Bastardy, the stuff of fears and also hopes, is always possible. And who can know for sure that any given child is not a changeling? Mothers and fathers can always find grounds to doubt or deny their children, and children can always find grounds to deny mothers and fathers. Likewise, that my lover may be my consanguineous kinsperson is a logical reality, and this merges with the oneirological nightmare that my lover is my consanguineous kinsperson. The particular family dissolves in the republic of dreams. The literal disappears in the figural.

This disappearance is the subject of jokes, but it is itself no joke. For belief in the difference between literal and figural kinship—in the possibility of knowing for sure who's who in the kinship system—is necessary to society if, as psychoanalysts and structural anthropologists generally aver, obeisance to the taboo on incest is a precondition for the continuation of society, or of society as we know it. This need to believe in the possibility of absolute knowledge of kin may be one reason so many people believe in it. Many thinkers assert, for example, that while the father-child bond is unknowable in an absolute sense, the mother-child bond is knowable.[3] Some even project onto a male god the certainty about kinship relations wrongly attributed, usually, solely to female human beings—sometimes going so far as to deny to women any essential role whatever in reproduction.[4] The desire to know who's who in the kinship system may also help to explain the attraction for literalists of figural standards of kinship that are more dependent on witnessed political rites than on biology, standards that at the same time include an incest taboo (as do some kinships by collactation).[5] Thus the fiction of knowledge of who one's literal parents are is matched by the actual knowledge of who one's figural parents are.

It makes as much or as little sense, in this view, to call "brother" the young man my sociological father thinks of as a "son" as to call anyone else "brother." King Claudius' command to young Hamlet—"Think of us as of a father"—makes Claudius, to all intents and purposes, Hamlet's biological genitor no less than Old Hamlet is Young Hamlet's father.[6] For it is, finally, no more or less fictive to say with the family literalist that "only my consanguineous parent's sons are my siblings" than to say with the universalist Christian that "all human beings are siblings"[7]—or with the particular nationalist that "all Frenchmen are siblings of the fatherland [*enfants de la patrie*]." The traditional distinction between literal and figural family, or between real and nominal kinship, erodes as family is conflated with nation—or with species.

It is partly the free-floating conditionality of kinship terminology that allows for the nationalist and universalist ideology according to which every person stands, or stands potentially, in relationship to any other person as a kinsperson. We are all siblings in this logic not because we are children of the same earth—though perhaps we are—but because consanguinity is ultimately unknowable and hence fictive. Put otherwise, if we are not all bastards or changelings, we may as well be: we are no more siblings than we are not siblings. In all cases (but one) our universal or national siblinghood can be counted on more than our particular familialhood.

SIAMESE TWINS AND CHANGELINGS

Friends, we would not have it known for the world, and I must beg you to keep it strictly to yourselves, but the truth is, we are no more twins than you are.

—Mark Twain, *Pudd'nhead Wilson*

The notion that we can know who are our kin is one of the great comforts of literary culture and religious cult. The Greek story of Oedipus, for example, proposes that kinship is ascertainable and depends, for its effective balance of fear and pity, on belief in that ascertainability. The spectator at Sophocles' version of the story, though discomforted by Oedipus's horror at learning who Oedipus is, is comforted by the idea that, since it was possible for Oedipus eventually to know who he was, it is at least possible for the spectator himself to know who he is and hence to avoid the incestuous fate of Oedipus. However, we have no absolute warrant to believe that Oedipus was actually the infant once consanguineously attached to Jocasta by the umbilical cord—except the Oracle's word. And as the wise Heraclitus says, the Delphic Oracle at the umbilical center (*omphalos*) of the Earth "neither speaks nor conceals, but [only] gives signs."[8] Similarly the Old Testament's tale of the Judgment of Solomon suggests that wise kings cannot know for sure who's who even when it is politically important to determine the "umbilical ancestress,"[9] as it is for the matrilineal Hebrew tribes. The sword-wielding Solomon does not figure out which of the two claimants is the literal mother, only which wants the child to live. Some commentators say that, after the Judgment, a heavenly voice announced that Solomon had, in fact, chosen the consanguineous mother; but most rabbis know well enough that such voices are to be distrusted, and the great rabbi Judah bar Ila'i said that "had he been present [in Solomon's court], he would have put a rope around Solomon's neck."[10]

The ambiguous interference of Greek Oracle and Jewish God in these questions of matrilineal descent suggests that the brief period of ocular certainty about matrilineal kinship afforded to human beings by the visibility of the *omphalos* is not entirely forgotten after the umbilicular scission (*Entzweiung*). (It is this brief period that distinguishes the epistemological situation of recently parturient mothers and their observers from that of most other adults and children.) Umbilicular certainty is erased only to be raised (*aufgehobene*) to the ideological plane of religious cults and legal or literary fictions.[11] For example, Christianity has its sect of Quietists devoted to navel-gazing (*omphalopsychita*), and it venerates umbilicular certainty in its cult of the Holy Umbilical Cord.[12] This corporeal relic, the shrivelled remnant of the umbilical cord that once consanguineously linked the earthly body of the Virgin Mary to the partly divine body of Christ Jesus, is a major focus for the Christian debate about whether Jesus was of the same essential substance as Mary or was merely the divine seed that God planted in her as in a prosthetic receptacle. "In his immortal and diviner part," writes Thomas Browne in a Christian tradition that applies to the Holy Family the ancient Greek notion that the woman has no essential role in procreation, "hee seemed to hold a nearer coherence, and an umbilicality even with God himself."[13] "The [Christian] deity," writes Bryant with a nod toward Delphi, "was worshipped under the form of a navel."[14]

What the idea of the Holy Umbilical Cord is to Christianity—an indubitable long-term sign of consanguineous affinity—the fact of Siamese twins is to secular Christendom. The Siamese twin is the only person who knows for sure who his or her kinsperson is, and American sideshows thus prominently displayed Siamese twins under banners like "Believe Your Eyes" and "Marvel of Marvels." (This quasi-cult role of the Siamese twin in secular Christendom was already foreshadowed in the Siamese twin–like aspect of the Holy Mother and Child. Mary is not only Jesus' umbilicularly attached mother, but also Jesus' conjoined sister, since Jesus and she are both children of God.[15] (According to the doctrine of the dormition, they are sometimes even of precisely the same age.)[16] The Siamese twin, the only real blood kin, is the letter in the figure of consanguineous kinship—the incarnation, as it were, of the true spirit of kinship. The argument that we cannot know for sure who our literal kin are—hence, that the fiction of assigning literal kinship matches the fiction in figural kinship—has to make exception of this person, who, by virtue of a literal attachment to her sibling, knows who her kinsperson is.

Mark Twain's America, with its interest in civil war and national unity, displayed Siamese twins in resorts of family "recreation." For example, an advertisement for Barnum's new Wood's Museum and Metropolitan Theatre (1868) read, "The long wants of a *Family Resort* now most satisfactorily supplied."[17] The theater announced the presence of "The Two," namely, Chang and Eng,[18] the twins from Siam whose freakish challenge to "normal" family bonds and boundaries "recreated" the family by defining it. During the Civil War ("brother against brother"), "The Two's" one body literalized the figure of consanguinity for the American nuclear and national families: the same blood ran through the veins of both brothers.[19] Barring fanciful surgery or imposture, the Siamese twins showed the public the one case where consanguineous kinship relations were ascertainable. (For them as for the Siamese felines in the Disney talkies, "We are Siamese if you please. / We are Siamese if you don't please.") The very fact of their existence opposed the otherwise universal indeterminability of kinship that threatens the ideological security of the consanguineous and national family. The Siamese twins' monstrosity, to which Americans resorted, helped to create the "normalcy" of the American family and nation.

"Two *me's* in one"—a motto of the Moliones, Siamese twins of Greek mythology[20]—was a conundrum of the nineteenth century.[21] Along with Chang and Eng, who continued to show themselves at freak shows in post-Civil War America, were Eliza and Mary Chalkhurst, the Kentish Siamese twins earlier memorialized in redware at the start of the Industrial Revolution,[22] and the Italian twins Giacomo and Giovanni Tocci, the subject of *Scientific American* essays written in the 1890s.[23] From the 1860s onward, Mark Twain focused on this conundrumed world of ordinary Siamese and conglomerate twins,[24] in his little tale *Those Extraordinary Twins* together with variations he worked on intermittently.[25] The tale goes some way toward summing up and explaining his century's remarkable concern with Siamese beings (see fig. 1). And Twain really completed *Those Extraordinary Twins* only in such forms as *The Prince and the Pauper,* with its exchange of identities, *The Mysterious Stranger,* with its co-natal "duplicates,"[26] and *The Tragedy of Pudd'nhead Wilson,* with its changelings and identical twins.[27]

In *Pudd'nhead Wilson* (1894) the "original" Siamese twins of *Those Extraordinary Twins* are recast farcically. They are cast first as identical twins who cannot be told apart for purposes of distinguishing an appropriate (i.e., nonincestuous) sexual mate from an inappropriate one, or a criminal from an innocent man; and they are cast also as interracial changelings who cannot be told apart for purposes of distinguishing a "black" from a "white" person.[28] Taken together, the Siamese twins in *Those Extraordinary Twins* and the identical twins and changelings in *Pudd'nhead Wilson* explore the philosophical and grammatical—and eventually also biological and racial—complexities of a binary dialectic involving species kind and familial kin.[29] To mark the one in twain is the game in Twain[30]—as it is in Platonic mathematical dialectic, where "both are two but each is one," or in musical duets.[31] But Siamese twins, as Twain knows, are physically united or conjoined not as in some such figure as Pauline marriage, where "the two shall be one flesh," but by their naturally conjoined blood-bearing tissues.[32]

By presenting the rare case where consanguineous kinship is ascertainable, Siamese twinning emphasizes the general inascertainability of consanguineous kinship that arises from the always possible existence of changelings and foundlings. Changelings and foundlings were not so rare as Siamese twins, however, either in American history or in America's fearful or hopeful popular mythologies. We find changelings and concern with changelings all the more frequently among the European aristocratic and American slave-owning classes dependent on nurseries and wet nurses. In these conditions—which we might call "coo coo"[33]—many a nurse mother from the lower classes seeks to have her son raised, like a new Moses, as the son of some Egyptian pharaoh's daughter. Or slave masters, like those in imperial Rome, fear that nurse mothers will seek such a substitution.[34] In these conditions, the number of real and imagined changelings and unwitting stepparents overshadows, at least in the popular imagination, the number of bastards and foundlings.[35]

Such "exchanges of infants in the cradle" as allow an American boy "with negro taint in his blood [to] substitute . . . for the legitimate white heir"[36] inform

Fig. 1. "I Thought I Would Write a Little Story," by F. M. Senior. In Twain, Those Extraordinary Twins, *Authorized Edition, pp. 216–17. (Widener Library, Cambridge, Mass.)*

the plot of Twain's *Pudd'nhead Wilson.* In this chronicle of an antebellum town, Twain combines the particular American fiction of racially pure blood with the general political fiction of the ascertainability of consanguinity. For the American rule of "descent" or of "statutory homogenization," according to which a man was either black or he was not, relied not only on the fiction that one can know paternity, or at least maternity, but also on the fiction that a specific generation makes a hugely consequential racial difference. In America, what made a man black might be a single drop of black blood.[37] (The notion that one drop of black blood turns an otherwise totally white person into a totally black person has roots in fifteenth-century Spanish Catholic anti-Semitism and has counterparts in twentieth-century German Nazism.[38]) Thus Twain writes of the changeling in *Pudd'nhead Wilson* that the "black" nursemaid Roxana's son "was 31 parts white and he too [like Roxana] was a slave and by a *fiction of law* and custom a Negro . . . [with] blue eyes and flaxen curls like his white comrade."[39]

Any proof of race in America depends on the ascertainability of consanguinity. Yet we cannot tell for sure whether a man is my kin or even my kind—my family or even my race—either by examining his looks according to the Aristotelean adage "Like father, like son" or by scrutinizing easily faked or mistaken birth papers.[40] Kinship is all the more unknowable in an America filled with bastardizing white slave owners having sexual intercourse (incestuously) with their own black slave-daughters. In Twain's antebellum South, in fact, the "white [people] were enslaving themselves, as it were, in the form of their children and their children's children."[41] The result of white slave-owners breeding with black slave women was that the "mixed" population increased rapidly. In 1795, one French visitor had already noticed that American people who were visibly "white" were nevertheless called "black" and on that account were enslaved.[42] A fictive biology lorded it over sociology.

It is no wonder in this racialist context that the desperate fiction that matrilineal descent is detectable (if not always detected) became near dogma.[43] The Judgment of Solomon, with its supposed changeling, was a favorite text for slave-owning lawyers and moralists. Few white Southern moralists noticed, however, that the loving *mater* who won her case in the Book of Kings might well not have been the biological genetrix.

The plot of Twain's *Pudd'nhead Wilson* centers on the Solomonic-like trope of black and white changelings, born on the same day and one "as handsome as the other." Their exchange is undetected by the white slave master, who "knows" his son only by the clothes he wears.[44] By the substitution of her child for her mistress's, the black slave Roxana thinks she has become, "by the fiction created by herself," her own son's slave.[45] And most readers of *Pudd'nhead Wilson* think so too—her knowledge is, as it were, the unquestioned proof in Twain's pudding. But can mother Roxana be certain that her biological son is now really her master? How could she know whether some other substitution did not occur some time before or after the exchange she herself managed?

Roxana thinks she knows which child is her biological child. She exclaims: "Oh. *I* kin tell 'em 'part."[46] Oedipus in Sophocles' punning play similarly thought he knew who were his kin and who weren't. But Oedipus the know-it-all turned out to be a know-nothing. And Oedipus's mother Jocasta did not recognize Oedipus as her own son despite his unusually marked feet.[47] (The foot printing carried out by obstetricians in modern hospitals is not more foolproof.) Kin mother Roxana know her kin any better than Oedipus and Jocasta? It's a wise woman who knows her own child. This is the case especially in certain demographic conditions, including American race slavery.[48] The puns of Roxana's dialect suggest the limits of maternal knowledge: "Dog my cats if it ain't all I kin do to tell t'other frum which."[49] Twain's puns tell of bestiality and miscegenation: dogs turn into cats just as white babies turn into black ones.

Roxana, Twain writes, is "the dupe of her own deceptions" insofar as she has become through habit her son's slave. But is she the slave of her black son? Most readers, accepting the sociologically needful and theologically sanctioned view that it is possible to know who's who, believe this is the case.[50] (Christian belief in the Virgin Mary's servitude to her masterful Son is an apt comparison.) Since there may have been substitutions in the cradle beyond Roxana's ken, however, Roxana may be the dupe insofar as she believes that it is her kin that is master.

There remains the possibility that there is, in any case, no essential filial or racial difference between "t'other" and "which." In some traditions, after all, kinship by consanguinity and kinship by collactation amount to the same thing.[51] "The milk of human kindness" that a black nurse mother would give to her white foster child thus would not only wholly "blacken" the previously white child—as some racialists averred—but also would make incestuous any sexual relations between that child and his nurse mother's consanguineous kin. Perhaps this incestuous quality in sexual relations between a foster son and his black nurse mother's kin is one reason that Southern white males sometimes thought of sexual relations with black women as extraordinarily exciting or disgusting.[52] Perhaps, too, a fear of this ambiguously incestuous quality in interracial sexual relations encouraged "aristocratic" white racialists to come to believe, as many did, that blacks were not really human: being nursed by an animal and having sexual intercourse with an animal would seem morally preferable to sexual intercourse with a human mother or her kindred.[53]

Just as the generational and hence racial "identities" of the changelings in *Pudd'nhead Wilson* can never be known with certainty, so that book's "identical" twins seem indistinguishable or interchangeable. *Pudd'nhead Wilson* poses this merger of identities as a juridical matter: the "identity [of the criminal twin] is so *merged* in his brother's that we have not been able to tell which was him. We cannot convict both because only one is guilty. We cannot acquit both because one is innocent. Our verdict is that justice has been defeated by the dispensation of God."[54] Since one cannot hang a literally identical twin without also hanging his sibling,[55] *Pudd'nhead Wilson* requires a dialectical scission of the one identity into two. This is similar to James Joyce's discussion of the death of Siamese twins, where Joyce refers to a "heated argument . . . regarding the juridical and theological dilemma in the event of one Siamese twin predeceasing the other. . . ."[56] The Midrash likewise interprets the Judgment of Solomon in the Book of Kings in relation to his separate determination of whether a two-headed human being has one legal identity or two.[57]

The "anatomical" separation of identical twins in *Pudd'nhead Wilson* is the job of the detective Wilson, known in the town as the man who once said that he wished he owned half a dog. "The idiot," remark the townspeople of this detective.

"What did he reckon would become of the other half if he killed his half. Do you reckon he thought it would live?"[58] Wilson eventually discovers who's who, or so it seems, by relying on the comforting hypothesis, born of Galton's ideas about fingerprinting,[59] that "one twin's patterns are never the same as his fellow twin's patterns. . . . There was never a twin born into this world that did not carry from birth to death a sure identifier in this mysterious and marvelous natal autograph." This rationalist thesis, which would mark the uniqueness of each "identical" twin, provides *Pudd'nhead Wilson* with its comedic conclusion.[60]

In the end, though, only "in-womb genetic markers" could determine absolutely who's who in changeling cases of the sort that occur in *Pudd'nhead Wilson*, with its interracial aspect, or cases of child abandonment of the sort that occur in *Oedipus the King*, with its incestuous aspect. The universal potential for changelings implies "no possible return to any point of origin."[61] So far as we now know, absolutely reliable in-womb genetic markers, long heralded in science fiction and in legal fictions, are in the early 1990s still "just around the corner" in the biological sciences.[62] Except in the case of genuine Siamese twins, only omniscient judges—the Oracle at Delphi, say, or God in the Judgment of Solomon—can testify adequately about the detection of kin and kind. Otherwise all people are changelings, or may as well be. They are interconnected both figuratively and literally as conglomerate children of the earth.

INCEST AND BASTARDY

> *What he sees often, he does not wonder at, even if he does not know why it is. If something happens which he has not seen before, he thinks it is a prodigy.*
>
> —Montaigne, "Of a Monstrous Child" (concerning a case of Siamese twins)

Many Americans, Sons of Liberty and Daughters of the American Revolution, would want to hold that a national fraternity or siblinghood of some divine or political sort is "the first objective, ethically . . . of the democratic way of life."[63] Thomas Paine's radical American assertion about the familial unity of humankind in *The Rights of Man* (1791), as it erases and rises above the line between consanguineous family and both nation and species, belongs to this universalist tradition that toes the line between literal and figural siblinghood:

> Every history of the creation . . . agree[s] in establishing one point, *the unity of man;* by which I mean that men are all of *one degree,* and consequently that all men are born equal, and with equal natural right, in the same manner as if posterity had been continued by *creation* instead of *generation,* [generation] being the only mode by which [creation] is carried forward; and consequently every child born into the world must be considered as deriving its existence from God.[64]

The ideal of American republican liberty involves this free association of siblings, or *liberi*. (The term *liberty* can mean both "son" and "free."[65]) But this liberty, as it conflates family with nation, collapses the ordinary distinction between incest and chastity, or between bastardy and illegitimacy.

The collapse is thus not without its discomforting political aspect. Already in the fifth century B.C. Plato had gone so far as to argue that the people of the ideal republic should think themselves member siblings of one family. (This is the "noble lie" of national kinship, which Plato compares with the fiction of common autochthony in Oedipus's Thebes.)[66] And if all people in the polis are essentially siblings, then every act of sexual intercourse with a fellow member of the polis must be incestuous.[67] Thus Plato, formulating a political ideology of kinship, raises for the first time in Western political theory the consequential question of the withering of the incest taboo and its connection with nationalism.[68]

As important to America as a Platonic concern with the incestuous implications of the idea of universal siblinghood was a similarly disconcerting Christian concern with how Jesus' injunction "All ye are brethren" might conflate family with species and hence make all sexual relations for human beings essentially incestuous. In this context all sexual relations are incestuous not only because, as in a dream, the lover figures the parent (as Freud might have it) but also because all people are really or essentially siblings. In the law of Roman Christianity it is therefore incestuous for a nun or friar—for a Sister or Brother—, who are equal Siblings in Christ, to have sexual intercourse with anyone.[69] (I here follow the usual custom of capitalizing kinship terms where they involve the Holy Family or the Catholic orders.)[70] The conflation of species with family means that only acts of sex with nonhuman beings—animals, extraterrestrial creatures, and gods—can be chaste, literally "nonincestuous." Where the species is the family, human beings will die out if they are afraid to breed in.

Brotherhoods and Sisterhoods in Christ imitate the transcendence of ordinary kinship relations that is the quintessence of the Holy Family in Christian ideology. In that Family, Jesus is the parent, spouse, sibling, and child of Mary, and He retains his role as father and son of Himself and as father and brother to all (other) human beings. The New Testament, recognizing this incestuous aspect of Christ's generation from Himself, includes in Jesus' genealogy the incestuous sons of Tamar, and it notes that the first outsiders to greet the newborn Jesus were magi—priests born of incestuous unions among a nation, the Persians, supposed to practice incest without guilt.[71] In this way, Christianity incorporates ordinary incestuous relations and raises them, as Stendhal suggests in *The Cenci*, to a spiritual plane where the incestuous backgrounds of such saints as Albanus, Julian, and Gregory make them all the more holy.[72]

By the nineteenth century, the extraordinary kinship structures that the religious orders' siblinghoods had figured (before the destruction of their houses in Elizabethan England and revolutionary France) were not so much destroyed as removed to a quasi-secular plane.[73] Both the sixteenth-century French notion of affinity by *alliance*, which moves toward an infinitely generalizable free friendship,[74] and the German romantic idea of the "elective affinity" of things to each other, not so much in the blood as in the spirit, would be ideologically important to the process of apparent secularization. The German idea likewise challenges both the primacy of kinship by consanguinity and the distinction between literal and figural kinship.[75] The influential cult of brother-sister love in romantic Christendom includes Lord Byron's depiction of sibling love in works such as "Cain," Hegel's consideration of like-unlike siblings in the *Phenomenology*, Vico's elaborate theorizing about sibling love in the *New Science*, and Karl Marx's premise that "in primitive times, the sister was the wife, and that was moral."[76] The premise figures the old idea of a perfectible siblinghood under God.

In republican America, too, the idea of a perfect siblinghood, or "liberty," did not lose its incestuous aspect. The quintessential practical American social experiment in this regard was John Noyes' nineteenth-century Perfectionist Society, a commercially successful society centered in Oneida, New York. Noyes takes "All ye are brethren" literally. He recognizes that a universal siblinghood requires either celibacy for the imperfect or incest for the perfect. (Noyes' followers were to be the latter: "Be ye therefore *perfect*, even as your father which is in Heaven is perfect.") Noyes explains his society's system of pantogamy—or universal marriage—by arguing that free sex without shame *is* possible within a holy community where "all things are lawful for me" (as Saint Paul said). And he stresses the corresponding view that free sex, including incest, is a sign of the "liberty" that grace confers.[77] He bases his argument for universal physical love, hence incest, on the same premise of universal siblinghood that the monastic orders used when they argue against any physical love whatever and celebrate the perfect life. "Love between the children of God," writes Noyes, "is exalted and developed by a motive similar to that which produces ordinary *family affection*."[78] John Ellis wrote in 1870 that "according to the doctrine of the Oneida Community, a man may have sexual intercourse with his grandmother, mother, daughter, sister, or with all of them, and be blameless. . . . At the Oneida Community [this] is regarded . . . as perfectly lawful and right."[79]

The Oneida Community was the most radical and commercially successful familist experiment of the century. There were, however, less radical familist and Fourierist communities, including New Harmony in Indiana and Brook Farm as presented in Hawthorne's *Blithedale Romance*.[80] (Fourier's "familism" included a communalist ideology based on the fraternal and sororal feeling supposed to exist between members of a more or less homogeneous family or tribe.)[81] And there were thinkers besides Noyes who were interested in the relationship of practical politics to the ideal of universal siblinghood and incest. George Lippard's best-seller *The Quaker City, or The Monks of Monk Hall: A Romance of Philadelphia Life, Mystery, and Crime* (1844)—published in "the city of brotherly love" and dedicated to William Hill Brown's bestseller incest-novel *The Power of Sympathy; or, The Triumph of Nature* (1795)—focused on a transitional family *ménage à beaucoup* which "seems all siblings and no parents."[82] In this book, with its evidently ambivalent title, Lippard describes how the Quakers' "Society of Friends" attempts to realize on earth "the great idea of Human Brotherhood."[83] Reciprocal friendship among the *liberi* was to make for liberty, just as *Freundschaft* in American-German idealist thought makes for *Freiheit*.[84] And in a late cinematic echo of the theme, the universalist spirit of the Quakers breathes life into the very American plot of *The Philadelphia Story:* there is "a Quaker spirit in the house" as near-brothers and near-sisters become free husbands and wives in their pursuit of happiness in republican remarriage.[85]

Herman Melville's *Pierre; or, The Ambiguities* (1852) moves from the commonplace topos of individual brother-sister incest, in terms of which it has often been considered, toward the incorporation and transcendence of incest and its taboo in a secularized universal siblinghood.[86] At the novel's outset, Pierre and his mother, Mary, whom Pierre calls "sister" and whom Melville compares to the Virgin Mary, express "a venerable faith brought over from France."[87] Their idyllic and class-conscious aristocratic family estate "seemed almost to realize here below the sweet dreams of those religious enthusiasts, who paint to us to a Paradise to come, when etheralized from all drosses and stains, the holiest passion of men shall unite all kindreds and climes into one circle of pure and unimpairable delight."[88] Yet the American Pierre, "a youthful Magian,"[89] falls in love at first sight with the servant Isabel, who has come to America from postrevolutionary Catholic France. The ambiguity of the possible family relationship between Pierre and Isabel constitutes the key element in the novel. Pierre fears that Isabel may be the unacknowledged bastard daughter of his revered dead father, but the novel insists throughout on the ultimate undeterminability—the ambiguity—of this quasi-orphan's parentage.[90] In this context it is not overdetermination in relation to kinship but nondetermination—not desire for incest but its actual inevitability for the universalist—that informs Melville's novel.

Smiles are "the chosen vehicle of all ambiguities" in *Pierre*,[91] and the ambiguous smiles of its fatherlike portraits—the "strange" closeted portrait that prepares Pierre to meet Isabel and the "stranger's head" portrait that hangs opposite the Cenci painting[92]—bear on the undeterminability of parentage. The closeted portrait speaks the words "I am thy *real* father" to Pierre, much as the ghost in Shakespeare's *Hamlet* says "I am thy father's spirit."[93] And like the ghostly spirit of Old Hamlet, the talking picture in *Pierre* emphasizes the question of who and what is real and nominal. It is a question especially for people who would just as soon not answer Tiresias' question to Oedipus, "Who was your father, son?" or for people who would answer that question with a slightly amended version of the Christian *Pater Noster*, "My Father, who art in Heaven."

Pierre and Isabel eventually divest themselves of mortal parents. "Henceforth, cast-out Pierre hath no paternity," the novel tells us. Says Isabel, "I never knew a mortal mother."[94] Now Pierre is a parentless child—like Billy Budd on the high seas in the ship called "Rights of Man." And like *Moby-Dick*'s abandoned Ishmael, Pierre is "driven out . . . into the desert, with no maternal Hagar to accompany him."[95] In the central love scene, Pierre demands that Isabel "call me brother no more! How knowest thou I am thy brother? . . . I am Pierre, and thou Isabel, wide brother and sister in the common humanity."[96]

The quality of the love between Pierre and his ambiguously illegitimate sister toes the line between the profane and sacred sexual relations in much the same way as bundling—sleeping together, undressed or dressed, on the same bed, without sexual relations. Apologists for bundling in the Dutch Protestant sections of New York State, where it was most common, justified the practice in terms of childlike or primitive "innocence" or in terms of "mortification of the flesh": some drew analogies with the sleeping customs of Amerindian tribes where whole families slept together and others discussed those Catholic orders where individual Brothers and Sisters sleep together. But bundling was also "the [courtship] custom of a man and woman, especially lovers"—which is how Webster defined it in 1864. And so it was frequently criticized as mere camouflage for premarital intercourse (between would-be husbands and wives) or incestuous intercourse (between brothers and sisters). American writers, countering English claims that bundling in New England was nothing more than a screen for incest, noted Julius Caesar's description of the Britons' "universal custom of promiscuous sleeping together," emphasizing Caesar's view that among the ancient Britons, "several brothers, or a father and his sons, would have but one wife among them"; and they claimed that the communalist sleeping custom of British families in the rural north was still merely the cover for incest and bestiality: "Pray, what term will you give to that promiscuous bundling of the father, mother, children, sons, and daughter-in-law, cousins, and inmates who call to tarry, and not infrequently stretch themselves in one common bed on the hovel's floor? / Nay, even, in some parts of your empire, the hogs and the cows join the group."[97]

Instead of calling his mother "sister," Pierre now calls his Madonnalike sister "wife." With his spouse and sibling Isabel, Pierre wavers between consummating and not consummating the love that is the principal ambiguity of *Pierre*.[98] His love cannot be contained within the confines of ordinary brotherly love, "the mere brotherly embrace." Rather, he loves Isabel as if she were a kind of nun, or Sister: "Isabel wholly soared out of the realms of mortalness, and for him became transfigured in the highest heaven of uncorrupted Love."[99]

It is in the commercialized "Church of the Apostles" that the brother-husband and sister-wife set up their Mettingen-like utopian household.[100] This church, renovated as a secular business center in New York City, recalls the community of Christian Apostles: it is the new Blackfriars.[101] Those who live at the church are "suspected to have some mysterious ulterior object, vaguely connected with the absolute overturning of Church and State, and the hasty and premature advance of some unknown great political and religious Millennium."[102] And Pierre himself begins to formulate a plan to further "the march of universal love"[103] with which the American Apostles are linked and which forms a keystone of their general ideology: "The great men are all bachelors, you know. Their family is the universe."[104]

Modeling himself on the figure of God and literalizing the meaning of "Isabel," or "Elizabeth," as "consecrated to God,"[105] Pierre begins to "gospelize the world anew."[106] He is himself the rock, the *pierre,* on which he plans to build a new church. ("Thou art Peter [*Pierre*] and upon this rock [*petrus*] will I build my church" [Matt. 14:17–19].) The doctrinal and practical basis of Pierre's church is the transcendence of the distinction between vice and virtue, a transcendence that involves erasing and rising above all distinctions between kin and nonkin. For Pierre all human beings are essentially autochthonous siblings "of the clod" and universalist "child[ren] of Primeval gloom."[107] From the unity of humankind in a common autochthony (like the one that the French national anthem "La Marseillaise" praises),[108] Melville figures into his novel the Platonic theme of a simultaneously spiritual and physical incest. Toward this end Pierre's tripartite familial *ménage* comes to include not only his ambiguously Sisterly sister (Isabel) but also his nunlike fiancée cousin (Lucy).[109]

Pierre would transcend the taboo on incest. (He is like Mohammed and the other holy and profane personages that Melville culls from the Western tradition—including Paolo and Francesca, Byron, the Aspasia-like Ninon de Lenclos, Semiramis, Cain, Enceladus, and the Cenci.)[110] Yet transcending the distinction between chastity and incest, or good and evil, means an end to being human as we know it.[111] In secular or commercial Protestant America, Pierre's libertine Catholic gospel is thus acted out as an individual fratricide: he kills his cousin. And his doctrine of transcendent neutrality to kinship is acted out as a suicidal neutering. Between perfect liberty and death, which the optimistic American revolutionary Patrick Henry set forth as comedic alternatives, there is, tragically, no essential difference—as probably there was not for Melville himself.[112]

At first blush, Melville appears to harness the general fear of committing incest in the interests of conserving social-class structure. Like much American literature, his *Pierre* suggests that if you marry outside your class you are likely to marry inside your family. People who are apparently of different classes, like Pierre and Isabel, can rise above that difference by recognizing their common descent—from the clod, say, or from Christ[113]—and by intermarriage. The common descent two people share can, however, rule out the possibility of chaste, or literally nonincestuous, sexual ties between them.

Marriage across class boundaries is therefore often represented in universalist literature as incestuous in the same manner as sexual liaisons with members of the Catholic orders: the offending pair who intermarry in the belief that people of all classes are brothers and sisters, in the universalist sense, turn out unwittingly to be blood relatives, in the consanguineous sense. In early American literature, masters who marry slaves and bourgeois factory owners who marry factory workers thus discover too late that their spouses are also their siblings.[114] As in Greek tragedy, the taboo against familial endogamy bolsters social-class exogamy. Where Oedipus in Sophocles' tragedy sees class exogamy, for example, Jocasta sees only familial endogamy. Oedipus says, "I at least shall be willing to see my ancestry, though humble. Perhaps Jocasta is ashamed of my low birth."[115] Where Pierre's mother sees only class exogamy, as if Pierre were just another Romeo, Pierre himself sees familial endogamy transcended in a new utopian community.[116]

BROTHER AGAINST BROTHER

I expected to find a contest between a government and a people: I found two nations warring in the bosom of a single state.

—Lord Durham, *Report on the Affairs of British North America* (1839)

The political figure of "We, the people," the voice of the American Declaration of Independence, first announces itself out of a bloody *Entzweiung,* or "divorce into two,"[117] within a single brotherhood.[118] It is a division of one group into two groups which, from its inception, has the appearance of being a struggle between two separate groups. Jefferson, the principal author of the Declaration, thus emphasizes the English colonialists' fraternal and political kinship with their English brethren in Great Britain. In a draft of the Declaration of Independence he refers to "our British brethren" and complains that "we have appealed to [the] *native* justice [of the British magistrates] . . . and . . . have conjured them by the ties of *our common kindred* to disavow these usurpations."[119] Similarly, in the Declaration of the Causes and Necessity of Taking Up Arms, the authors refer to their common forefathers and foremothers;[120] and in the Declaration and Resolves of the First Continental Congress, they emphasize that "Our ancestors, who first settled these colonies, were at the time of their immigration from the *mother country* entitled to all the rights and liberties and immunities of free and natural born subjects within the realm of England."[121] In the same vein, an early draft of the Declaration of Independence complains that now the *mother country* permits its magistrate "to send over . . . soldiers of *our common blood* . . . to invade and deluge us in blood" and "to impress our fellow citizens . . . to the *high seas* to bear arms against their country to become the executioners of their friends and brethren or fall themselves by their hands."[122]

Unlike the English and French in Upper and Lower Canada, to whom Canada's founding father, Lord Durham, addressed his *Report on the Affairs of British North America* about twin nations in conflict in the bosom of one continent,[123] the American colonists thus founded their one nation with an annulment of the bonds of kinship that had connected them with Britons. On "the high seas" of the mind, beyond the jurisdiction of any merely familial or national authority, Jefferson declared a new family or nation. (In a manner of speaking, Jefferson was piloting a "ship of the mind"—called "The Rights of Man"—with the foundling child Billy Budd.) "Manly spirit," he said, "bids us to renounce forever these unfeeling brethren. We must endeavor to forget our former love for them and to hold them as we hold the rest of mankind, enemies at war, in peace friends."[124] "You know a kingdom knows no kindred," wrote Queen Elizabeth some two centuries earlier.[125]

In the moment of divorce, as in that of marriage, a person both is and is not kin to the other party. Likewise in the moment of fraternal scission, as in that of national constitution, people both are and are not akin. In the moment of American foundation, the nation in America thus foundered between civil war, which is endogamous (brother against brother), and international war, which is exogamous (brother against other).

The rhetoric of brotherhood and otherhood that informs this moment of foundation is not empty figuration. It is no more figural, in an absolute wise, for Americans to call the British their brethren than for any individual to call another person brother. (Nor is it less figural.) What are brethren in a political context but people who we think of as brothers? Already in the Renaissance period to which Melville alludes in his *Pierre,* Alberico Gentili had written in his *Law of War* that "an agreement to be brothers, although it does not make men brothers, surely has some effect." Gentili's statement is true enough, so far as it goes, though it is misleading insofar as it suggests that there is something different from an "agreement to be brothers" that might "make men known to one another as brothers." Agreement in this context is all there is. National siblinghood depends upon agreement or belief just as siblinghood in the family does. As a national family can be split apart by agreement, moreover, so can it be extended. Thus the Romans, to whom the French and American revolutionaries looked as republican models, agreed to call the Haeduans and the Batavians "brethren"; in this simple manner, Haeduan and Batavian "others" became Roman "brothers," as Gentili says.[126] Jeffersonian rhetoric moves from praising "the ties of our common kindred" to "renounc[ing] forever these unfeeling brethren" as the American people is founded in the slide of the British from their status as "brothers" to that of "others."

Where brothers so easily become others the dream of liberty for all American people was bound to falter. American Hegelians like Denton J. Snider and Henry C. Brokmeyer speculated with good reason that a second tempering by blood of the itinerant national spirit was inevitable, because "divorce does not a nation make." The blood of Abel would again cry out from the earth. In the 1860s "Northerners" and "Southerners" fought a bloody and intemperate war. This conflict, like the earlier one, was an intranational or "civil" war between political brethren. (Mark Twain and others sometimes directly compared the American "Civil War" to a contest between Siamese twins, much as Montaigne had done during the French civil war of the sixteenth century.)[127] And it was also an international war between two brotherhoods or states. By the so-called "War Between the States," which name the Southerners naturally preferred, the South and North might have become two states under God—much as the Thirteen Colonies and England had done and as French and English Canada may still do. By "The Civil War," which name the Northern victors preferred, the American national siblinghood emerged as one nation.

With the Emancipation Proclamation and slavery's end, all human beings, recognized by the Declaration of Independence as self-evidently created equal, stood undivided as one national family, one brotherhood. Well, almost undivided: if all humans are created equal, then we are compelled to regard as other than human those beings whom "We, the people" do not happen to recognize as having been created equal. Black people, for example, whom Martin Delany called a "nation within a nation,"[128] had been unequal by policy before the Declaration of Independence. (In 1789 Clermont-Tonnerre had used the same phrase—"nation within a nation"—to vilify French Jews; the Maranno Cardoso previously had called the Jews "a Republic apart.")[129] John C. Calhoun, among others, had argued that blacks were not included among the beings whose equal creation the *Declaration* certified.[130] Calhoun's view was opposed by Lockean theorists who looked to a theory of natural rights as opposed to civil rights. But it was, unfortunately, not always non-American to argue that blacks were not human beings. Thus Cartwright, in *The Prognathous Species of Mankind,* and Nott, in *Types of Mankind,* insisted that blacks were a separate species, and George Fitzhugh claimed in *Cannibals All!* "that the Negro was something less than human."[131] Racialists put into question whether blacks had the same humanoid blood as whites; the 1833 essay entitled "Are the Human Race All of One Blood?" thus had suggested that the blood pools of whites and blacks were separate. Melville, knowing only too well the terms of this false debate, added in small print to an inscription in his *Mardi,* "In-this-re-publi-can-land-all-men-are-born-free-and-equal," the bitter words—"Except-the-tribe-of-Hamo."[132] Abraham Lincoln, remarking that the Declaration of Independence is meant to apply to the "the whole human family," said that "the Republicans inculcate . . . that the negro is a man; . . . The Democrats deny his manhood."[133]

Remarkably, most proponents of abolition and equal rights in the United States did not oppose the racialist (and Christian) conflation of species with family. They tried instead to use the idea of the siblinghood of humankind to their own ends; they appealed to a common generation, or racial descent, for all human beings black and white. They hearkened back to the American argument, already venerable from the sixteenth century and still alive nowadays in the speeches of Martin Luther King.[134] For universalist American abolitionists and Christian preachers, the favorite biblical passage was that God "made of one blood all nations of men for to dwell on the face of the earth."[135]

A few abolitionists, however, recognized that this conflation of species with consanguineous family—like the worshipful myth of the famously native American *alma mater* Pocahontas ("we by descent from her, become a new race, innocent of both European and all human origins—a race from earth . . . but an earth that is made of her")[136]—ultimately could not protect despised creatures from exclusion from the human family. They argued that what really binds people together in the American nation is not biological descent but autochthonous descent (as in Greece), religious consent (as in Christian "rebirth in Christ"),[137] or national regeneration (as in revolutionary France). For some American abolitionists, national regeneration in America, like the spiritual *régénération* promised to the Jews of France by Abbé Grégoire and Napoleon,[138] represented the hope of full political status for everyone regardless of racial generation or nativity. The "umbilical cord" of consanguinity, with its tribal and racial divisiveness, was to be overcome by what Lincoln, believing like Hawthorne in the "electricity of human brotherhood,"[139] calls the universalist "electric cord."[140]

However, calling black men and women less than human or treating them as such was a factor in America's political foundation—even in its sexual history—which distinguished it from the foundation of modern France.[141] Many Americans blamed the massacre of white people in Haiti on such arguments as Condorcet's well known claim that the revolutionary doctrine of "liberty, equality and fraternity" makes every black man a human brother to every white man.[142] They were threatened by the gist of the cosmopolitan French and English abolitionist motto "Am I not a man and brother?"[143] And they were discomforted by such potentially coalescent lyrics as those of "The Rainbow," which, in the 1849 edition of Montgomery's *Songs on the Abolition of Negro Slavery,* ring out

> Black, white, and bond, and free,
> Castes and proscriptions cease;
> The Negro wakes to liberty,
> The Negro sleeps in peace;
> Read the great charter on his brow,
> "I AM A MAN, A BROTHER *now*."[144]

Many white racists, sure or fearful that blacks could be no brothers of theirs, said contemptuously that the doctrines of the French and English abolitionist movement were merely "the sentiments of man and brotherism."[145]

Perhaps denying to American blacks the status of brother humans had the conservative effect of maintaining, even after the 1860s, America's familial vacillation between endogamous and exogamous conflict—familiar already from the scission where one nation became two (the War of Independence) and the twinning where two nations became one (the Civil War). The role of "brother becoming other" or of "other becoming brother" in the speculative theater of American ideology—a role once played by the distant British and then by the neighboring Northerner or Southerner—could now be played out, in the tradition of Twain's *Those Extraordinary Twins* and *Pudd'nhead Wilson,* by the part of "blacks becoming white" or "whites becoming black."

Consider here the importance to American ideology of the legal fiction of the "statutory homogenization of the races," a fiction according to which every person is simply either white or nonwhite. (It was in the tradition of this fiction that a grand jury before the Emancipation, deliberating the expulsion of "free colored" people from South Carolina in the late 1850s, argued that "we [Americans] should have but two classes, the Master and the slave, and no intermediate class can be other than immensely mischievous to our peculiar institutions.")[146] No satire, certainly not Mark Twain's *Pudd'nhead Wilson,* could do better than such legal rulings and fictions. For by classifying a person of "mixed race" as a Negro, the court was "denying that intermixture had occurred at all."[147] (Louisiana courts held in the 1970s that a person with 1/32 "black blood" was legally a "Negro.")[148] The white racialist idealist insists that there are only two terms to describe people, brothers ("whites") and others ("nonwhites"), and he bolsters his view by the legal fiction that white blood, though it can be wholly "blackened," cannot be partly diluted. In fact, in the United States, there were rarely mediating terms between white and nonwhite—terms of the sort one does find in Canada and the West Indies[149]—just as in universalist thinking generally, there is no mediating term between brother and other. And let us here recall the view of Thomas Jefferson, who in his *Notes on the State of Virginia* makes the usual segregationist, though not necessarily dehumanizing, Enlightenment analogy between species and race: "Will not a lover of natural history, then, one who views the gradation in all the races of animals with the eye of philosophy, excuse an effort to keep those in the department of man as distinct as nature has formed them?"[150]

Who had better grounds, in such a monstrous historical context, to demand the immediate abolition of slavery in America than those people who really recognized the practical consequences of the doctrine that all men are equal brothers? Who, indeed, better than the radical Perfectionist John Noyes? Noyes claimed to understand the radical implications of such brotherhood as the American Declaration seemed to espouse. His Perfectionist experiment, as we might expect, demanded abolition along with both liberty and that guiltless incest which he and other communalists believed that *libertas* required. (Liberal nationalists, fearing the apparent similarity between the "nationalism" they pretend to espouse and the "racism" they pretend to hate, are always quick to attribute to pure racism the tendency toward incest—which is easy enough to do, thanks to such explicitly proincest racialist ideologists as Joseph Gobineau and Richard Wagner—but they are slow to attribute to national liberalism the same tendency.)[151] Noyes, through his monthly paper *The Perfectionist,* made converts to his ideas about the universalist brotherhood of mankind and racial equality.[152] Among the more prominent subscribers to Noyes' views on abolition were Edmund Quincy, the Quaker sisters Sarah and Angelina Grimké, Henry C. Wright, and other abolitionists, including William Lloyd Garrison, who writes eloquently of Noyes in his biography. In a letter to Garrison, Noyes describes his "hope of the millennium beginning . . . At The Overthrow Of This Nation."[153] It was the overthrow of all nations, as of all families, that Noyes wanted. Instead of a world of many nations, Noyes hoped for the one family nation of humankind—what John Gower, in fourteenth-century England, had called the primal "man's nation."[154]

KINSHIP AND KINDNESS

Much that is left unsaid here about the relationship between literary figures and kinship terminology, and about the political implications of the idea of national kindred, will be considered in subsequent sections of this book. But three problems should be specified briefly at this point. The first concerns the idea, to which we will return in chapters 2 through 4, that it is a common genitor, not quite one of us, who makes us all siblings. That is, we are often able to call ourselves brothers and sisters only by assuming a common parent. "Alle Menschen Brüder werden! [All men are, or become, brothers!]," writes Schiller in the great *Ode to Joy*. But people become brothers only because, "Above the stormy canopy / There must dwell a loving father"—a parentarchal God, perhaps, or a national mythology of parental founding or autochthony.[155]

The second problem, to which we will return mainly in chapters 5 through 7, is that it is only our opposition to another group of siblings, who are not quite us, that makes us siblings. Here the uncompromising conflation of species with family has made universalist nationalism a dangerous and cruel ideology. For the universalist ideology of love and kinship leads—has led—inexorably, to actions of hatred and unkindness. From the position "all men are my brothers" it comes to follow easily that "only my brothers are men, all others are animals." When only my siblings are human, all others are not human. *Volk* is conflated with species. In much the same way, confusing species and family becomes the basis for the institution of a particularized fraternity ("we men are brothers, they are others"), not for the institution of a universalized siblinghood ("we are all siblings"). In many languages the word for "human being" and that for "fellow tribesperson" are consequently one and the same.[156] The universalist confusion of species with family thus becomes an effective ideological basis for the institution of particularist siblinghoods defining one nation against another by means of specific exclusionary tactics like misogyny, racism, anti-Semitism, and, as we shall see, earthlingism. Muslims and Jews in sixteenth-century Spain and African Americans in nineteenth-century America came to play out the role of the other species, for example, while in revolutionary France, the Platonic ideal of sexual and propertal communalism—of *égalité*—was obscured, as women were excluded conceptually and politically from the human species and that *alma mater* "Lady Liberty" was won, if at all, only at the expense of a Rousseauist sororal oppression.[157]

The third problem is that deniability of kinship, taken apart from other conditions, means neither that kinship will be acknowledged nor that kinship will be denied. I can acknowledge that my child is mine, *thank God*, even as I know that it is deniable that my child is mine; and I can deny that my child is mine, *God help me*, even as I recognize that my child probably is mine. Children thus fantasize that they are the offspring of royal parents, men hope that they have unknowingly fathered children or fear they have not fathered the children that they call theirs, and mothers deny that their children are theirs. There are various demographic as well as psychological and political factors at work here—all cultured, to some extent, by a skeptical focus on the impossibility of absolute knowledge of kin. In some societies, for example, there may be an unusually high rate of sexual intercourse whose participants either know that they do not know who their partners are (as in the Amazons' anonymous matings in Greek myth)[158] or believe wrongly that they know who their sexual partners are (as in the bed tricks of Elizabethan and Jacobean drama). And where bastards, foundlings, or changelings are known or believed to be widespread, there will often develop, by way of socially needful compensation, specific types of figural kinship. The large number of foundlings and oblates in medieval Europe, for example, helped to set the stage for the political fictions of premodern familial nationalism in the sixteenth century. In eighteenth-century France the foundling d'Alembert was aware that the growing number of parentless children and orphans was tending to make Frenchmen equally kin or nonkin;[159] revolutionaries generally were concerned with establishing a *liberté* in *fraternité* where everyone would be equally legitimate and illegitimate; and Rousseau, theorist of *liberté*, sent off to the foundling hospital his five illegitimate infant children partly in order that he might recognize them thereafter, not at all as particular consanguineous children of his own, but only as fellow multiple twins of the republic and equal children, or *liberi*, of the nation.[160]

2

FROM COEXISTENCE TO TOLERATION

OR

Marranos (Pigs) in Spain

How could a society as tolerant as Castile, in which the three great faiths of the West had coexisted for centuries . . . how could a clergy that had never lusted for blood except in war (Queen Isabella thought even bull-fighting too gory), gaze placidly upon the burning alive of thousands of their fellow Spaniards?

—Henry Kamen, *Inquisition and Society in Spain . . .*

THE END OF *CONVIVENCIA*

For hundreds of years, Muslim Spain was the most tolerant place in Europe. Christians, Muslims, and Jews were able to live there together more or less peacefully. The three religious groups maintained a *convivencia,* or coexistence, thanks partly to a twofold distinction among kinds of people that was essential to the particularist doctrine of Islam influential in Spain. Islamic doctrine distinguishes first between Muslim and non-Muslim peoples and second between those non-Muslims who are, like Muslims themselves, "peoples of the Book" (i.e., Christians and Jews) and those non-Muslims who are "pagan." These two distinctions, taken together, could amount to the difference between life and death. For example, Muslim courts ruled on the basis of the Koran that those "others" who were "peoples of the Book" could not legally be put to the sword for refusing to convert to Islam while those "others" who were pagan could be. Christians and Jews had to be put up with, and usually were.[1]

Spanish Islam's limited tolerance toward religious heterogeneity and toward national differences was something that Spanish Christendom, when it conquered

Spain from the Muslims, was generally unwilling and perhaps ideologically unable to maintain. With the conquest of Spain from the Muslims, Christian Catholicism came to constitute the basis for a radically exclusionary definition of Spanish Christendom. (The word "catholic" comes from the Greek *kata,* meaning "according to," and *holos,* meaning "the whole.") For just as the Islamic division of humankind into particular groups encouraged a limited *convivencia,* the Christian union of all humankind into a single brotherhood encouraged a certain intolerance. In one version of the categorizing process I have outlined, the crucial Christian doctrine "All men are brothers"—or "All human beings are siblings"—sometimes turned all too easily into the doctrine "Only my 'brothers' are men, all 'others' are animals and may as well be treated as such."[2] The politics and metaphorics of this transformation involving kin and kind is the subject of this chapter.

The interconnected historical hypotheses here are: first, that in Spain there was a basically Islamic particularist ideology of several siblinghoods according to which some people are siblings and some are others, and that this ideology allowed for a coexistence grounded in protection for *dhimmis,* or non-Muslim residents of Muslim states, as human beings; and second, that the Islamic ideology was followed by a basically Christian universalist ideology of one siblinghood where all people are siblings and none are others, and that this ideology allowed for an intolerance grounded in the exclusion of nonsiblings from full humankind.

These hypotheses are not without historical complexities beyond our present purview. After all, Islamic rule did not always foster coexistence. During the Almohad terror in the latter part of the twelfth century, for example, Jewish communities that refused to convert were sometimes put to the sword—as memorialized in the poetic lament by the twelfth-century Jewish scholar Abraham ben Meir ibn Ezra of Toledo.[3] However, "there are no more than half a dozen [instances of the forced conversion of Jews to Islam] over a period of thirteen centuries"[4]—a remarkable record when compared with the history of Christian proselytizing. Some scholars complain that the Pact of Umar treated Christians and Jews with less dignity than brother Muslims. However, the Pact established laws protecting the two groups, admitted them to the polity as human beings, and guaranteed them a generally dependable protection. Similarly, although some interpreters of the Koran say that Sura 2:256 ("There is to be no compulsion in religion") means that the Muslims were not so much tolerant toward other peoples of the Book as resigned in the face of obdurate belief,[5] the result was still coexistence instead of compulsion and murder.

Likewise, Spanish Christendom was not always intolerant. Yet to the extent that *convivencia* ever existed in such places as fourteenth-century Aragon,[6] it was largely a short-term holdover from previous Islamic law codes or a practical strategem for dealing with large Muslim presences in traditionally Christian-ruled states and previously Muslim-ruled ones:[7] the final Christian "reconquest" in 1492 thus marked the effective end of any pretense at *convivencia.* In any case, Christian *convivencia* never existed in anything like the way many Spanish nationalists have described it. In 1311 James II carried out the explicit orders of Pope Clement V when he prohibited the *çala* (the public prayer ritual in Islam that is mandatory for all Muslims) under pain of death. And while Muslim law categorized Christians, however bothersome, as "fellow human beings" and treated them as "a people of the Book," Christian legal codes, which were linked to a religious univeralism that could not easily recognize the existence of "others" who were not animal, "frequently classed Muslims in the category of 'slaves, mules, donkeys, cows, or other animals.'"[8]

How the various peoples of Spain defined and treated one another critically influenced modern European Christendom's understanding of caste and race.[9] And Spain at the time of the reconquest came to serve as a model for how the doctrine that "All human beings are siblings," in its merging of the usual distinction between the human species and the family, negates concepts, like "people of the Book," which mediates kind and kins and thus affect the politically sensitive definition of "nation."[10] The reconquest of Spain during the seven centuries leading up to the expulsion of the Jews in 1492 and of the Muslims in 1502 was *the* nationalist event in Spanish history. (On the very day in 1492 that Christopher Columbus set sail from Palos for what turned out to be the New World, he noted in his log the shiploads of Jews and *conversos* leaving their Old World home of a millennium under threat of death.) The expulsion of the Muslims consolidated a brutal ideology of who was in the one-family nation and who wasn't. The official view became that Christians with only Christian ancestors were Spanish nationals and that all others were not. There was to be no such mediating concept as fellow "people of the Book."

The crucial events in the gradual historical evolution of the exclusivist definition of the modern Spanish nation probably occurred during the hundred years between the mid-fifteenth and mid-sixteenth centuries. First, there was the introduction of the famous Statutes of the Purity of the Blood (*limpieza de sangre*) in Toledo in 1449—and elsewhere a little earlier or later.[11] These statutes distinguished between original Christians and *conversos*—those people who ostensibly had converted from Islam or Judaism or whose ancestors had converted—on the basis of blood lineage.

The statutes were at first denounced by the pope.[12] The Roman Catholic creed, after all, traditionally stresses essentially not kinship by consanguinity but rather rebirth and kinship through Christ.[13] Many powerful people with *converso* ancestry somewhere along the line argued for this tradition, as did the Dominican cardinal Torquemada in his *Treatise Against the Midianites and Ishmaelites.* But there were also racists influenced to some extent by the Christian doctrine of Arianism, according to which Jesus was not or not entirely consubstantial with God his Father, though he was consanguineous with Mary his mother. King Leovigild's sixth-century Gothic Christianity did not disappear from Spain with the so-called Spanish Conversion to the Christian orthodoxy of his sons Hermenegild and Reccared I. Under Muslim rule, in fact, Gothic Christians kept their old law code—according to the Pact of Umar.[14] And they maintained the old Arian heresy according to which the Son, though He had a likeness (*homoiótés*) to the Father, was not of absolutely the same substance (*homousios*). It would seem to follow that kinship in Christ does not fully transcend consanguinity, and that spiritual religion is not all that matters: blood counts.[15] (Similarly, the sixth-century Spanish Catholic Saint Isidore of Seville was interested in the potentially nation-forming opposition between *brotherhood-german* and *brotherhood-spiritual*. Orphaned as a young child and raised in a monastic brotherhood, Isidore was the brother-german of Brother Leander. Both loved their common sister-german Florentina and wanted her to become a Sister-in-Christ; Isidore's *Regula monachorum* was adopted by many Spanish Catholic Brothers in the seventh and eighth centuries.)[16] But even if Spanish Christendom were to have forgone entirely the Roman Catholic notion that Christianity transcends blood kinship—as Gothic Christians and so-called tribal pagans may have wanted Spanish Christendom to do—how then would it be a crime to be a Jew? After all, the Mother of God and all the apostles were Jewish—as the distinguished jurist Alfonso Díaz de Mantalvo put it.[17] And surely the Church was properly the home of the Jews, and the Gentiles were the outsiders who had been invited in—as argued the Bishop of Burgos, Alonso de Cartagena, in *Defensorium unitatis christianae* and Bishop Alonso de Oropesa in *Lumen ad revelationem gentium*.[18]

Despite the powerful arguments against the blood statutes that focus on the polar opposites of spiritual kinship (Isn't a convert a brother in Christ?) and consanguineous kinship (Wasn't Christ a Jew?), a nationalism of exclusion finally became dominant in Roman Catholic Spain in the latter part of the fifteenth century. The myth of pure blood (*sangre pura*), unmixed with Muslim or Jewish blood, took hold. The joint sovereigns Ferdinand and Isabella benefited, perhaps, as they unified Spain into a nation of one blood, from a *Germania* or "union of siblings-german."[19] In later centuries, the Spanish myth of pure blood traced a tribal bloodline to a Gothic or Teutonic ancestor, Tubal, from the twenty-second century B.C.![20] The Spanish kept their fixations on blood purity even after there were in Spain virtually no more Jews and Moors. By 1788, the term *limpieza de sangre* had come to refer to class difference, the upper class maintaining its "purity" by refusing to do manual work because it was beneath their dignity and honor. Thus transformed to suit contemporary ideologies of social class, pure blood eventually informed the rhetoric of German Nazis and Spanish and Italian Fascists.[21] The latter sought to cut off blacks and Jews from pure-blooded Italian "Aryans" (see fig. 2).[22] And pure blood is still cherished among modern Spaniards. In 1988, the "blue-blooded" president of a Madrid-based institute to promote cultural exchanges between Spain and its Muslim neighbors asserted that the Spanish "take pride in our *sangre pura*, pure blood. No Catholic wants to face the thought of Moors on the family tree."[23]

The final consolidation of Spanish nationhood followed on the strict enforcement of the blood regulations under Philip II, in the latter half of the sixteenth cen-

Fig. 2. Cover of the journal La Difesa della Razza *for August 20, 1938. (Library of the Jewish Theological Seminary of America, New York)*

tury.[24] One drop of "Jewish blood" might make a person non-Christian in Spain just as, in parts of the United States in the 1800s, one drop of "black blood" made a person nonwhite. "If it were proved that an ancestor on any side of the family had been penanced by the Inquisition or was a Moor or Jew," writes Kamen, "the descendant could be accounted of impure blood and disabled from office."[25] The official Instructions of 1561 thus stipulated that "all the [penitential garments imposed by the Inquisition to bring shame on the wearer] of the condemned, living or dead, present or absent, be placed in the churches where they used to live . . . in order that there may be a perpetual memory of the infamy of the heretics and their descendants."[26] Juan Escobar de Corro later argued in his *Treatise on Testing for Blood Purity and Nobility* that "purity" and "honor" are exactly synonymous and that any stain on an impure lineage was ineffaceable and perpetual.[27] Costa Mattos wrote that "A little Jewish blood is enough to destroy the world."[28]

Cervantes' tale in *Don Quixote* of people of Arab "race" whose families had converted under Christian rule to Christianity (Moriscos) may help to illustrate the dilemma. The tale concerns a young Morisco woman who learned her Christianity from her mother just as she had sucked milk from her. "Mamé la Fé catolicá en la leche." In terms of spiritual kinship she was a sister in Christ, and in terms of collactaneous kinship she was a Christian daughter in milk. But according to the statutes of pure blood, she was no Christian. The Roman Catholic Inquisitors—and some of her consanguineous Muslim kinspersons as well—claimed that her Christian belief was a mere fiction (*invención*). The young Morisco was thus expelled from Spain due to the "crime" of a nation (*nación*) to which she felt she belonged only by fiction. And her Christian lover suffers an imprisonment in Algiers like that of Cervantes himself.[29] Cervantes says in the "Prologue" that he created the book as out of an imprisonment and that he stands in relation to its hero, Don Quixote, not as a father to a consanguineous son who is "like" his father and part of his father's nation, but as a stepfather (*padrasto*).[30] Cervantes knows the folly and horror of such attention to pedigrees as Don Quixote discusses in "One of the Most Important Chapters in this History": "From all this I wish you to infer, my dear sillies, that the subject of genealogies is a most confused one."[31] Fernando de Rojas, whose background was *converso,* wrote his famous *Celestina* a year after the expulsion of the Jews. There, he seems similarly to "attack the concepts of external honor and purity of blood (always behind the mask of a servant or prostitute). In this he lent his voice [as Stephen Gilman says] to the protest of his fellow conversos whose blood was not pure and who, like the prostitute Areusa, demanded that honor be attached to deeds and not to the distinction of birth."[32] She paraphrases the complex historic proverb "When Adam delved and Eve span / Who was then a gentle man?"[33]

But the blood statutes and the Inquisition's peculiar attention to genealogy "triumphed and became the law of the realm in Spain and later in Portugal. They spread their rule over other races as well (black African, Chinese, and Moors) and into the Iberian colonies," with dire consequences to the so-called indigenous populations of the New World and peoples in the Orient.[34] The boundary of the Christian "nation" became no more than race and genealogy, shorthand reports of which gave Spaniards easy access to a breederlike knowledge of who was in the *Germania* and who wasn't. Blood now defined the nation: national kinship was literalized as consanguineous and consanguinity itself was upheld as ascertainable (even as fears of bastardy and of foundlings increased).[35] Diego Laínez was right to denounce the cult of blood purity as "the national humor or error."[36] And Ignatius Loyola, the founder of the Society of Jesus (in 1540) who refused to associate himself with most racialist aspects of the Inquisition, also said that the Spanish cult of blood purity was "the Spanish characteristic."[37]

Spain, formerly the European model of *convivencia* and intellectual progress, had become in the sixteenth century the least tolerant place in all Europe, and soon it was to become one of the most backward. Once there had been an influential Muslim ideology where there were human others (fellow "peoples of the Book" and pagans) as well as human brothers (fellow Muslims), and some of those others had to be lived with, no matter how distressing their existence might be. But now, with the rise of specifically Spanish Christendom, all humans were brothers and all others were animal or may as well have been. Spain fulfilled its national aspiration in the reconquest only at the loss of any specifically human term mediating between Christian human beings and other creatures.

THE TAUREAN NATION

From this loss came the nationalist ideology of the bullfight. The rise in sixteenth-century Spain of the unique prominence of the bull festival, rightly called Fiesta Nacional or Fiesta Brava—not a mere sport (*deporte*)—corresponds to the rise of the ideology of the modern Spanish "nation." The killing of bulls in urban arenas was developed to its still-present zenith during the last years of the reconquest, which is associated with "the idea of the growth of some form of Spanishness."[38] *Taurofilia* and bullfighting in Spain express a national ideology of otherhood and brotherhood.

The Spanish fighting bull, like the creature in the ancient Minoan ritual of bull-leaping that some ethnographers say grounds the bullfight historically,[39] helps to define the difference between humankind and animalkind—who is a bull and who is merely a taurean human being—and between human kin and nonkin—who is in my family, hence lovable only in the chaste way, and who is not? The Minotaur, son of the bestializing Pasiphaë and her bull-lover, straddles the line between human and nonhuman kind: he has a head at once human and bovine. And the Minotaur's family straddles the line between family kin and nonkin: his "sister" (or half-sister) Phaedra has an incestuous passion for her "son" (or stepson) that would be nonincestuous only if her family kin were not her species kind.

Campos de España writes that "in the bullfight the Spaniard has found the most perfect expression for defining his human quality."[40] The bullfight helps to fix ideologically the difference between national and nonnational. What is the unique *nación* of the *toro bravo* and the quality of its treatment by the *nación* of Spanish Christians? *Nation* means "a particular class, kind, or race," not only of persons but also of animals.[41] *Bravo,* the term that the Spanish use to describe the nation they admire, means "wild."[42] Yet the *toro bravo* is not a "game" animal, like a deer, or a "domestic" animal, like a llama, or an animal *sauvage,* like a mountain lion. The *toro bravo* is distinguished among animals both domestic and wild in that, according to a long tradition of breeding, it is actually cultivated so as to be or become "wild," or artificially natural. The bullfight itself, the great national festival of Spain, is merely a *desbravando.* The animal, bred artificially to be *bravo,* is "civilized" in the ring. The bullfighter's technical term *rompiendo,* or "breaking," is thus appropriate, as is Hemingway's remark about the matador's "educating" the bull almost as if it were a man.[43] And so the bull, in a *corrida* that will "break" it as on the rack, is given a humanoid name.[44] The bullfighter follows definite regulations to torture the bull, "outmaneuvering" it with the aid of painful harpoonlike pikes (*banderillas*) in the neck muscles and the *picador*'s bloodletting spikes in the enlarged hump on the neck. He "prepares" (*lidiar*) the bull for butchering as a sacrificial victim.[45] (Easter Sunday is when the bullfighting season begins.)[46] The bullfighter butchers or sacrifices the bull according to prescribed rites and with prescribed implements; if he should fail in this respect, the spectators call him *asesino,* "murderer," as if the dead bull had been a human king and the matador a mere "assassin," a term that derives from the Arabic *hashshashin.*[47] Strikingly, parts of the dead bull—principally the "ears"—are distributed as awards, and several parts are eaten.[48] Thus Spanish Christians, who are theoretically omnivorous—for them, all food is legally edible—transform the Minoan rite of bull-leaping into a national festival incorporating such regulations regarding butchering and such restrictions regarding what can legally be eaten as generally characterize only particularist religions like Judaism and Islam.

In the sixteenth century, when the bullfight truly became the national festival, a universally proselytizing bullish Christendom freed itself from the discomforting burdens of any sort of Muslim *convivencia* or Jewish tolerance. No longer was there in Spain a specifically human intermediate term between national kin and national nonkin, between Christian brothers and others who were not Christian. Islam had had such a term in its notion of a "people of the Book" that is neither Muslim nor pagan, neither brother nor other. And the ancient Jewish Commonwealth had its notion of a "strange people in a strange land," nevertheless protected as a human nation by distinctly humane laws and promised, as by Rabbi Moses ben Maimon of Córdoba (Maimonides) in the twelfth century, a share in the world to come.[49] But Spanish Christendom, in its unwaveringly universal aspect, had no such term—except in its peculiarly inhumane reworking of "pagan" bull-leaping. The Inquisition and its secular arm burnt alive those it called crypto-Jews and -Muslims, and those it did not burn to death it either proselytzed or expelled wholly out of Spanish existence—much as Shylock is expelled from Venice in act 4 of *The Merchant of Venice.* (In 1594—a few years before Shakespeare wrote the play, with its themes of forced conversions, racial difference, and relations between the three peoples of the Book—Roderigo Lopez, court physician to Queen Elizabeth, born a Portuguese Jew but a convert to Christianity, was executed before a festive crowd which laughed at his dying assertion that "hee loved the Queene as well as Jesus Christ.")[50] Ferdinand the Bull, more taurine than humane, helped the Inquisition transform the idea of human others who are to be treated humanely into an ideology where all others are not fully human and must be either Christianized or—since the statutes of blood purity often made Christianization impossible—destroyed. At the well-known Festival of the Christians and Moors in Spain, even today the expulsion of the Muslims (1502) is reenacted annually in the same Spanish cities and villages where the bullfight plays its part in telling Spanish nationals who they really are.[51]

AN AMSTERDAM OF RELIGIONS

T. S. Eliot got it right: "The Christian does not want to be tolerated."[52] He cannot tolerate difference without also wanting to sublate (*aufheben*) that difference.[53]

The systematic political philosophy of religious toleration toward all men arose prominently in the seventeenth century thanks partly to certain thinkers' recognition that Christendom, unlike some polities or religions, requires an extra and perhaps extraneous theory of toleration, or "policy of patient forbearance in the presence of something which is disliked or disapproved of."[54] The New Testament says "All ye are brethren," but the politically needful policy of toleration would have to recognize that there are not only brothers in the world but also others who should be tolerated as they are, no matter how much they or their existence discomfort us. John Locke's treatises on the idea of religious toleration confront this tradition of a potentially intolerant Christianity—one that slips from the proposition "All men are my brothers," to the proposition "Only my brothers are men." Locke, who as a young man had rejected the idea of taking holy orders in the Church of England, gave expression in his treatises on toleration to the political dilemma inherent in any polity of universal brotherhood; his ideas are crucial to the development of pluralist and liberal toleration in the modern world.

The ideology of national toleration also has roots in the experience of the Iberian Marranos who fled the Spanish and Portuguese Inquisitions. Many Spanish crypto-Jews went to Portugal, where they were called and called themselves "The Nation," as also in France and Amsterdam.[55] This "Nation" was caught between a world of Christians who derogatorily called them *marranos*, or "pigs," and a world of Jews, who called them *anusim*, or "compelled" and even "raped."[56] Yosef Yerushalmi explains that "the novelty of Marrano apologetics and polemics goes far beyond the relative degree of its Christian learning. The knowledge which these writers had of Christianity was derived not merely from books, but from their own personal experience of Christian life, ritual, and liturgy. They are thus the first body of Jewish writers *contra Christianos* to have known Christianity *from within*, and it is this which endows their tracts with a special interest."[57] Among such tracts is Isaac Cardoso's *Philosophia libera*, one of the first works of general philosophy published by a professing Jew specifically for a generally European audience.[58] And there is also Cardoso's *Las Excelencias de los Hebreos*, a treatise published in 1679 in Amsterdam, a haven for Marranos.[59] Having been raised as a Spanish Catholic and living as a devout Italian Jew, Cardoso describes in this controversialist work the Spanish claims that the Jews are cruel and inhuman, and he counters those claims with an argument that the Spanish merely project onto others the faults they fear in themselves. In this context he describes how the Spanish kill both men and animals for pleasure at their Fiesta Agonal (or bullfight); how they regard the dead with exhilaration and joy; and how they sacrifice men to their gods, throwing them to the so-called wild beasts.[60] In the same work, Cardoso emphasizes the need for political toleration toward Jews not only because Jews are a loyal, industrious, and hence politically useful people—which was the usual argument for toleration—but also because Jews constitute "a Republic apart."[61]

Leo Strauss, in a work researched in the Weimar Republic, writes in reference to the Marrano philosopher Uriel da Costa—a Catholic who, after relinquishing Christianity and converting to Judaism, relinquished Judaism and took up a general critique of all religion—that "the situation of the Marranos favored doubt of Christianity quite as much as doubt of Judaism."[62] (One might think here of the anti-*converso* pamphlet of 1488 that spoke of the monstrous animal "which carried Mohammed on his back from Jerusalem to Mecca and which, like the *conversos*, belonged to no known species.")[63] Certainly, the unique philosophical stance of the Marranos, which was skeptical and liberal, helped to mark and make for a new sort of toleration. It is a stance whose proponents include the Marrano skeptic philosopher Francisco Sánchez in his *Quod Scitur Nihil* [That nothing can be known; 1581] and Sebastien Châteillon in his *De haereticis* [1554], where he criticizes Calvin for helping the Inquisition to persecute the Spanish anti-Trinitarian Michael Servetus and eventually to burn him at the stake. "We know in part," writes Châteillon, "that Socrates was right, that we know only that we do not know. We may be heretics quite as much as our opponents."[64] It is a stance that includes Montaigne's free-thinking essays, written by the son of a Marrano at a time of brutal religious conflict in France, where Catholics persecuted Protestants as if they were members of another race or even species.[65] And it includes Pierre Bayle's skeptical and tolerant writings, as well as Spinoza's treatment of freedom of thought and speech in his anonymously published *Theologico-Political Treatise* (1670), which, with its celebration of the domestic liberty of Amsterdam, has been called the first philosophy of democratic liberalism.[66]

Amsterdam was a haven for political radicals and religious outcasts from Europe, including such proponents of toleration as Henri Basnage de Beauval, Pierre Bayle, and John Locke.[67] No political refugee living in Holland in the seventeenth century needed reminding that Christianity in practice did not live up to its claims of universal love. Spinoza, haunted by the same memories of Spanish cruelty as many Hollanders, was no exception. "I have often wondered," he writes, "that persons who make a boast of professing the Christian religion, namely, love . . . and charity to all men, should quarrel with such rancorous animosity, and display daily towards one another such bitter hatred, that this, rather than the virtues they claim, is the readiest criterion of their faith."[68] Much of Spinoza's political thinking starts from analyzing the link between the preaching of universal love based on universal kinship and the practice of persecution. Sometimes Spinoza flatters the majority of his readers by appearing to agree with them that Matthew's famous claim about the Jews—that they believe in the doctrine "Love thy neighbor and hate thine enemy"[69]—is correct. But indirectly, he points out both that the Jews were bidden to love their fellow-citizens as themselves and that there is an inevitable conflict between the requirements of universal love and those of politics: "Though the Jews were bidden to love their fellow-citizens as themselves (Lev.

19:17–18), they were nevertheless bound, if a man offended against the law, to point him out to the judge (Lev. 5:1, and Deut. 13:8–9).''[70] Even as Spinoza pretends for heuristic purposes to agree with Matthew's assessment of Jewish hate and Christian love, he demonstrates that the link between the teaching of the Sermon on the Mount and that of Mount Sinai is one of polar opposition. He shows the connection between those teachings. For according to Spinoza, the difference between the commands ''Hate thine enemy [the foreigner],'' which Matthew attributes to the Jews, and ''Love thine enemy,'' which Matthew attributes to the Christians, is due exclusively to the changed political circumstance of the Jewish people in the Diaspora. Moses could think of the establishment of a good polity, whereas Jesus—like Jeremiah and Isaiah—addressed a people that had lost its political independence.[71] Spinoza shows that, since ''religion has always been made to conform to the public welfare,''[72] Christianity and Judaism are political refractions of the same doctrine.

A word about Spinoza's ''indirection'' in argument is in order. Some historians have said that Spinoza wrote in convoluted fashion in order to hide an atheism that would have troubled his readers and hence interfered with his purpose. (In 1671 Spinoza wrote a moving letter about this matter to the Portuguese-born Isaac Orobio de Castro, categorically denying the charge that he ''with covert and disguised arguments [taught] atheism.'')[73] Be that as it may, Spinoza wrote in such a way as not to offend his readers unnecessarily. He did not want to jeopardize his larger political purpose, which was the support of free philosophical inquiry.[74] Spinoza's ''esotericism'' was pedagogic and political: he sought, according to the principle of ''economy,'' to speak to different men at their own planes of understanding.[75] Spinoza remarks that Saint Paul was ''to the Greeks a Greek and to the Jews a Jew''; Spinoza himself was something of an ideological Machiavelli.[76]

Amsterdam, with its famous domestic tolerance, provides Spinoza with his purpose. The phrase ''an Amsterdam of religions'' meant something like ''a universal [domestic] toleration.''[77] And Spinoza praises Amsterdam as a place within which ''men of every nation and religion live together in the greatest harmony''—more as in a Canadian mosaic, perhaps, than in an American melting pot. In Amsterdam, says Spinoza, a man's ''religion and sect [before the judges] is considered of no importance'':[78]

> Now, seeing that we have the rare happiness of living in a republic, where everyone's judgment is free and unshackled, where each may worship God as his conscience dictates, and where freedom is esteemed before all things dear and precious, I have believed that I should be undertaking no ungrateful or unprofitable task, in demonstrating that not only can such freedom be granted without prejudice to the public peace, but also, that without such freedom, piety cannot flourish nor the public peace be secure. Such is the chief conclusion I seek to establish in this treatise.[79]

Spinoza wanted religious tolerance for all men. And surely he would have applauded the efforts of his former teacher, the French-born, Lisbon-raised Amsterdam Rabbi Manasseh ben Israel, to convince Oliver Cromwell to allow the return of the Jews to England. (In England, the question of the readmission of the Jews was mooted under the growing desire for religious liberty; such works appeared in the English language as Manasseh's *Vindiciae judaeorum* in 1656 and Spinoza's *Theologico-Political Treatise* in 1689.)[80] However, the sort of religious tolerance that many of his well-meaning liberal contemporaries desired was not all that Spinoza had in mind. Such tolerance had existed already in the world—as in the old Spanish Islamic *convivencia.* (In 1930 Germany, Franz Rosenzweig wrote, in *The Star of Redemption,* that "in a certain sense, Islam demanded and practiced 'tolerance' long before the concept was discovered by Christian Europe. And on the other hand love of neighbor could lead to consequences such as religious wars and trials of heretics—not aberrations but legitimate developments which will simply not fit into any superficial conception of this love.")[81] This limited tolerance by Islamic Spanish *convivencia,* which earlier I idealized for heuristic purposes, and which the tolerant Lessing idealized, during the Enlightenment, in such works as *Nathan the Wise,* was admirable.[82] But what Spinoza sought was not an ideal if limited freedom of religion based on theological principles, but rather a separation of philosophy from theology. This break would mark an end to the terrors of religious inquisition and guarantee a safe place in the world for freedom of philosophical inquiry.[83]

John Locke, who knew both the political and Cartesian writings of Spinoza,[84] himself lived in political exile from 1684 to 1689 in Amsterdam. Already interested since the 1660s in the limits of human understanding and the question of toleration,[85] Locke attended a debate there in the 1680s between the Marrano Orobio and the Remonstrant Protestant theologian Philip van Limborch. One of his first publications was a lengthy review of this debate, appearing anonymously in the Remonstrant Jean Le Clerc's *Bibliothèque universelle et historique.* (The debate is the subject of Limborch's *De veritate religionis Christianae amica collatio cum erudito Judaeo,* and it influenced his *Historia Inquisitionis,* a massive critique of the Inquisition.)[86] Locke's review is connected ideologically both to his *Fundamental Constitutions of Carolina* (1669), which emphasizes the principle of religious toleration, and to his *Letter on Toleration,* addressed to Limborch in 1689. The liberal Locke's various letters and essays on toleration[87] and his critique, in *Two Treatises of Government* (1689), of the notion in Filmer's *Patriarcha* (1680) that "Nations" are merely "distinct Families"[88] are among the earliest systematic nonuniversalist arguments for specifically religious toleration in the Christian West. As we shall see, Locke and his liberal contemporaries introduced a new particularism into the debate concerning toleration.

TOLERATION

To belong to this omnipresent shepherd, it is not necessary for the entire flock to graze on one pasture or to enter and leave the master's house through just one door. It would be neither in accord with the shepherd's wishes or conducive to the growth of his flock. Do you wonder why some people deliberately turn these ideas upside down and purposely try to confuse them? They tell you that a union of religions is the shortest way to that brotherly love and tolerance you kind-hearted people so earnestly desire.

—Moses Mendelssohn, *Jerusalem* (1783)

Anti-Semitism is the Jewish aspect of Christianity—so goes the claim. The accusation that racism and anti-Semitism in Christendom are fundamentally Jewish—an accusation encountered with reference not only to the period of the Spanish Inquisition[89] but also to European history overall[90]—generally boils down to the claim that, since Judaism is supposed to heed consanguinity and tribal affiliation and Christianity is not, racist or anti-Semitic Christians are fundamentally Jewish. This charge has been refuted for myriad historical circumstances, including those of sixteenth-century Spain.[91] That people continue to make the allegation, citing everything from the curse of Ham[92] to the rules concerning monetary interest, is not the fault of those who have refuted it. Yet students of politics and religion have been slow to emphasize that Jewish particularism heeds tribal difference in such a way that it can become precisely the basis for a realistic tolerance.

The particularism of Judaism can encourage tolerant coexistence insofar as its ancient Hebrew Commonwealth had rules recognizing that there are not only Jewish siblings but also other human beings. Those "others" have specific legal and political rights as human beings.[93] Judaism is not essentially a proselytizing religion; it provides a clear standard of goodness independent from being Jewish.[94] Good human beings who are other than Jewish—they run the gamut from "righteous gentiles" to "primitive idolaters"—are protected under the laws of the ancient Hebrew Commonwealth so long as they obey the Noachic covenant. This was recognized by such Quakers as William Penn, author of *The Great Case of Liberty of Conscience* (1671) and the *Constitution for the Colony of Pennsylvania* which guarantees "religious freedom," who wanted to see a Christian Commonwealth with tolerance toward "heretics" as well as non-Christians, and argued that Christendom should emulate the coexistence promulgated by the Hebrew Commonwealth.[95]

If it is not Jewish particularism that leads inevitably to religious discrimination and racial intolerance among Christian universalists, then what does? When we try to see past local issues—like the myths of tribal autochthony that might allow underprivileged social classes to think that they are "unpolluted children of the earth" even if they not pure-blooded Spanish noblemen (*hidalgos*)[96]—a major factor would be the doctrinal absence, essential to the universalist dogma of Christianity, of the Old Testament category of "human beings who are other than siblings." Christianity, indeed, gains its fundamental New Testament mediation between humankind and God (in the person of the man-God Jesus) only as a trade-off for the Old Testament mediation between sibling human being and nonhuman other. The absence of the category of "nonsibling human being," expressed by the formulation "all human beings are siblings, none are others," is of the essence of

Christianity. (By the same token, that absence serves to specify Christianity's general rationale for tolerating bothersome creatures: not that we should put up with human others—there are none such—but that some apparently nonbrothers may turn out to be brothers, or some apparently nonhuman creatures turn out to be human.)

The doctrine that all men are brothers was a frequently cited New Testament text in sixteenth- and seventeenth-century defenses, both Protestant and Catholic, of specifically religious toleration.[97] For example, the German freethinker Sebastian Franck insists that "anyone who wishes me good and can bear with me by his side is a good brother, whether Papist, Lutheran, Zwinglian, Anabaptist, or even Turk, even though we do not feel the same way until God gathers us in his own school and unites us in the same faith. . . . Even if he is Jew or Samaritan, I want to love him and do him as much good as in me lies." And in his 1530 translation of a Latin *Chronicle and Description of Turkey* by a Transylvanian captive, which had been prefaced by Luther, Franck added a pro-Islamic appendix holding up the Turks as in many respects an example to Christians.[98] The French Protestant Châteillon, attacking the Calvinists for their notorious persecution of Servetus in the mid-sixteenth century, quotes Franck's saying that "my heart is alien to none. I have my brothers among the Turks, Papists, Jews, and all peoples."[99] Luis de Granada, one of the few Spanish Catholics to plead openly for religious toleration, writes in 1554 that "Christian charity and zeal for the salvation of souls oblige me here to say a word in warning to those who, out of a mistaken zeal for the faith, believe that they do no sin by inflicting evil and harm on those who are outside the faith, be they Moors or Jews or heretics or gentiles. They deceive themselves greatly, for these too are brethren."[100] Likewise the Socinians' unitarian confession of faith—a Catechism published at Rakow, Russia, in 1605 that was based on the unitarian teaching of the Italian Sozzini that Jesus was not divine—defines toleration in terms of brotherhood: "In so far as we are concerned, we are all brothers, and no power, no authority, has been given us over the conscience of others. Although among brothers some are more learned than others, all are equal in freedom and in the right to affiliation."[101]

Christians often conflated species with family, as we have seen, so it is not surprising that the argument that we should tolerate others' religious views because they are our kin, or "brothers," should sometimes take the form of a claim that we should tolerate their views because they are our kind, or "human beings." During the Thirty Years' War, Hermann Conring based his appeal for toleration on the premise that "Protestants are human; they are human beings like everyone else."[102] And in the American colonies Roger Williams wrote, "I speak of Conscience, a persuasion fixed in the mind and heart of a man, which enforceth him to judge and to do so with respect to God, his worship. This Conscience is found in all mankind, more or less: in Jews, Turks, papists, Protestants, pagans."[103]

Benevolent people, then, used the rhetoric of universal brotherhood as part of an attempt to bring about a beneficent tolerance. However, their idea of universal brotherhood often constituted for them an entire politics or antipolitics, so they were generally blind to or uninterested in the totalitarian and intolerant tendency of the universalist fraternity they praised. For the traditional universalist argument that we should tolerate bothersome humanoid creatures for the reason that all human beings are brothers does not allow for conceiving a creature as being at once nonkin and kind and thus encourages us to treat as nonhuman those we might already regard as nonkin.

In any event, the creed "All human beings are siblings" is difficult to live up to. Politically speaking, one nation, or siblinghood, defines itself against another, and probably needs to. Psychologically speaking, a universal siblinghood seems to lead either to celibacy (as for the traditional Saints) or to incest (as for the heretical Corinthians).[104] So most universalists, even as they uphold in some idealist fashion the view that "all human beings are siblings," come to live as though they accept the particularist view that requires such attention to blood lineage as allows for national definition and for sexual reproduction without incest. Thanks to their peculiar combination of ideal universalism and actual particularism, however, benevolent and would-be tolerant universalists may fail to understand the multifaceted character of the category of "nonsibling human." We have already seen why they should fail. The Old Testament category of a being mediating between brother and other contradicts too discomfortingly the cherished ideals both of universal siblinghood and of a being mediating between man and God. So universalists often fail to consider what sort of political rights, if any, "human beings who are other than siblings" should be accorded. In Christianity—if not in Christendom—there are supposed to be no such beings.

In his *Jerusalem,* Moses Mendelssohn tried to veer the ideology of a universalist Enlightenment ("all men are brothers") off what he took to be its probably inevitable course toward barbarism ("only my brothers are men, all others are animals"). In the Germany of his day, Jews were pressured to renounce their faith in return for civil equality and union with the Christian majority.[105] The pressure was kindly, but it was also a form of intolerance toward nonkin. So Mendelssohn attempted to insinuate, between the two ordinary categories brother and other, a mediating term that would allow long-term "strangers" the status of human beings: "Regard us, if not as brothers and fellow citizens, then at least as fellow men and co-inhabitants of this country."[106] And Mendelssohn tried to warn his contemporaries against the sort of person who "outwardly . . . may feign brotherly love and radiate a spirit of tolerance, while secretly [and perhaps unbeknownst to himself] he is already at work forging the chains with which he plans to shackle our reason so that, taking it by surprise, he can cast it back into the cesspool of barbarism from which you have just begun to pull it up."[107]

John Locke, in his *Letter on Toleration,* writes that "it is not the diversity of opinions (which cannot be *avoided*) but the refusal of toleration to those that are of different opinions (which might have been *granted*) that has produced all the bustles and wars that have been in the Christian world upon account of religion."[108] Locke may be overstating his case. Diversity of religious opinion probably can be avoided—at least in the public sphere. After all, it is the universalizing impulse of Christianity precisely to homogenize diversity of religious opinion by converting non-Christians and Christian heretics. And, where conversion proves difficult, it is the tendency to void, or empty, Christendom of these elements by whatever means.

But *should* Englishmen put up with non-Christians? Those who said "no" bolstered their view by claiming that the ancient Hebrew Commonwealth did *not* tolerate idolatry and that the Christian polity ought to be like the ancient Hebrew Commonwealth. Locke strengthened his argument for toleration first by contradicting this claim. He points out, correctly, that although the ancient Hebrew Commonwealth did compel Jews (brothers) to observe the rites of the Mosaic law, it did not compel non-Jews (others), even idolatrous strangers. "In the very same place where it is ordered that an Israelite that was an idolater should be put to death [Exod. 22:20, 21], there it is provided that a stranger should not be vexed nor oppressed."[109] Locke then argues that the ancient Hebrew Commonwealth is, in any case, an inappropriate model for actual states. That Commonwealth was distinct from actual polities of Europe in that it was "an absolute theocracy" exhibiting no "difference between that commonwealth and the church."[110] (The same point is made by Spinoza.) And, according to Locke, "there is absolutely no such thing under the Gospel as a Christian commonwealth." Although certain states have "embraced" Christianity, all maintain an older form of government with which Christianity per se does not meddle. In this way, Locke rejects the aspect of the ancient Hebrew Commonwealth that conflates religion and politics, while at the same time using the rhetoric of Jewish particularism to bolster the practice of toleration toward non-Christians and Christian heretics or schismatics.

Events in the Middle East in the early 1990s remind us that an Islamic or Judaic particularism with a tendency toward universalism, however much it may provide one precondition for a tolerant society in a state where one religion or another clearly dominates, is not in all historical contexts gentler than a universalism with a tendency toward particularism. (It is this quality of tolerance as a merely paternalistic noblesse oblige that Kant criticizes as "haughty" in his 1784 essay "What is Enlightenment?")[111] Locke himself does not extend toleration to all groups. He excepts atheists and philosophical free thinkers from "religious" toleration.[112] And more significantly—in view of the suggestion of his English translator Popple that Locke believes in "absolute liberty"[113] as well as modern day critiques of so-called "pure tolerance"[114]—Locke also specifically excludes from toleration those persons who commit acts associable with child sacrifice and incest:[115] not only are such acts criminal under English law but even when displaced to the symbolic level of cult and ritual, they reflect or encourage a potentially intolerant catholicism. (Blood from the sacrificed Son provides the extraordinary substance of communal siblinghood to which, according to Catholic doctrine, all men essentially or potentially belong; and incest or celibacy of one sort or another is always the sign of an ideology of universal siblinghood.)[116]

Locke calls tolerance "the chief characteristic of the true church."[117] He would not seem to require it for free inquiry, as would Spinoza. His influence on its legislation was at first disappointing.[118] And, as we have seen, fraternal liberalism already had such inherent problems as are suggested by the history of the idea of brotherhood in the United States, with its connections with race slavery and civil war, and by the Dutch imperialists' cruel treatment of Africans and Indonesians.[119] Yet the larger sixteenth- and seventeenth-century discourse on toleration did guide the separation of church and state informing, we suppose, the efforts of some modern democracies to thwart religious inquisitions and witch hunts, and to respect what we call "the rights of others." In this sense the influence of the exploration in Amsterdam of religious and national toleration probably extends to present liberal democracies that concern themselves a little less with the circus tragedy of the bull, so discomforting to Queen Isabella, and a little more with the agonies of humankind.

3

THE FORKED TONGUE

OR

The Road Not Taken in Québec

Mon pays,
Ce n'est pas un pays,
C'est l'hiver.

—Gilles Vigneault, from his song "Mon Pays"

BOTH SIDES AGAINST THE MIDDLE

I expected to find a contest between a government and a people: I found two nations warring in the bosom of a single state: I found a struggle, not of principles, but of races; and I perceived that it would be idle to attempt any amelioration of laws and institutions until we could first succeed in terminating the deadly animosity that now separates the inhabitants of Lower Canada into the hostile divisions of English and French.

—Lord Durham, *Report on the Affairs of British North America* (1839)

Popular ways of speaking about language do not differ much from those about race or even species. Not only are the linguistic and natural historians' terms for genus and species often the same, but, more important, as Charles Darwin observed, "the proofs that [different languages and distinct species] have been developed through a gradual process are curiously the same."[1] In its dialectics of universal-particular and terminus-origin, the ideology of linguistic historiography (Grimm) differs little from that of species historiography (Darwin) or racial historiography (Gobineau).

A case in point is the rhetoric of the universal in natural history and linguistics. Many natural historians hypothesize a human "monogenesis"—a single genetic origin of all presently living human beings—and sketch a family tree that illustrates a supposed divergence of humankind from a single DNA stock. The hypothesis of

"Eve, mother of humankind" in the logic of this natural history,[2] like the apostle's claim that "God made of one blood all the peoples of this earth" in the logic of Christian kinship,[3] is both comfortingly unitarian ("All men are my consanguineous kin") and critically divisive ("Only my consanguineous kin are brothers, all others are animals"). Many historical linguists likewise hypothesize a single original source or locale for all human languages, some claiming to have "reconstructed the ancestor of all living languages"[4]—the pre-Babel Ur-language that seventeenth-century theorists called "Adamic"[5] and modern linguists call "Nostratic." Belief in the historical existence of this unitarian language is "a kind of religion [that] emphasize[s] the unity of humankind and the need of brotherhood,"[6] yet here, too, the ideal ("All human beings speak variants of the same language that we speak") turns all too easily into the particularist political dogma ("Human beings are the creatures that talk our language, all others are animals"). *Nostratic* means "not yours" as much as "ours."

Exactly how language mediates race or nation depends on whether the state or *polis* is conceived as essentially unilingual or multilingual. On the one hand, there are legally unilingual countries, like the United States, where citizens have political rights as individuals rather than as members of one or another particular linguistic or racial group. In the United States, for example, every American citizen has the same universal right to go to school or to argue in court, but there is no guaranteed right in the United States to attend school or to plead in court in the language of one's choice (if one's choice is not English). When American courts do grant permission to plead in a non-English language (as sometimes happens where there are many people who speak Spanish), and when differences frequently occur between the meaning of the law as written in English and its meaning in the non-English translation, the courts make their disposition according to the "original" English.

On the other hand, there are legally multilingual jurisdictions in North America where citizens have rights as members of particular racial or religious groups. In Québec, for example, citizens have group rights based on the division, according to the terms of the constitutional British North America Act (1867), of most of the people into Catholic and the Protestant groups or more precisely, into the English and French Catholic group and the English and French Protestant (or Huguenot) group. The written law exists—indeed, must exist—in equally valid or "original" form in both English and French.

The French and English Québécois have a historical right to attend school in their "native" religion. (The British North America Act established two state-supported religious schools boards, the Protestant board overseeing separate schools for English and French Protestants and the Catholic board overseeing separate schools for French and English Catholics. There are no secular or universal schools of the American sort.)[7] From the group right to attend school in one's native religion evolved the group right to attend school in one's native language. For a century and more after the signing of the British North America Act, Protestant church leaders thus argued for English-language rights, basing their pronouncements on the historical religious rights of the Anglo-Saxon minority—that is, the rights of the group of persons who are British "stock," or *britannique de souche*—and Catholic leaders similarly argued for the French-language rights of the *Québécois de souche.*[8]

While the United States heeds equal rights of individuals as members of an ideally unilingual nation, generally subordinating the status of citizens as members of particular linguistic or racial groups, Québec heeds the rights of the English and French groups or the Protestant and Catholic groups, generally subordinating both the individual rights of citizens and the group rights of non-English and non-French linguistic or racial groups. (Exceptionally, the United States grants to some individuals as members of racial or linguistic groups an anomalous treatment under the law: The fact that blacks and Hispanics are covered by various affirmative-action rulings tends to lessen the difference in American ideology between race and language, as suggested by the occasionally racially ambiguous term *Hispanic*. There are parallels to the lessening of the difference between religion and language in Québec.)

What happens to those people in Québec who are, by consanguineous or linguistic generation, neither French nor English, or neither Catholic nor Protestant?[9] Québec politicians in the late 1960s and 1970s followed the policy of the traditionalist Union nationale, an appropriately named political party of earlier decades, in assigning to every person either a French or English mother tongue (*langue*). This meant that the state had to determine the language (*langue*) in which a particular citizen felt him- or herself to be fluent or actually was so—in order to assign the appropriate language of school instruction or to designate the proper agency of school tax collection. Since determination of the mother tongue generally also meant determining the national or racial origin of the parent (mother), the ideology of language policy in Québec was conflated explicitly with issues of consanguineous "nation" and race.

In the United States an immigrant becomes an American citizen in the civic ritual of "naturalization" by becoming a member of the one group of people reborn, or regenerated, as Americans when the Founding Fathers declared their independence from Great Britain. In Québec, on the other hand, the immigrant becomes a member of one people or the other (English or French). In the 1980s, for example, immigrants from non–English-speaking countries (allophones) and from most English-speaking countries (anglophones) were classified as "French" (francophone) for educational and taxation purposes. Thus, a person with Greek-speaking, Greek Orthodox parents was called "French." By the same legal fiction, an English-speaking person from Singapore who was not *britannique de souche* was classified as French even though he might speak only English. (This racialist aspect of Québec politics, linking linguistic assignation with descent from a particular genetic stock, matches a discrimination already inherent in the British North America Act itself.)

According to a similar fiction, most Jews in Québec were classified as "English Protestants" even when they were, like the Moroccan *sefarads*, "native" Arabic- or

French-speakers. The fact that the Jews of Québec, whether *ashkenaz* or *sefarad*, both constituted something like a single nation as a group (with a single sacred written language) yet spoke various native languages besides French or English as individuals, tended to challenge the thesis—dear to European linguistic nationalists—that a common spoken language is the main distinguishing characteristic of nationhood.

It is a peculiar characteristic of Québec—one that helps to clarify its often misunderstood political and linguistic ideology—that its struggle over language and politics concerns not only immigrants who have no legally recognized language (i.e., who do not have English or French) but also true bilinguals (i.e., who have both English and French). Questions about assigning nationality to people who are, in linguistic terms, perfectly bilingual—or who have, in racial terms, one English and one French parent—hold the same position in the speculative theater of Québécois ideology as do questions about Siamese twins and interracial changelings in a United States concerned with problems of consanguineous and national identity. Moreover, linguistic miscegenation is often debated in Québec much as racial miscegenation is debated in the United States. (Some French Canadians called themselves Canadian *nègres*.)[10] For example, in 1972 at the French-language Université de Montréal a debating union resolved that "English in twentieth-century Québec is a politically oppressive language." Similarly, in 1973 it was resolved at the English-language McGill University that "English is an essentially oppressive language" (as opposed, say, to the language of an imperialist or colonizing political oppressor, like French in twelfth-century England). Finally, it was resolved at Concordia University that "the historical interaction between French and English make them essentially one language with a shared future terminus." In this last resolution one topic would be "FrAnglais" (French turning into English) or FrEnglish (English turning into French).[11] Another would be "bilingual publicity"—the signs of the times.

SIGNS OF THE TIMES

Controversies about the language of public advertising loomed large in the headlines of Montréal's newspapers when, in 1955, the federalist Canadian National Railroad decided to call Montréal's new downtown station/hotel complex The Queen Elizabeth in honor of the English monarch, and the provincial Ligue d'Action nationale collected two hundred thousand signatures calling for the use of a French name such as Le Château Maisonneuve in honor of the French founder of Montréal.[12] Pierre Laporte, in a well-known essay published in the French journal *L'Action nationale* entitled "'Queen Elizabeth?' . . . Never!," complained that "our cities are plastered with English names."[13] I was eight years old at the time, living with my parents and siblings on TransIsland Road in a duplex situated between French Catholic and English Protestant neighbors.

In 1959 the popularist Société Saint-Jean Baptiste de Montréal was arguing that English should never appear publicly without French. "It is necessary," wrote the Société's chairman, "to prohibit English unilingualism on everything that reaches the public: signs . . . billboards, menus, instructions, etc."[14] By the mid-1960s the linguistic face of most of the island of Montréal was becoming distinctly French. For example, a major downtown department store, Morgan's, became La Baie/The Bay; most new public buildings had French names; and there were serious discussions about laws, eventually enacted in the early 1970s, to require bilingual consumer contracts and company names.[15] In 1974 the Liberal party premier of Québec, Robert Bourassa, oversaw the legislative passage of Bill 22, which declared French as Québec's "official language" and required that all billboards and public signs include French.[16]

The momentous alternatives facing Québec in the mid-1970s was whether to be bilingually French and English—as ancient Babylonia, say, had been both Akkadian and Sumerian—or unilingually French. The desire or need to choose between bilingualism and unilingualism was the key element of the Québec scene when I returned there in 1973, after eight years of study elsewhere. Like a visual historian of a disappearing popular culture, I photographed hundreds of bilingual signs; like a writer for a litterateur's academic journal with political pretensions, I penned an essay entitled "La publicité bilingue au Québec: une langue fourchue" (as it appeared in the *Journal canadian de recherche sémiotique*) or "The Forked Tongue: Bilingual Advertisement in Québec" (as it appeared in *Semiotica*).

In the mid-1970s the broad outline of the unilingual "solution" to Québec's language problems was already unmistakable. In 1963, after all, the Société Saint-Jean Baptiste had argued that private companies and corporations in Québec should only have French names.[17] (The terrorist Front de Libération du Québec, or FLQ, was founded in the same year.) And in the mid-1970s there was pressure from an increasingly "separatist" populace—represented by the new Parti Québécois—to pursue a unilingual policy.

The Parti Québécois came to power in 1976, and in 1977 one of the party's first acts was Bill 101—called "The Charter of the French Language"—which proclaimed French as the "public language of Québec."[18] The Parti Québécois minister, Camille Laurin, claimed that Bill 101 was a "law of the people coming from the core of our collective history."[19] He explicitly appealed to the francophone *Québécois de souche*—those who claimed to trace their bloodline to the "original family" of seventeenth-century immigrants. Promising his constituents that "the Québec we wish to build will be French in essence,"[20] Laurin called for a Québécois "reconquest of French."[21] To match the military conquest of New France by England in 1759, there would be a linguistic reconquest of French in 1977 that would reclaim the "genuine" French of old Québec and assimilate Québec's allophone and anglophone populations.[22] Peoples of English blood stock—*britannique de souche*[23]—would be encouraged to emigrate, and indigenous Amerindian and

Inuit populations, which might have made good claims to being the true "original" settlers of the land, were now called *autochthone,* partly in order to distinguish their nativity from the fictive originality of the Québécois.[24]

Laurin's call for a Québécois "reconquest of French" often relied on the traditional social fiction of mother tongue—hence mother nation—in much the same way that sixteenth-century Christian rhetoric calling for the reconquest of Spain depended on the notion of blood purity. In Québec the rhetoric of reconquest and expulsion—of bloodlines and origin—influenced political discourse for a decade. The nationalist *Québécois de souche,* calling for others' "francisization" or expulsion, remembered the brutal expulsion by the British of the Québécois' Acadian kin two centuries earlier.[25] (Just at the time that The Queen Elizabeth Hotel controversy hit the Montréal newspapers in 1955, twenty thousand descendants of the exiled Acadians were gathering in the Canadian maritime provinces to commemorate the two hundredth anniversary of the Acadian Expulsion.[26] English Governor Charles Lawrence's 1755 proclamation calling upon Acadian francophones to meet at the church in Grand Pré, pronounced in both English and French, was a memorable sign of the times.)[27] Seeking to avoid a cajun-like extinction of language, and fearing bilingualism as the antechamber of cultural annihilation, the new Québécois chose to use the language of expulsion against others rather than to be assimilated themselves.[28] "I am a two-pawed singer / who yaps his beautiful songs / for a race on the way to extinction" wrote the poet Jacques Michel.[29] During the crisis of October 1970, when Prime Minister Trudeau declared the infamous War Measures Act, the FLQ *Manifesto* proclaimed that "the day is coming when all the [English-speaking] Westmounts of Québec will disappear from the map."[30]

Camille Laurin, a psychiatrist-turned-politician, figured that Bill 101's language provision was the appropriate treatment for Québec's national sickness. That disease, he said, was "linguistic degeneration" and all that it entails. Degeneration was an old theme in Québec's political discourse,[31] generally involving various quandaries about which dialect to use: Canayen, FrAnglais, the French of France, or *joual.* (*Joual,* sometimes called *cajun,*[32] was at the center of the ideological controversy. The term derives from local pronunciations of *cheval,* meaning "horse"; the dialect it designates is sometimes compared with American "jive-talk" or "gumbo.") Focusing on the language issue, Hubert Aquin had complained in 1962 in *Liberté* that Québec writers' "fatigue culturelle" was nearly incurable. "As soon as I began to write," Gérald Godin had written in *Parti Pris* in 1965, "I realized that I was a barbarian, i.e., a foreigner, according to the etymological meaning of the term. My mother tongue was not French but *franglais.* I have to learn French almost as a foreign language."[33] Gaston Miron, in a 1973 essay, had discussed the linguistic schizophrenia and alienation that informed the diseased cultural life of Québec; briefly considering the "debilitating effects" of bilingual signs on the "purity" of the "French language," he had called for a new "linguistic decolonization."[34] Similarly, in 1975 Jacques Godbout had claimed that the "ideology" of *joual* was an "infantile disease of nationalism."[35] It was out of this sociopsychiatric diagnostic tradition focusing on language that Laurin's colleague, Gaston Cholette, came to stress "the psycho-cultural importance" of unilingual signs. The eradication of bilingual signs, by purifying the language of the tribe, would rehabilitate the self-esteem of the "essential nation."[36]

The Parti Québécois's legislation against bilingual *publicité* in Bill 101 was aimed not only at curing the ills of the Québécois "nation"—at turning its *hiver,* in Vigneault's words (cited as the epigraph to this chapter), into a *pays*—but also at "converting" to the French lingua franca both the historically anglophone and the various allophone communities. René Lévesque, the Parti Québécois premier, remarked on the effect of bilingual *publicité* on these nonfrancophone communities in stating that "in its own way, each bilingual sign says to an immigrant: 'There are two languages here, English and French; you can choose the one you want.' It says to the Anglophone: 'No need to learn French, everything is translated."[37] With a few exceptions mandated noblesse-oblige style by Premier Lévesque, the Parti Québécois's "message" to both the historically anglophone and the immigrant communities was that French was *the* language of the province. The message had the executive backing of the Commission for the Protection of the French Language, which pursued and prosecuted violators of the law. Anglophone and allophone critics called its detectives "language police"; the Montréal-born novelist Mordecai Richler, writing from the United States, called them "tongue troopers."[38]

In 1985, with the defeat of the Parti Québécois, the Liberal party's Bourassa reassumed the premiership of Québec. Trying to arbitrate the conflicts among the various linguistic communities, Bourassa hinted that he would create bilingual "territories" on the island of Montréal, where the English and French languages would appear together again legally on the same sign.[39] But in the wake of nationalist reaction to the Canadian Supreme Court's denial of the right of Québec courts to outlaw any bilingual signs, Bourassa instead passed through the Québec legislature a bill authorizing unilingual French signs on the outside of buildings and permitting bilingual signs with "clear French predominance" on the inside. Not surprisingly, the ambiguity of such "territorial" plans led to general apprehension. Who would distinguish *le dedans* from *le dehors,* and how?[40]

Viewed from the vantage point of the early 1990s, the *visage* of Montréal is more French than ever, and the number of anglophones and *britanniques de souche* continues to decline. ("De plus en plus, le Québec se francise," writes Uli Lochner.)[41] The Meech Lake Accord, though it lacks specific provisions about language and culture,[42] now stipulates at the federal level that Québec is a "distinct society." Yet French fear of anglicization remains. Some language nationalists complain about banners like "Bâtir ce pays sur le rock" (Build this country on rock and roll);[43] others complain about the debilitating effect of "anglicized French" on public signs[44] or argue that all bilingualism "deforms" Québec's "collective unconscious."[45]

It is not surprising that the problem of specifically bilingual public signs should remain: the essentially political dilemmas that underlie such signs endure. A bilingual agenda, sometimes based upon the individual rights of all persons (e.g., freedom of expression) and at other times based upon the group rights of the historically anglophone community (the modified British North America Act), is still pitted against a unilingual agenda informed by the national rights of the *Québécois de souche*.[46] "No compromise will satisfy both Alliance Québec and the Société Saint-Jean Baptiste," wrote an astute Dion in *Le Devoir* in 1989.[47] Yet despite the continuing face-off—and sporadic events in the past decades that range from name calling to murder or assassination—there remains in Québec a genuine political decency and linguistic tolerance.[48]

VOLKSWAGEN BLUES

We step and do not step into the same rivers; we are and are not.

—Heraclitus

The essay on bilingual advertising that follows was originally delivered in 1973 as a university lecture. At that time people in Québec were reaching beyond questions of individual free speech and toward questions of tribal or linguistic nationalism and economic class. Québec was vacillating between national separation from and federation with the rest of Canada. Separation was endorsed for nationalistic reasons by a labor movement that assumed, in an era of pseudosecularization, an anticapitalist rhetoric reminiscent of the older church.[49] One symptomatic debate was whether non-French signs and bilingual signs should be allowed in the public sphere.

"The Forked Tongue" infuriated both parties in this debate because it pitted both sides against the middle. The editors of the French journal *Liberté* and of the English journal *Our Generation* criticized "The Forked Tongue" in diametrically opposite ways. The former argued that I should have taken the side of nationalist "labor," which wanted to allow only French signs, and the latter argued that I should have taken the side of antinationalist "capital," which wanted to allow all signs. In fact, "The Forked Tongue" mainly called for a tolerant bilingualism that might have helped somewhat to reveal—through the intensive form of public translation that is intermediated bilingual publicity—the then potentially fruitful political reciprocity of French and English in Québec. Coleman argues that "language policy through the mid 1970s served mainly to legitimate capitalism in Québec by taking the foreign (i.e., English) face off of it."[50] (It is an indication of the political fury of those days that after the publication of "The Forked Tongue," an anonymous writer threatened to kill me if I wrote again about bilingualism.)

We can't step into the same river twice, at least not without ambiguity, but that doesn't mean we should let sleeping dogs lie. Academic journalists with political predilections often prefer to let the dismembered past lie for them. (An extreme example would be Paul de Man writing in a bilingual Belgium during the fury of the Nazi period.)[51] But Québec in the 1990s is still living out the political promises and contradictions of the 1970s. Interlinguistic mediation, though now generally removed from the public *visage* of Québec, beguiles its literary writers and critics all the more. There has been an increasing fascination, in recent years, with the diglossia in Anne Hébert's *Kamouraska*;[52] a renewed scholarly respect for *joual;* a rejuvenated interest in the theory of translating into English French texts that contain English words and translating into French English texts that contain French words;[53] much concern with the role of "transfugee" writers;[54] a burgeoning movement to increase the number of French anthologies of English writers;[55] and a renewed focus on publishing bilingual journals.[56] Despite all this, most social and literary critics ignore or obscure the epigraphic, literary quality of bilingual public advertisement; they mask its bilingual aspect with bureaucratic neologisms and often endorse outright censorship in the name of "the nation's health." In this way cultural criticism is denied its highest political purpose and becomes its own victim.

In the 1990s the tribal brotherhoods of Québec still search nostalgically for a national *régénération* from the ashes of the decisive Battle of the Plains of Abraham. Every Québec schoolboy and schoolgirl, both French and English, can recount the fateful day in 1759 when the opposing generals, Montcalm and Wolfe, died within hours of each other, almost as if the generals were spoofing—recapitulating in grand Hegelian historical style—the simultaneous deaths of the Oedipal twin brothers Polyneices and Eteocles.

In Jacques Poulin's Québécois novel *Volkswagen Blues* (1984), with its memorable linguistic code-switching,[57] language remains, as Heidegger is quoted as saying, "the house of being." But the house of language in *Volkswagen Blues*—the place where we are *chez nous* (at our own home) in the blues of America—is a metaphorical ferry on the move: a Volkswagen ("nation's car"). In Poulin's novel it is the character named Saul Bellow—a Jew from Montréal who long ago wandered to the United States—who has the drollest words about the tribal quality of the politics of language:

> "When you're looking for your brother, you're looking for everybody!" *Il traduisit lui-même sa phrase dans un français hésitant:*
> *—Quand vous cherchez votre frère, vous cherchez tout le monde!*[58]

BILINGUAL ADVERTISING IN QUEBEC (1974)

> *Translation ultimately serves the purpose of expressing the central reciprocal relationship between languages. It cannot possibly reveal or establish this hidden relationship itself; but it can represent it by realizing it in embryonic or intensive form.*
>
> —Walter Benjamin, *Illuminations*

An advertisement is a trope which turns our attention toward a particular word or object. This trope and the word or object toward which it turns our attention are rhetorically as interdependent as a riddle and its solution, or as an epigraph and that on which it stands or might properly stand.[59]

In the modern world, most advertisement turns our attention to brand names or to commodities. Modern advertisements, indeed, seem to constitute a unique language of commodities, the international use of which informs the modern world. The influence of this language is most apparent where bilingual and multilingual advertising is practiced.

Bilingual advertisement turns our attention toward a word or object in two languages at once. Like interlinguistic translation, bilingual advertisement depends on and itself expresses the reciprocity between two languages. Such reciprocity is most apparent in written advertisements, such as those in bilingual and commercial Québec. The ordinary exchanges of linguistic meaning in these signs are symptomatic of extraordinary exchanges in the language of commodities and in the political economy in general.

There are two principal kinds of mediators between English and French messages in the bilingual signs of Québec: objects outside the sign and semiotic units within the sign.

Objects Outside the Sign

Most bilingual signs in Québec present separate English and French messages which are mediated by the one object to which they both are supposed to refer. This object, on which the messages may or may not stand, is essential to the complete signification of the bilingual sign.[60] The sign "PONT/BRIDGE,"[61] for example, is supposed to refer to an object (a bridge) which mediates between "PONT" and "BRIDGE." Signs which depend on similarly objective mediators include "CUL-DE-SAC/DEAD END," "COURSES AU GALOP/THOROUGHBRED RACING," and "MCGILL UNIVERSITY/UNIVERSITÉ MCGILL" (see fig. 3-i). In these signs, one message is a complete or independent translation of the other. The possibility of such translation depends on the supposed existence of objects (e.g., a bridge, road, racetrack, or campus) which the words on the sign are supposed to represent. These objects, outside the sign, are mediators or translative agents.

Semiotic Units Within the Sign

Some bilingual advertisements in Québec contain linguistically different messages connected not (only) by objects outside the sign, but (also) by semiotic units within or on the sign. In the sign "UNIVERSITÉ <u>MCGILL</u> UNIVERSITY,"[62] for example, the French message, UNIVERSITÉ MCGILL" and the English message "MCGILL UNIVERSITY" are connected by "McGILL," which mediates semiologically between and participates in both messages. In Québec there are several kinds of semiotic mediators: (1) pictures; (2) numbers; (3) literal intersections; (4) words; and (5) brand or corporation names. A semiotic mediator, as we shall see, turns or "adverts" the reader's attention away from the (other) words on the sign (which, however, it may help to explain or modify) and toward itself as a semiotic unit (e.g., "McGILL") or toward the object (e.g., the campus of McGill University) which that semiotic unit is supposed to represent.

1. *Pictures.* In the road sign "PEDESTRIAN X PIÉTONS" (see fig. 3-ii), X is a pictorial or diagrammatic representation (of a highway intersection) which mediates between and participates in both English and French messages at the same time. Similarly, in the sign "PONT BRIDGE" is an interlinguistic mediator, whether the reader supposes to be an interpretative aid for reading the words of the sign, or to be, like these words, a signifier of the object (a bridge) to which the sign directs our attention.

2. *Numbers.* Things apparently signified by pictures and diagrams often vary from one language group to another. There is less variance in the signification of numbers, which are among the most common mediators in Québec advertisements. In the sign "GABARIT 9' CLEARANCE," the number "9" and the diacritical mark "'" mediate between the English message "9' CLEARANCE" and the French message "GABARIT 9'." The mediating number "9" is printed in the prominent color red, and is located centrally after "GABARIT" and before "CLEARANCE," the correct location for a number in the syntactically different French and English languages. The sign "TAUX/RATES" (see fig. 3-iii) expresses similarly syntactic reciprocity between French and English. Located in the middle of the sign and printed in red, cardinal numbers signifying monetary units mediate between two linguistically different columns of ordinal numbers signifying chronological units. The numerical mediators in "VITESSE MAXIMUM 25 MAXIMUM SPEED," "À LOUER 381–3373 FOR RENT," and "NEW BRUNSWICK RBC-000 NOUVEAU-BRUNSWICK" also connect and complete French and English messages.

3. *Literal Intersections.* In bilingual advertisements French and English words sometimes intersect at common letters. In Figure 3-iv, for example, "ICE" and "GLACE cross at "C." The problem of whether the French or English word predominates visually is, to some extent, resolved by printing the French message horizontally (the way we read) and the English message vertically, and by coloring "C" in a color identical to that of the other letters in "ICE" but different from that of the other letters in "GLACE." Figure 3-v illustrates a similar crossword puzzle, in which the letter "O" connects and literally participates in the French message "2 POUR 25" and the English message "2 FOR 25."

4. *Words.* There are two kinds of ordinary verbal mediators: a word which appears alone on the sign and a word which connects an English and French message.

A word which appears alone on the sign: The *Dictionnaire FrAnglais Dictionary* states that "more than 8,000 words are alike, or almost the same, in English and in French."[63] The orthographic (if not phonetic) similarity between the English and French languages makes possible the employment of the same verbal unit as a sign to both English and French readers. "TELEPHONE" (see fig. 3-vi), "PROVISIONS," and "SERVICE" are examples of signs which express this similarity between French Canadian and Canadian English, although signs such as "ESSENCE" suggest how

i. *MCGILL UNIVERSITY/UNIVERSITÉ MCGILL* (*entrance sign*)

ii. *PEDESTRIAN X PIÉTONS* (*pedestrian sign at crossing*)

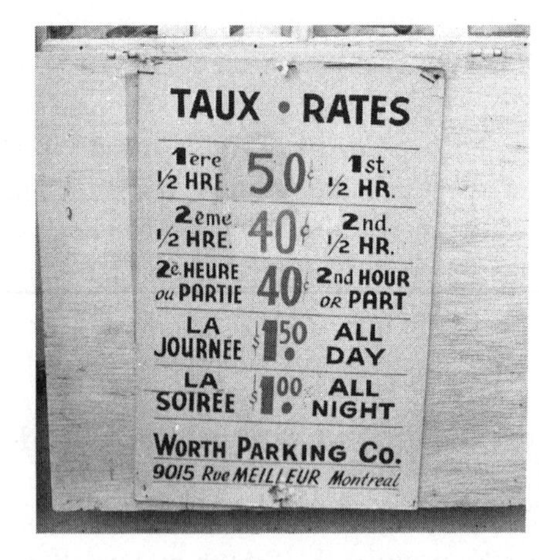

iii. *TAUX/RATES* (*parking lot sign*)

iv. *ICE/GLACE* (*ice-dispensing machine sign*)

v. *2 POUR 25* (*balloon man sign*)

vi. *TELEPHONE* (*telephone booth sign*)

Fig. 3. Bilingual signs. Montréal, 1973.

vii. *UNIFORM BOUTIQUE D'UNIFORMES* (commercial sign)

viii. *AUTOBUS EXC. BUSSES* (no parking sign)

ix. *EDIFICE CIL HOUSE* (entrance sign at corporate headquarters)

x. *YOUR TELEVISION IS INCOMPLETE WITHOUT CABLE TV LTD/VOTRE TÉLÉVISION EST INCOMPLÈTE SANS CABLE TV LTD* (billboard)

xi. *NETTOYAGE À SEC CAPRICE INC. DRY CLEANERS*, (proprietor's sign)

xii. *APT MEDARD APT*

the same verbal unit may signify to French readers one thing (gasoline) and to English readers another thing (lofty substance).

A word which connects an English and French message: FrAnglais mediators between two linguistically different messages may be nouns, adjectives, prepositions, or verbs. Typical examples of mediation by nouns include "EXCELLENT CANADIAN CUISINE CANADIENNE PAR EXCELLENCE," "MONTREAL OSTOMY CENTRE D'OSTOMIE DE MONTRÉAL," "SNOW AND ICE FALL FROM ROOF DANGER FONTE DE NEIGE ET DE GLAÇONS," and "UNIFORM BOUTIQUE D' UNIFORMES" (see fig. 3-vii). Such signs suggest syntactic and lexical similarities and differences between French and English. The last two, for example, show how the French possessive is adjustable to English syntax.

French nouns usually precede and English nouns usually follow adjectival connectives. One curious example is "GIBEAU ORANGE JULEP." In "VITESSE MAXIMUM SPEED," a verbal mediator modifies two nouns, and in "VITESSE MAXIMUM 25 MAXIMUM SPEED" a numerical mediator is in apposition to the two modified nouns. AUTOBUS EXC. BUSSES (see fig. 3-viii) employs an exceptional adjectival mediator which both the English and the French words precede. The English reader may interpret this sign as "BUSSES EXC." and the French reader as "EXC. AUTOBUS," so that the same semiotic unit ("EXC.") operates as a noun (in the French "exception d' ") and as an adjective (in the English "excepted"). The difficulty of reading such signs, which are supposed to be comprehensible even to speeding bilingual motorists, is often like that of reading the book of FrAnglais poetry entitled *Mots d'heures: gousses, rames,* in which French verses reproduce phonetically the English verses of *Mother Goose Rhymes.*[64]

5. *Brand or Corporation Names.* In the bosom of a single continent, two nations war about the linguistic affiliations of brand name mediators between linguistically different messages on bilingual signs. Some nationalists argue, for example, that the supposed economic oppression of French Canadians by English Canadians is reflected in the widespread use of brand name mediators affiliated with English. In "EDIFICE CIL HOUSE" (see fig. 3-ix), for example, CIL both is an interlinguistic mediator and represents the initials of a series of words in English, one of the languages between which it mediates (English and French). This sign, perhaps, does reflect the imprudent refusal of "Canadian Industries Limited" to accommodate French readers. This accommodation is often practiced by international corporations, such as the "International Civil Aviation Organization" or "Organisation d'Aviation Civile Internationale," the sign of which, "ICAO OACI," shows two series of letters in reverse order. The more typical signs of national companies, however, do not so tamper with their initials or names. "PHARMACIE KAREN'S PHARMACY" makes impossible any accommodation of the name "Karen's" to the French language, which cannot easily absorb the English possessive. "FARINE *FIVE ROSES* FLOUR," which could have been written internationally as "FARINE 5 ROSES FLOUR," is a similarly English brand name mediator. Such signs probably would have the subliminal

political tendency of convincing speakers of both languages that English brand names can mediate French and English languages, and so solve the linguistic and perhaps, in some ideal fashion, also the economic problems of Québec. Some signs, such as "YOUR TELEVISION IS INCOMPLETE WITHOUT CABLE TV LTD/VOTRE TÉLÉVISION EST INCOMPLÈTE SANS CABLE TV LTD" (see fig. 3-x) and "LA CIE GENERAL SUPPLY DU CANADA LIMITÉE / THE GENERAL SUPPLY COMPANY OF CANADA [LIMITED]," would seem to accord to the English brand name the power to bridge (language differences) and to complete (sentences).

Not only English but also French and FrAnglais mediators are enhanced by their role in bilingual advertisements. Examples of such mediators include "SERVICE DE RENOVATION METROPOLITAIN HOME SERVICE," "NETTOYAGE À SEC CAPRICE INC. DRY CLEANERS" (see fig. 3-xi), and "POSTE DE LA SALLE/STAND FOR LA SALLE." The political implications of these and other brand name mediators can hardly be determined from their linguistic origins or etymologies, which do not consider the peculiar movement of brand names into all ordinary languages. "CIL," for example, has now become a word in Canadian English (in which "CIL" was once merely the initials of a series of words) and in French Canadian (in which it was once linguistically meaningless). Sometimes company directors purposefully change the names of their corporations in order to mask their linguistic origins, and so placate linguistically nationalistic consumers. "Libby's," for example, changed its name in French to a French idiom. The directors of one international corporation chose its quintessentially supernational name, **EXXON**, when advertisers argued that it was a phonetically meaningless sound, unobjectionable to most linguistic groups. **EXXON** was supposed by zealous advertisers to have the tendency of inscribing internationally the supposedly blank slates (*tabulae rasae*) of the spirits of nations not only with a new letter (**X**) but also with a new rhetorically powerful superword.

The study of the political power of brand names and of their peculiar definition of genus and species must focus not only on their movement into ordinary languages, but also on their movement, even in a unilingual country, from the linguistic to the supposedly superlinguistic level. In bilingual Québec, the superlinguistic aspect of the language of commodities and the masquerade of brand name mediators as being superior to English and French have specific ideological tendencies. We have seen that the semiotic mediator in a bilingual advertisement is a nucleus or focal point of the advertisement, especially when readers are bilingual (and so read both messages), or when the mediating symbol is centrally located or more brightly colored or larger than the other semiotic units on the sign. Bilingual advertisement often tends to enhance the mediator with the intellectual privileges of presenting the most important part of the message of the advertisement and of conjoining two ordinarily separate languages. The mediator "serves as a measure which makes linguistically different messages commensurable and so reduces them to equality."[65] A mediator is a universal equivalent standing in a relation to the words it translates like the relation of money to the commodities it exchanges. The translating mediator as semiotic unit (in addition to the thing which this unit is supposed to signify) is a kind of equalizing agent among languages. The curious sign "APT MEDARD APT" (see fig. 3-xii), for example, would equalize subliminally "appartement" and "apartment." Similarly, "ENJOY COCA-COLA RAFRAÎCHISSANT" would equalize "enjoy" and "rafraîchissant," words both linguistically and syntactically different. Such mediators are supposed by many (as we have suggested) to rise above and solve (by an act of translation or linguistic equalization) a political problem implied by the antagonism between two linguistically different messages. In bilingual Québec, the capacity of the language of wares so to equalize (on the ideal plane) tends to create a society in which "the only comprehensible language which we [English and French human beings] can speak to each other or which can mediate between us, is [not that of ourselves, but] of our commodities in their mutual relations."[66] Or so it seems. Even more in a bilingual than a unilingual country, it is apparently not we who speak with each other about commodities, but rather ventriloquistic commodities that speak a unique and alien *Warensprache* through us.[67]

Commercial advertisement employs a verbal trope (e.g., a translating mediator from one language to another) in order to precipitate a commercial exchange (e.g., a transfer of commodity from seller to buyer). The relation between tropic exchanges of meaning and economic exchanges of commodities implies a major social role for advertisement. Unlike the epigram, which turns our attention to contemplating a thing, commercial advertisement turns our attention to purchasing a thing. The advertisement pretends that it is a call to action, but purchase, the action to which it directs us and which it presents as a salve for the wound of desire, is an unproductive exchange. The most insidious ideological tendency of modern commercial advertising is not the witting lie, but rather the often unwitting tendency, tellingly emphasized in bilingual advertisement, to suggest that linguistic and political problems are solvable by economic consumption.[68] This tendency, often praised by advocates of "consumer society," ignores the fact that language and advertising (which is a part of language) are more properly ideal political productions than commodities.

The interlinguistic mediator in bilingual advertising in Québec is a telling symptom of the ideological tendencies of the language of commerce and advertisement in general. Nowadays, however, mediated bilingual advertising is infrequently practiced. Most new signs contain linguistically separate, unilingual or nonverbal messages, despite the fact that mediation distinguishes a brand name and economizes on space. Many forces may help to explain this demise of the commercial and noncommercial FrAnglais mediator. So-called separatism, for example, associates national liberation with linguistically separate or unilingual messages; and consumerism associates consumer protection with wholly independent or unmediated messages.[69] Federal studies of biculturalism, which might be expected to encourage bilingual advertising, have failed even to notice interlinguistic mediation, and so seem, like reports from nineteenth-century colonial administrators, to certify or approve a supposedly unavoidable war between two nations.[70] The language of commodities continues to monopolize ever more immediately the life and thought of both English and French Canadians. It is sad that interlinguistic mediation, which reveals as well as conceals that monopoly, continues to be ignored or attacked by those who do not understand that the forked tongue of advertisement in Québec is not bilingual but rather commercial.

4

FROM DORMITION TO NATION

OR

The Sinful Soul of England

It is, or it is not, according to the nature of men, an advantage to be orphaned at an early age.

—Thomas De Quincey, *The Caesars*

INTRODUCTION

Elizabeth's childhood, so goes the story, must have been unhappy. She disappointed her father (Henry VIII) by not being born a boy, was bereaved when her mother (Anne Boleyn) was beheaded for sibling incest, was declared a bastard by her father, who eventually exiled her from the court, and had four stepmothers (Jane Seymour, Anne of Cleves, Catherine Howard, and Catherine Parr). After her father died, the orphaned Elizabeth's ten-year-old half-sibling (Edward, son of Jane Seymour) became king. Her onetime stepuncle (Thomas Seymour) became her stepfather by marrying, in indecent haste, one of her stepmothers (Catherine Parr). This uncle-father, for his subsequent seduction of Elizabeth, or for an attempt to marry her, was executed by his own brother (Edward Seymour, "Protector of England"). Elizabeth had seen a good deal of sin and suffering by the time her manuscript entitled "The Glass of the Sinful Soul" (1544) was published in Germany as *A Godly Medytacyon of the Christen Sowle* (1548). Elizabeth was then fourteen years old. She had completed the manuscript when she was eleven.

In the next few years, the Protestant monarch Edward died. Elizabeth's other half-sibling (Mary Tudor) became the Catholic monarch and arranged for the eighteen-year-old Elizabeth to be sent to prison at the Tower of London and then to custody at Woodstock. Written with a diamond on her window at Woodstock some time in the mid-1550s are the lines: "Much suspected by me, / Nothing proved can be, / Quoth Elizabeth prisoner."[1] And yet, despite the insecurity of this existence, in 1558 at the age of twenty-five, Elizabeth became Queen of England; although her legitimacy was never legally established, she became, so goes the historical account, one of the most powerful and influential rulers that Europe would ever see.

This brave tale of triumph over adversity is well known. (The unseemly incestuous details are usually omitted from the elementary school texts, however, along with any mention of the "Glass.") Do not book reviewers in the United States, mired in the Horatio Alger tradition, still praise those books about "Good Queen Bess" which their publishers patronizingly target for the "eleven year old marketplace" precisely because such books provide their readers—or their readers' parents—with a moral role model for "juveniles" growing up in unsettled circumstances?[2]

The comforting explanation of the accomplishments of Elizabeth's mature years in terms of the precariousness of her early years is not, of course, without some scholarly justification. And that explanation has informed books about Elizabeth targeted for adult scholars. Here the aesthetic requirements of biographical and historical narrative seem to have encouraged authors to emphasize, often *in vacuo,* the tension between adversity and triumph in order to clarify, or seem to clarify, the connection, crucial to the analysis of national politics in Britain, between Elizabeth's personal life and her public presence.

In this context there arises, as we shall see, a need for a reexamination of the family of the brilliantly educated and precociously intelligent preadolescent Elizabeth in relationship to the subsequent politics of nationhood in the Elizabethan era. But how, exactly, are the circumstances of the young Elizabeth's private "adversity" linked to her later public "triumph"? This is no easy matter to discern. We are not father confessors or psychoanalysts—at least not Elizabeth's—who look into the souls of our subjects.

But even so, we can still consider the speculum of the sinful soul that Elizabeth herself provided during these formative years. For Elizabeth's work, understood in historical context, mirrors how, for that monarch, ordinary kinship, which is the precondition for what she calls "fornication" and incest,[3] might be transcended by one or another kind of extraordinary spiritual or political kinship.

Elizabeth had known ordinary kinship, or its legal figurations, in situations where one person is both sibling and lover/spouse (as Anne Boleyn was to George Boleyn or Catherine of Aragon to Henry VIII) or both parent and lover-spouse (as Thomas Seymour to Elizabeth herself). She would know extraordinary kinship and its apparent transcendence of the incest taboo from its Christian formulations. These formulations, which she expresses in her speculum, would include the doctrine of universal siblinghood, according to which, as I have described it, all human beings are siblings so that every act of sexual intercourse is incestuous; and it would also include the doctrine of the quadruply affined *sponsa Christi,* according to which—as Elizabeth suggests in the letter to her stepmother Catherine Parr that she attached to the 1544 manuscript—one human being is at once the parent, sibling, child, and lover-spouse of another being. Elizabeth would also know—and perhaps, as we shall see, foreknow—extraordinary kinship in the developing British and generally European doctrine of nationhood. The interconnection of Christian kinship and nationhood, as well as Elizabeth's own life and work, constitutes the subject of this chapter.

THE GLASS OF THE SINFUL SOUL

If your first spring and auther
God you view
No man bastard be,
Unles with vice the worst he fede
And leveth so his birthe.

—Elizabeth's translation from Boethius, *Consolation of Philosophy*

In 1531 and 1533, Marguerite d' Angoulême, queen of Navarre, devout and free-thinking sister to King Francis of France, published a remarkable religious meditation of some seventeen hundred lines entitled *Le Miroir de l'âme pécheresse.*[4] Marguerite had been an acquaintance of Anne Boleyn since 1516.[5] And in 1534–35, after Anne, Henry VIII's mistress since 1527, married the English king in 1533 and gave birth in the same year to Elizabeth Tudor, Marguerite renewed her association with the well-educated and reform-minded English queen.[6] At about this time, it seems likely, Marguerite sent Anne a copy of her book.[7] Anne was beheaded in 1536. But in 1544, her daughter Elizabeth, then eleven years old, made an English translation of Marguerite's *Miroir,* most likely from a 1533 edition of that work that she found in her mother's collection.[8] Elizabeth called her translation "The Glass of the Sinful Soul."

Elizabeth sent the manuscript together with a covering letter and an elaborate needlework cover she had embroidered to her stepmother, Queen Catherine Parr, as a New Year's gift for 1545.[9] Catherine, an author of religious meditations like *Lamentation of a Sinner* (1547) and *Prayers, or Meditations* (1546), may have amended the manuscript, as Elizabeth had asked her to do in her covering letter, and probably added some new material of her own.[10] Catherine then sent the manuscript and revisions to John Bale, her friend—and Elizabeth's.[11]

In his capacity as nationalist scholar and theologian, Bale "mended" some of Elizabeth's "Glass of the Sinful Soul," adding to it a long "Epistle Dedicatory" and "Conclusion," and prudently retitling the work *A Godly Medytacyon of the Christen Sowle.* It was published in 1548, just a few months after the death of Henry VIII. In his edition, Bale included a few multilingual biblical translations by the princess and a woodcut depicting her kneeling before Christ.[12]

Queen Elizabeth wrote a fair number of literary works. Most have been published, some in good scholarly editions. Yet remarkably there was no readily available edition of the "Glass" until 1993.[13] Scholars of previous generations could have learned much from the "Glass." Published several times during Elizabeth's lifetime, it had considerable literary and political influence, and its study sheds light on historiographical matters such as handwriting, the education of women, and devotional and translational literature by and for women in the Elizabethan era.[14] The occasional efforts to explain the dearth of attention paid to the work by stressing that Marguerite's original is a "poor" poem and Elizabeth's translation "inaccurate," that works by "mere" eleven year olds are not worth studying, and so forth, have not been convincing. Much more convincing as an explanation is the expression in Elizabeth's work of an ideology both important and discomforting in its personal and historical aspects.

The "Glass"'s treatment of bastardy and incest, for example, has potentially disconcerting ramifications for ideas of liberty and politics generally, likewise illuminating the historical rise of the English nation and the biographical role of Elizabeth herself. For the most profound themes of the "Glass" involve the reworking and expansion in nationalist and secular terms of such medieval theological notions concerning kinship as universal siblinghood, where all men and women are equally akin, and dormition, where the Virgin Mary plays at once the role of mother and daughter as well as wife.[15] Above all the "Glass," whose French original had the subtitle *Discord étant en l'homme par contrariété de l'esprit et de la chair,* concerns the transmutation of the desire for, or fear of, physical incest into the desire for, or fear of, spiritual incest. It thus reflects, as we shall see, the beginnings of a new ideal and real political organization which—partly out of the concerns of England's great monarch with incest and bastardy and partly out of political exigencies of the time—were introduced by her as a kind of "national siblinghood" to which she was simultaneously the mother and the wife.

The "Glass" is a reflection of Elizabeth herself. (She wrote to Catherine Parr that "the part that I wrought in it [was as] well spiritual as manual.")[16] Interpretation and contextualization of that glass helps to elucidate—in terms both of individual psychology and of national politics—not only how a preadolescent young woman of 1544 formed her spirit but also how that spirit informed the political identity of the English nation (as Bale predicted it would) and participated in producing the modern nation.

INCEST, BASTARDY, AND THE BIRTH OF A NATION

C[o]elum Patria ("Heaven [is my] Fatherland")

—embroidered by Elizabeth on the cover of a New Testament bible

In 1544, the year she wrote the "Glass," the eleven-year-old Princess Elizabeth had a fourth stepmother, Queen Catherine Parr. Her first stepmother, Queen Jane Seymour, had died giving birth to Elizabeth's half brother, later King Edward VI. The marriage of her second stepmother, Anne of Cleves, had been declared null and void. Her third stepmother, Queen Catherine Howard, had been executed on the charge of adultery. "Mother, mother, mother," says Shakespeare's Hamlet—and also Melville's Pierre.[17]

Anne Boleyn had risen to the place of queen thanks to Henry VIII's memorable charge that Queen Catherine and King Henry VIII were living in adultery and incest, and that their marriage therefore ought to be declared null. The charge, which recalls the complexities of the liaison between King Claudius and his sister-in-law Gertrude, in *Hamlet,* was momentous in the English Reformation. Catherine was the widow of Henry VIII's brother Arthur; she was Henry's sister-in-law. Should Henry have married? Legally, could he? On the one side of this debate stands the law of the Levirate, according to which a man must marry the childless widow of a deceased brother.[18] On the other side stands the doctrine of carnal contagion, according to which it is incest to have sexual intercourse with one's sister-in-law.[19] Henry VIII himself took the view that sexual intercourse with Catherine was incest.[20] Thus began a series of specifically English charges of incest in the royal family.[21] Such charges are germane to the foundation of the English Reformation, and, like Thomas More's Romish claim that such Brothers or ex-Brothers as Martin Luther—as well as Elizabeth's tutors and mentors John Bale and Bernardino Ochino—commit incest when they marry, they are part of a general Renaissance revaluation of profane and sacred sexual liaisons.[22]

Elizabeth's mother, Anne Boleyn, not only rose to power by means of a charge of incest (against Catherine); she also fell from power partly as a result of one. We can gain a fuller grasp of the ramifications of the charge by examining how it affected Elizabeth's legal status and hence, of course, the English people's natural concern with problems of succession. For the people had reason to wonder whether a princess conceived in adultery or incest was legitimate.[23] Elizabeth was deemed a bastard on several counts, five of which are worth pursuing in the present context.[24]

First, Elizabeth's *pater,* Henry VIII, had claimed publicly that she was a bastard and that her uncle Lord Rochford, her mother's brother, was her consanguineous genitor.[25] Just as Anne was accused of having had sexual intercourse with her brother Lord Rochford, Elizabeth was declared a bastard by a 1536 act of Parliament.[26]

Second, Sir Thomas More argued that the union between Henry VIII and Catherine of Aragon was not incestuous, hence that both the divorce of Catherine and the marriage to Anne Boleyn were null. It follows from this that, whether Henry or Rochford was Elizabeth's genitor, she was in any event a bastard.

Third, Henry VIII and Anne Boleyn were married barely nine months before Elizabeth's birthday, in suspiciously speedy and secret circumstances. Even if the marriage to Anne were legitimate (which More said it was not) and Henry were the genitor (which Henry himself said he was not), Elizabeth might seem to have been at least conceived out of wedlock.

As if this were not enough to cause the English people to doubt Elizabeth's legit-

Fig. 4. Illuminated Book of Hours showing the Annunciation. The couplet is in Anne Boleyn's handwriting: "Be daly prove you shalle me fynde/to be to you bothe lovying and kynde." (Courtesy of the Trustees of the British Library)

imacy, there was a fourth allegation—that Elizabeth's mother was also her sister, or, put otherwise, that Anne Boleyn was not only Henry's loving wife but also his kind daughter.[27] (Below an illumination of the Annunciation—the angel Gabriel's intimation to Mary of the divine incarnation in her womb—in a Book of Hours that Anne Boleyn gave Henry, she inscribed the couplet: "Be daly prove you shalle me fynde / to be to you bothe lovyng and kynde"; see fig. 4.)[28] Though this allegation of incest is false, we ought not to dismiss it as altogether frivolous. For in the context of the Christian religion, children of incestuous unions—including the annunciated god (Jesus) and several saints (Gregory the Great among them, as we have seen)—come to assume powerful places in both profane and sacred institutions.

Finally, there was a fifth claim concerning Elizabeth's illegitimacy. Elizabeth's consanguineous aunt, Mary Cary (née Boleyn), had been her father's mistress either before Anne or at about the same time (probably 1527–28). Thomas Cranmer, Archbishop of Canterbury and Elizabeth's godfather, relied on the doctrine of carnal contagion and the parallel 1536 act of Parliament. According to the doctrine and the act it was nominated incest to sleep with the sister of one's mistress ("flesh of my flesh"). Cranmer declared both that the marriage between Henry and Anne was incestuous and that Elizabeth was a bastard.[29]

As we shall see, the doctrine of carnal contagion, and the charges of incest and bastardy that go along with it, affected Elizabeth's sense of identity, and it helps to give us some access to what one biographer has called her spiritual "girlhood." Its implementation in law during the 1530s and 1540s affected the very foundations of the new England. Its redefinition of the taboo on incest in the most generalizable terms helps to define the origin of the English nation state and its new Anglican institutions.

CARNAL CONTAGION

He that comforts my wife is the cherisher of my flesh and blood.

—Shakespeare, *All's Well That Ends Well,* 1.3.46–47

English men and women used the Doctrine of Carnal Contagion to claim that Henry VIII's marriages to his sister-in-law Catherine of Aragon and to Anne Boleyn were incestuous. If the king's brother (Arthur) had slept with a woman (Catherine of Aragon), it was argued, then she was the king's kin and his marriage to her was null and void. (Any offspring such as Mary Tudor, later Queen Mary I, would be illegitimate.)[30] By the same token, if the king had slept with a woman (Mary Boleyn) then that woman's sister (Anne Boleyn) was the queen's sister and there stood between them a diriment impediment to marriage. (Any such offspring as Elizabeth Tudor, later Queen Elizabeth I, would be illegitimate.) The argument that the siblings of a sexual mate become one's own siblings tended to make Elizabeth both legitimate, as it would nullify Catherine's marriage with Henry, and illegitimate, as it would nullify Anne's marriage with Henry.

The doctrine of carnal contagion involves the spread of blood kinship, as if it were a disease. As J. H. Fowler summarizes the medieval theologian Rabanus Maurus, "there is something like a communicable disease metaphor involved in early medieval notions of sexuality. If one sleeps with a woman who sleeps with another man who sleeps with another woman who sleeps with me, then whether I will it or not my flesh is inextricably bound up with the flesh of that first man's. A term which continually shows up in these canons and letters to describe fornication is *contagio carnalis*—carnal contagion."[31] Thus the English *Jacob's Well* states that "whan a man hath medlyd wyth a womman, or a womman wyth a man, neyther may be wedded to otheres kyn, into the fyfte degre, ne medle wyth hem; for if thei don, it is incest."[32] Fornication not only leads to venereal disease and to incest through illegitimacy, it also leads to incest through the secret spread of kinship by contagion of the flesh. This contagion, which involves a general teleology of all sexual activities, leads to views such as Jonas of Orleans' strident argument that "all illicit carnal relations are incestuous."[33] Promiscuity gives rise to incest insofar as one becomes kinsperson to all the kinspersons of one's marital or extramarital partner and record-keeping becomes all but impossible.

During Elizabeth's time, potentially universalist "figurative" kinship structures of this sort were replacing "literal" physical ones. It would seem, especially to one in a position like Elizabeth's, that all sexual liaisons were, or were likely to be, incestuous. This appearance, which seems to inform Elizabethan "sexual nausea," was a fulfillment of earlier Church history. For example, the archaic doctrine that cousinship even to the seventh degree makes a sexual liaison incestuous—indeed cousinship to any known and verifiable degree even so far back as Adam![34]—means that most all sexual liaisons in a small town are incestuous, and in a large town without perfectly accurate genealogical record-keeping it makes all liaisons potentially incestuous.

In carnal contagion cases, moreover, the diriment impediment to marriage involves keeping the books not only on sexual unions within wedlock but also those without. According to the "great mystery"—or quasi-philosophical dialectic—of Pauline Christianity, a man "shall be joined unto his wife, and they two shall be one flesh."[35] The Church, taking marriage to be the essential telos of all sexual intercourse, easily extended the marital union to also include the conjunction of fornicator and fornicatrix. "Know ye not," asks Paul, "that he which is joined to an harlot is one body? for two, saith he, shall be one flesh."[36] Thus sexual intercourse, whether marital or extramarital, spreads kinship by bringing the relatives of each party into the kindred of the other.[37]

The Christian notion of the growth of kinship relations without as well as within marriage casts doubt on the old, or the Old Testament, distinction between legal and illegal sexual relations, between marital and extramarital relations, and between sexual relations between people who are related "in-law" and those who are not. It puts into question the crucial distinctions between incest and endogamy (whether one marries or not is now essentially immaterial) and between endogamy and exogamy.

By conflating extramarital with marital sexual intercourse, the doctrine of carnal contagion undermines and *transcends* the ordinary notion of kinship, which looks to marriage, as to incest, as a definitive institution. As we shall see, the conflation seems also to allow for the transformation of the Catholic *sponsa Christi,* queenpin of Christian kinship systems (she is linked to God in a chaste and incestuous relationship that is at once marital and extramarital) into a Virgin Queen such as Princess Elizabeth, already in her eleventh year, seems to comprehend. John Bale, in introducing us to Princess Elizabeth's translation, writes of this first public accomplishment of the young princess that "such noble beginnings are neither to be reckoned childish nor babyish, though she were a babe in years that hath here given them."[38] The child is mother to the woman; a full and accomplished childhood can tell much about a life.

SPECULATION ON PRINCESS ELIZABETH

Edidpus [Oedipus] busy serche did wrap him in most harmes;
for whan of him selfe he axed as he no Corinthe wez,
but Guest, he met with Laius, who after kild he had,
and mother his owne in mariage tok, with whom he got kingdom,
with dowary hers, whan than happy he thoght he was,
Againe he questioned who he was, whiche whan his wife wold Let
more earnest he, the old man as gilty he wer rebukd;
Omitting no good menes to make bewrayd al that was hid.
Than whan suspect herof his mynd had moche distract
And old man had skrigd out, "O worthi me whom nide to spike constrains;"
yeat kindeled and vexed with Curiositis stinge made answer,
"Compeld to heare, yeat heare I must."

—Elizabeth, translation from Plutarch, *De Curiositate*

The eleven-year-old Princess Elizabeth was in the position of Sophocles' Antigone. She was caught between horror at sexual transgression and pious duty to a family constituted by such transgression.[39] A brilliant and brilliantly educated young woman, blessed or cursed with the knowledge of her origin,[40] what might she have felt about herself and about her consanguineous and national families? Did a perhaps too curious young Elizabeth feel that she herself had, as in some Oedipal drama, "the seed of incest" for which her aunt-mother Anne and uncle-father Lord Rochford had been executed, and of which her father Henry VIII himself stood accused?[41]

Many biographers aver that for her aunt-mother, Anne Boleyn, Elizabeth never displayed any posthumous affection. That would be understandable enough in the circumstances. But what about the adage "like mother, like daughter"? Did Elizabeth see, or fear to see, her mother in herself, as in a glass? Is that why Elizabeth adopted as her own the badge of the sinful Anne Boleyn with its inscription *Semper Eadem* ("Always the Same")?

Elizabeth's own familial liaisons were tinged by incest. Elizabeth's uncle-father

Thomas Seymour seduced the thirteen-year-old, or tried to. Thomas was Elizabeth's uncle because Thomas's sister, Jane Seymour, who had been Lady in Waiting to Queens Catherine of Aragon and Anne Boleyn, had married Henry VIII in 1546 only one day after the execution of Elizabeth's mother. And he was her stepfather because in 1547 he married, "in indecent haste" (the funeral of Henry VIII was barely done with, much as Old Hamlet's when Gertrude married Claudius), the queen dowager Queen Catherine Parr, Elizabeth's fourth stepmother.[42] (Catherine Parr followed the triad of stepmothers Jane Seymour, Anne of Cleves, and the beheaded Catherine Howard: "mother, mother, mother.") Some time after Catherine Parr died, Thomas Seymour tried to marry Elizabeth herself.[43] (In a letter to Edward Seymour of January 28, 1549, Elizabeth writes "Master Tyrwhitt and others have told me that there are going rumors abroad, which be greatly both against my honor and honesty (which above all other things I esteem) which be these; that I am in the tower; and with child by my Lord Admiral. My Lord, these are shameful slanders.")[44] Was it a childhood memory about Thomas Seymour that somehow effected the aged Elizabeth's infatuation half a century later with the young Robert Devereux (Second Earl of Essex)? He whose maternal great-grandmother, Mary Boleyn, was Anne Boleyn's sister?[45] When in 1598, fifty years after John Bale published her "Glass of the Sinful Soul," Elizabeth translated Plutarch's warning against an Oedipal curiosity about familial origins, did she reflect upon the publicly acknowledged inascertainability of her own paternity and the never entirely suppressed charge, often leveled by recusant Catholics, that she was conceived in sinful incest? ("Compelled to hear, yet hear I must.")

We might speculate that the young Princess Elizabeth was attempting to deal with problems involving both paternity and incest.[46] Of Elizabeth's interest in incest we may be sure. She chose to translate a book about incest, *Le Miroir de l'âme pécheresse,* probably from her supposedly incestuous mother's copy—a book written by and probably given to her mother by Marguerite of Navarre, an author known for her spiritual libertinism and love for her brother.[47] Marguerite of Navarre was also known to the Princess as one of her potential "mothers" and "aunts": Henry VIII had once entertained the idea of marrying Marguerite, which would have made her one of Elizabeth's stepmothers,[48] and he had also tried to arrange a marriage between the thirteen-month-old Princess Elizabeth and Marguerite's nephew, the young Duke of Angoulême, third son of Francis I, which would have made her one of Elizabeth's "aunts."[49]

Moreover, the young Elizabeth did not fail to remark in the "Glass" how the general theme of Marguerite's work involves sensuality and incest. Marguerite's *Miroir* depicts, as Elizabeth puts it, a young woman "bound by her concupiscence," having "a body ready and prompt to do all evil" and subject to "my enemy, my sensuality (I being in my beastly sleep)."[50] Elizabeth, in the New Year's letter to Catherine Parr that accompanied her translation, notes that the "Glass" concerns how a "naughty" woman, for whom it was once a sin to be related to a being as both his daughter and wife, can become affined guiltlessly to another being as "mother, daughter, sister, and wife."[51] In the "Glass," Elizabeth discovers and explores a way to rise above the taboo on ordinary incest, deriving partly from the freethinking and spiritual libertine French tradition for which there is a kinship by *alliance* that supersedes kinship by blood (where some people are brothers and some are others) and looks to universalist standards of kinship (where all people are equally brothers and others).[52] Such universalist standards make all sexual intercourse equally chaste or unchaste, literally incestuous. They eventually even require the conflation of a libertine whore with the virginal mother. (Shakespeare's Cranmer reminds us of Elizabeth's reputation for virginity when, in *Henry VIII* [5.4.60–61], he predicts that his godchild Elizabeth, daughter of the whorish libertine Anne, will die "yet a virgin,/ A most spotted lily.")

The question of ordinary paternal legitimacy, which dogged Elizabeth throughout her life, is transcendible through a Roman or Hellenist "cosmopolitan" or Christian "universalist" view. This view would make every human being equally a child of the earth or of God, say, or essentially an "orphelyn of fadyr and modyr" (Chaucer's Boethius).[53] In her translation of Boethius' *Consolation of Philosophy*, where Christianity prescribes Jesus as the divine Father of everyone—in much the same way that Rome prescribed Caesar—Elizabeth might be consoled by the idea that she was no more or less bastard than anyone else:

> All humain kind on erthe
> From like beginninge Comes:
> One father is of all,
> One Only al doth gide.
> . . . .
> What Crake you of your stock
> Or forfathers Old?
> If your first spring and Auther
> God you view,
> No man bastard be,
> Vnles with vice the worst he fede
> And Leueth so his birthe.[54]

In the words of the Elizabethan poem called *The Lord's Prayer*, "Our Father, which in heaven art, / . . . hast [made] us all one brotherhood."[55] When Adam delved and Eve span there was no gentleman—or, at least, everyone was equally a gentleman and nongentleman.[56] All are essentially children of the same genitor. The Boethean appeal to universally equal kinship suggests that the distinction between illegitimacy and legitimacy is transcendible or irrelevant.

At first blush such a sentiment might seem able to console one such as Elizabeth, whom many people called bastard. No one should call Elizabeth a bastard any more because in God's eyes there are not bastards and legitimate children but only children of God, say, or of the Earth. But, on second thought, the sentiment for universal kinship must also have been unsettling to Elizabeth, and to Bale. If it is not our particular family's blood that distinguishes one person from another, one might ask by what right does any British monarch rule? (Elizabeth and Bale think of Elizabeth as potential monarch.) England was not, after all, an elective or constitutional monarchy (like Hamlet's Denmark) or an empire where adopted or adrogated sons inherit the throne (like Caesar's Rome). In England, for good or ill, the right to the throne had to be defended in terms of blood.[57] Thus Elizabeth's supporters needed to establish not merely that legitimate British princesses could inherit the throne (a task that Bale, countering the proponents of the Salic Law, sets for himself in his fine catalog of women rulers of Britain)[58] but also that Elizabeth was legitimate in terms of blood (a nearly impossible task). Or Elizabeth's supporters needed to take a polarly opposite tack, replacing familial blood as the standard for kinship and for political inheritance with some such quality as "nobility of spirit" (a task that Bale seeks to accomplish in his Epistle Dedicatory).

Replacement of blood with nobility of spirit would itself be dangerous to any would-be ruler. It would bring into disrepute the familial reverence ordinarily due to earthly parents. ("Honor thy father and thy mother" says the Old Testament.) And hence, by common analogy, rejection of blood as the standard for kinship would destroy the political reverence due to the king. ("The king is the father of his people" says the old adage.) Transcendence of consanguinity as the standard for kinship would tend to transform the very idea of family and nation, as Bale suggests in his ancillary essays.

THE *SPONSA CHRISTI*

Then is God in us, and all we are in Him, and He in all men. If we have Him through faith, then have we a greater treasure than any man can tell.

—Elizabeth, "Glass of the Sinful Soul"

In the case of physical incest, two people who are consanguineously related to each other act as though they are not. A consanguineous sister, for example, acts as a spouse, and that is called bad, even absolutely profane. In the Holy Family, however, the Virgin Mary is at once the spouse, sibling, child, and mother of God the Father and/or Son, but her obviously incestuous sexual intercourse with Him (or Them) is called "good," even absolutely sacred. The term *taboo* means both "sacred" and "dangerous, forbidden."[59] And the term *sponsa Christi* [spouse of Christ], which defines Mary's relation to God, emphasizes a woman's marriage with Christ in a union at once extramarital (Mary is married to Joseph) and incestuous. Mary is the human female parent of God as Son (she parented the Son); she is the spouse of God; she is the sibling of God (the Son and she are children of the same Father); and she is the child of God (the Father's child). There is no denying that Mary's relationship to God, which is the model for perfection in Roman Christian life, amounts to incest of a kind. ("Is't not a kind of incest, to take life / From thine own sister's shame?" Shakespeare's Isabella demands of Claudio. Isabella, the would-be Sister, is sister to Claudio, and her name, which means something like "consecrated to God," is cognate with *Elizabeth.*)[60] As such the Virgin Mary is the archetypal *sponsa Christi;* and her mysterious puzzle has occupied many a church father and pope from the earliest to the latest periods of Christian theology.[61]

Incest, the centerpiece of political controversy during Elizabeth's girlhood, has long been associated with puzzles, as in the case of the riddle of the Delphic oracle, or the Theban sphinx whose terms Oedipus cannot solve but is compelled to enact. Consider the riddling of this late medieval epigraph inscribed in the exact middle of the collegial church of Écouis, in the cross aisle:

> Here lies the child, here lies the father,
> Here lies the sister, here lies the brother,
> Here lie the wife and the husband,
> Yet there are but two bodies here.[62]

The solution to this riddle involves a local story: "The tradition is that a son of Madame d'Écouis had by his mother, without knowing her or being recognized by her, a daughter named Cecilia, whom he afterward married in Lorraine, she then being in the service of the Duchess of Bar. Thus Cecilia was at one and the same time her husband's daughter, sister, and wife. They were interred together in the same grave at Écouis in 1512."[63] The woman here has a relationship to her man like that of Mary to God and of any nun who, as *sponsa Christi,* is similarly related to Christ. Similar epigraphs appear in other churches in Europe:

> Here lies the son, here lies the mother,
> Here lies the daughter with the father;
> Here lies the sister, here lies the brother,
> Here lies the wife with the husband;
> And there are only three bodies here.[64]

Shakespeare begins *Pericles, Prince of Tyre* with an equally familiar riddle about a kind of incestuous self-consumption:

> I am no viper, yet I feed
> On mother's flesh which did me breed.
> I sought a husband, in which labour
> I found that kindness in a father.
> He's father, son, and husband mild;
> I mother, wife, and yet his child:
> How they may be, and yet in two,
> As you will live, resolve it you.[65]

And nearly identical puzzles inform such writers as Gower and Twine as well as the sixteenth-century Navarre-born Spanish poet Julian Medrano.[66] The free-thinker Rabelais, in his *Gargantua* and *Pantagruel,* which were published in the 1530s, writes of people on the Island of Ennasin that "they were so related and intermarried with one another that we found none of them who was a mother, or a father, an uncle, or an aunt, a cousin or a nephew, a son-in-law or a daughter-in-law, a god-father or a god-mother, to any other; except indeed for one tall noseless man whom I heard calling a little girl of three or four, Father, while the little girl called him Daughter."[67]

What else is Elizabeth's "Glass," in this context, except a kinship riddle? Elizabeth herself teases out the matter thus: "I am sister unto Thee, but so naughty a sister, that better it is for me to hide such a name."[68] Certainly Elizabeth had incest on the mind when she wrote these words. Her father Henry VIII had committed incest with Catherine of Aragon, and her deceased stepmother Catherine Howard had committed adultery with Catherine's cousin Thomas Culpepper. Elizabeth's genetrix Anne Boleyn had been accused of being sexually "handled" by Elizabeth's uncle (Rochford) and by Elizabeth's father (Henry VIII). Elizabeth's uncle-father (Thomas Seymour) was soon to be accused of "handling" Elizabeth herself. To father or Father, to brother or Brother, to son or Son, Elizabeth's narrator calls out in the "Glass," "O my father, what paternity! O my brother, what fraternity! O my child, what delectation! O my husband, O what conjunction!"[69]

Lest we miss the extraordinary quality of the poet's love, the speaker asks of her unnamed fourfold kin, "Is there any love that may be compared unto this, but it hath some evil condition?" (see fig. 5). For it is ultimately Father, and not father,

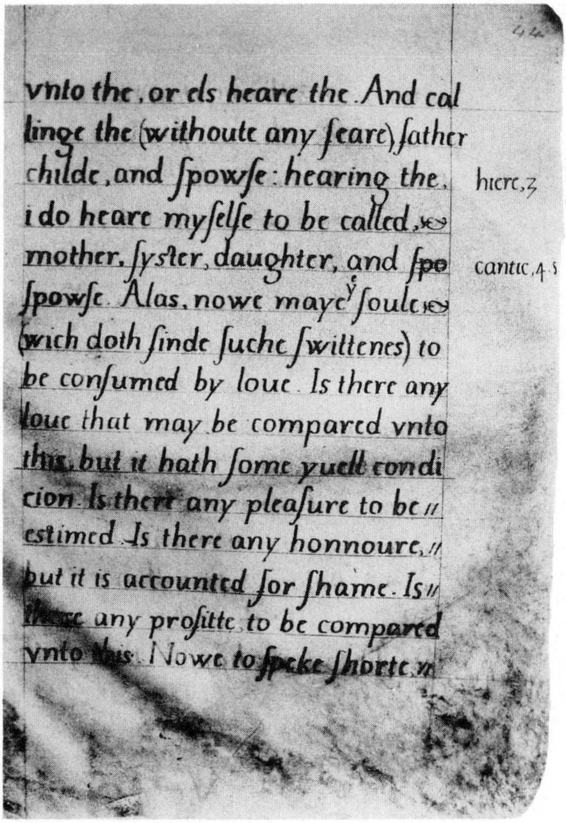

Fig. 5. Autograph page in Elizabeth's hand: "And calling Thee (without any fear) father, child, and spouse, hearing Thee I do hear myself to be called mother, sister, daughter, and spouse. Alas! now the soul may (which doth find such sweetness) [to] be consumed by love. Is there any love that may be compared unto this, but it hath some evil condition? Is there any pleasure to be esteemed? Is there any honor, but it is accounted for shame? Is there any profit to be compared unto this?" Elizabeth, Glass of the Sinful Soul. Cherry 36. (Bodleian Library, Oxford)

44

vnto the, or els heare the. And cal
linge the (withoute any feare) father
childe, and ſpowſe: hearing the, hiere,3
i do heare myſelſe to be called,
mother, ſyſter, daughter, and ſpo cantic,4 5
ſpowſe. Alas, nowe maye ͤ ſoule
(wich doth ſinde ſuche ſwittenes) to
be conſumed by loue. Is there any
loue that may be compared vnto
this, but it hath ſome yuell condi
cion. Is there any pleaſure to be //
estimed. Is there any honnoure, //
but it is accounted for ſhame. Is //
there any proſitte to be compared
vnto this. Nowe to ſpeke ſhorte,

who handles the young girl. And so, to Jesus, she cries out: "Thou dost handle my soul (if so I durst say) as a mother, daughter, sister, and wife. . . . Alas, yea, for Thou hast broken the kindred of my old father, calling me daughter of adoption."[70] In the end the narrator comes to recognize that on one's own, one can do nothing to overcome the sinful desire for physical incest. Only through the grace of God can profane incest be converted to sacred. The soul must look into herself as into a

Fig. 6. Mosaic depicting the Virgin's Dormition, with Christ holding the childlike soul of His Mother. La Martorana, Palmero, Siciliy, A.D. 1143. (Alinari-Art Resource, New York)

mirror, and "(beholding and contemplating what she is) doth perceive how of herself and of her own strength she can do nothing that good is, or prevaileth for her salvation, unless it be through the grace of God, whose mother, daughter, sister, and wife by the scriptures she proveth herself to be."[71]

The "Glass" is about the conversion of a soul from sensual or physical sin to a kind of spiritual incest, or about that logical or psychological-spiritual metamorphosis in which such opposites as incest and chastity become each other. (By the same token the "Glass" is about a conversion from betrayal, defined in terms of fourfold kinship, to a union in faith.)[72] Bale, in consideration of such conversion, writes in his Conclusion: "And though the facts be as the purple, yet shall they appear as white as the wool." From this viewpoint, the "Glass" is a "spiritual exercise of her [Princess Elizabeth's] inward soul with God"[73] and the figure of fourfold kinship, far from being a mere "oddity" in the poem, is central to it.[74] For in this exercise it is not Elizabeth, or her immediate family members, who occupy center stage as "sexually abused persons" or as "sexual abusers." It is God who is "familiarly commoned with."[75] God is treated as though he were a human being (which Jesus partly was) and a family member (which Jesus is to the *sponsa Christi*) (see fig. 6). God is kind and kin.

QUEEN MARGUERITE OF NAVARRE

Elizabeth's life was a continuation and fulfillment of the promise of Margaret's.

—Percy W. Ames

The full title of Marguerite's poem states that its central theme is the place of God as spouse—*Le Miroir de lame pecheresse, auquel elle recongnoist ses faultes et pechez, aussi les graces et benefices a elle faictz par Jesu-christ son espoux* [The mirror of the sinful soul, where the soul recognizes its faults and sins, as well as the graces and benefices made to her by Jesus Christ her spouse; 1531, 1533]. And, indeed, the central issue of the *Miroir* is the transmutation by the sinner of her profane desire for, or fear of, ordinary physical incest into a sacred desire for, and love of, that extraordinary incest which informs the Holy Family.

The same issue is present in some other works of Marguerite.[76] The central story in the *Heptaméron* (1559), for example, concerns a young man who unknowingly has sexual intercourse with his mother and then marries the offspring of this union—his sister, daughter, and spouse. His mother had chosen for this young son "a schoolmaster, a man of holy life"; but "Nature, who is a very secret schoolmaster taught him a very different lesson to any he had learned from his tutor."[77] Neither the son nor the daughter ever learns of their blood kinship, and for them (if not for their knowingly incestuous mother) the tale ends happily: "And they [the son and daughter] loved each other so much that never were there husband and wife more loving, nor yet more resembling each other; for she was his daughter, his sister and his wife, while he was her father, her brother and her husband."[78] In *Le Miroir de l'âme pécheresse,* the same sin of earthly incest reappears as the blessing of heavenly incest.[79] The work as a whole is informed almost entirely by the topos that a woman mystically in love with God is involved with him in a fourfold incestuous relationship and that this relationship, by virtue of its being spiritual, is not only bereft of its horrid and profane quality but actually made sacred. The *Miroir* and the *Heptaméron* are thus polar opposites, containing both thematic parallels (the former focusing on spiritual incest, the latter on physical incest) and verbal parallels ("mother, sister, daughter, and wife").

The protagonist in the *Miroir* is a woman who compares herself with the Virgin Mary—the mother and sister of God the Son, and the daughter and spouse of God the Father. She acknowledges that her wicked desire for physical sex, even incest, can be overcome only by a liberating, graceful raising of the physical into the spiritual. Without God, according to the *Miroir,* fleshly desire will turn to naughty action. Marguerite of Navarre's *Heptaméron* says of the woman who knowingly slept with her son that "she must have been some self-sufficient fool, who, in her friar-like dreaming, deemed herself so saintly as to be incapable of sin, just as many of the Friars would have us believe that we can become, merely by our own efforts, which is an exceedingly great error."[80] The woman's presumption was trusting to her individual power to overcome lust "instead of humbling herself and recognizing the powerlessness of our flesh, without God's assistance, to work anything but sin."[81] It is for Marguerite of Navarre as it is in Shakespeare's Navarre, in *Love's Labour's Lost,* where the members of the "little academe" are unsuccessful in their attempt to live a life like that of celibacy because "every man with his affects is born, / Not by might master'd, but by special grace."[82] The *Heptaméron* suggests that it is the manner of men to commit incest; incest of one kind or another is inevitable, because without grace repression of incestuous desire is bound to be unsuccessful.[83]

Marguerite's reformist and spiritualist work involved her in conflicts with the traditional Catholic and Protestant movements. That the *Miroir* itself contains no mention of male or female saints, merits, or any purgatory other than the blood of Jesus, was noted as a dangerous theoretical tendency, and students of the College of Navarre acted a comedy in which Marguerite was represented as a Fury of Hell.[84] Marguerite's biblical commentaries, moreover, were condemned by the censors at the Sorbonne, who ordered her books to be burned. (They were saved by express order of King Francis, her beloved brother.)[85] Marguerite's apparent support of the antinomian and pantheistic "Spiritual Libertines" was attacked by Calvin in his strident pamphlet *Against the . . . Libertines* (1545).[86] Many Libertines believed that everything is a manifestation of the spirit of God. For them, this meant undoing the distinction between "good" and "evil" acts, since nothing could be truly outside God. And they laid out almost in anthropological terms how both a spiritual libertine and a traditional nun, in imitating the Virgin Mary, ought to make of God a father, husband, brother, and son.

SOCIAL ANTHROPOLOGY OF UNIVERSALIST ORDERS

We see, then, that these savages have an unusually great horror of incest or are sensitive on the subject to an unusual degree, and that they combine this with a peculiarity which remains obscure to us of replacing real blood-relationship by totem kinship. This latter contrast must not, however, be too much exaggerated, and we must remember that the totem prohibitions include that against real incest as a special case.

—Freud, *Totem and Taboo*

Totemic tribes that enjoin exogamy (marriage outside the tribe) and allow for the existence of other totemic tribes can thereby avoid incest. But a tribe that believes its totem to be universal and all other human beings to be part of itself—or, teleologically speaking, potential converts to its universalist doctrine—makes exogamy impossible and all intercourse incestuous. Christianity calls for the establishment of such a tribe in its motif of the universal brotherhood of man ("All ye are brethren") and, in its proselytizing character, it claims to think of and treat as brothers even those who believe themselves to be non-Christians.

"All ye are brethren"—the question may arise here as to whether Jesus really meant what he said. For the implication of being faced with a choice between universal incest and universal annihilation—either to love one another equally or to die out—is a heavy burden. Some theologians, therefore, claim that Jesus did not really mean what he said—or did not really mean it for everyone. They fall off from the words of the New Testament and claim that not all people were meant to become "perfect." To be celibate, they say, is merely a "counsel"—which is "one of the advisory declarations of Christ and the apostles, in medieval theology reckoned as twelve, which are considered not to be universally binding, but to be given as a means of attaining greater moral perfection."[87] Thus Thomas Vautrollier, in his *Luther's Commentarie ypon the epistle to the Galatians* (1575), following out the tendency of ideas about "perfection" such as those of Bale in his "Epistle Dedicatory" and "Conclusion" to Elizabeth's *Godly Medytacyon of the Christen Sowle* (1548), dismissed the controversy as to whether all people should treat all others as siblings: "the Papists divide the gospel into precepts and counsels. To the precepts men are bound (say they) but not the counsels."[88] A few English Renaissance thinkers, however, take Jesus to mean what he says in the New Testament, and argue that *Perfection,* which refers to "the austerity of monastic life, monastic discipline," was the only nondegenerative form of life: Archbishop John Hamilton, in his *Catechism* (1552), writes that "matrimonye was degenerat fra the first perfeccion."[89]

Whether or not Jesus meant people to be perfect, the Christian monachal and libertine sects are microcosms of a perfect and potentially cosmic universal Siblinghood in which everyone is a Brother or Sister in Christ. Freud remarks that terms like "Sisters in Christ" have analogues in societies where kinship terms "do not necessarily indicate any consanguinity, as ours would do: they represent social rather than physical relationships."[90] But the monachal use of such terms, we shall see, assumes more than this replacement of "physical" relationships by social ones: it assumes the conflation of all social and theological relationships with biological relationships where a Sister or nun who violates the taboo on incest by marrying the Son of her Father is both sacred and taboo.[91]

The Old Way: *siblings* Becoming *Siblings*

The rule of the Old Testament is that a man or woman must leave his mother and father and marry in order to fulfill the ancient commandment to be fruitful and multiply. The New says, on the contrary, that a man or woman must give up entirely his or her old kinship ties—even hate his mother and father—in order to replace those ties, not with new human ones (to replace a father with a husband), but with divine ones (to replace an entire family with Christ).[92] These divine ties would make every human being a child of adoption to God. Who then are my family? "Ceux qui feront le vouloir de mon Père, / Mes frères sont, et ma soeur, et ma mère" (They who do the will of my Father, they are my brothers, and my sister, and my mother).[93] For such a child of adoption to God, all physical sexuality is as incestuous as it would be incestuous for consanguineous brothers and sisters to sleep together. Thus Sisters (nuns) and Brothers (friars and monks) are barred from having sexual intercourse with any human being by rules not only against fornication, but also against incest.[94]

In theory, the love of blood relatives might combine with the fear of earthly incest and make for an individual's decision to join a monastic order or even to found one. Courtly love may be a similar way to love one's sibling *in extremis* without violating the taboo against physical incest. Indeed, "in the middle ages a sister was not infrequently the object of courtly love, partly, it appears, because the presence of the incest barrier served to re-inforce the knight errant's resolution to adhere to the ideal of chastity."[95] Historically, earthly sibling love and heavenly Sibling love have often been joined in the same persons. In the sororal families or nunneries of Europe, women gave up all consanguineous ties (for them, he who had been a consanguineous brother was now essentially the same as any other man) and modeled their new family on Mary's relations with God (who was alike to her as brother, husband, son, and father). For women who adopted this sacred fourfold relationship to God, it was an act of absolutely profane incestuous sexual intercourse to have sexual intercourse with any person at all. Because they were no longer consanguineous sisters to any person and forever Sisters to all persons equally, sexual intercourse with a man who had been a consanguineous brother was now no better or worse than sexual intercourse with any other man.

Sigmund Freud, in his discussion of how some people "find happiness . . . along the path of love . . .. by directing their love, not to single objects, but to all men alike," calls Saint Francis of Assisi the man who "went furthest in exploiting love for the benefit of an inner feeling of happiness." Franciscan "readiness for universal love of mankind," says Freud, is, "according to one ethical view, . . . the highest standpoint which men can achieve."[96] Yet Francis's love for every being universally seems inextricably linked with love for his Sister in particular, as his remarkable poem "Brother Sun and Sister Moon" suggests.[97] What Francis and Clare could never permit in physical relations becomes a blessing in spiritual relations. (In *The Soul's Journey into God,* the great Franciscan thinker Bonaventure likewise put the balance of all the soul's relationships into its one, supposedly whole, relationship with God as Christ—a spiritually incestuous relationship, since the soul becomes the daughter, spouse, and sister of God.)[98]

In Christian hagiography generally, a saintly person's intense earthly sibling love is often followed by an extraordinary Sibling love of all human beings, just as if each and every human being had become a Brother or Sister. The *Acts of the Saints* (Acta sanctorum) includes more than 150 men and women who were brother and sister as well as Brother and Sister.[99] Sibling celibates appear from the very beginning of Christian monachism: Saint Anthony, traditionally the first Christian monk, placed his sister in a nunnery when he left the world for the ascetic life.[100] More strikingly, brother-sister liaisons played an important role in the historical beginnings of the Christian orders, for the sister of each of the three great cenobitical founders (Saint Pachomius, Saint Basil, and Saint Benedict) helped to preside over a community of nuns that followed an adaptation of her brother's rules for monks.[101]

A few examples of earthly sibling and Christian Sibling love in the lives of great saints with doctrines influential in pre-Dissolution England might be useful. Saint Benedict, founder of the order in which Thomas More was educated, visited his sister, Saint Scholastica, once a year. On the last of these visits, according to Saint Gregory the Great's biography, Scholastica entreated Benedict to stay the night. When he adamantly refused, she fell to prayer until a sudden storm arose, so that she had her way. The consummation of that night, spent all in spiritual conversation, could be seen as the incorporation and transcendence of any earthly attraction, physical or otherwise, that might have existed between the brother and sister.[102]

Legends about Gregory the Great's own life involve incest and its atonement. As Hartmann von Aue tells the story, Gregory was the child of a brother-sister union and unknowingly married his own mother. When he became Pope he forgave his mother's incest and his own, restored Benedictine discipline, and enforced the rule of celibacy for the clergy.[103] This Christian solution—repentance and atonement—to the oedipal situation suggests that the Catholic orders made possible an atonement for the desire for incest or unchastity, even when the actual act was not in question. The Holy Family atones for the earthly one by making all even.

Some brother and sister saints voiced explicit concern for their siblings' sexuality in terms that border on identification and possessiveness. In his *Book on the Institute of Virgins and on the Contempt of the World,* Saint Leander exhorts his sister Florentina as Sister to marry the Son and enter the religious life. "Ah, well-beloved sister, understand the ardent desire which inspires the heart of thy brother to see thee with Christ. . . . Thou art the better part of myself. Woe to me if another take thy crown."[104] And he identifies the virginity of his sister as Sister with the goal of the entire Church: "Christ is already thy spouse, thy father, thine inheritance, thy ransom, thy Lord, thy God."[105] Florentina's earthly crown will be Leander's as much as any man's. In an epitaph composed for the tombstone of his sister Saint Irene, Saint Damasus expressed a similar proprietary interest: "A witness of our love (our mother) / Upon leaving the world, / Had given thee, my blood sister, to me as a pure pledge."[106] Nor was such concern with a sibling's chastity confined to men, as in the case of Saint Lioba and her cousin Saint Boniface.[107]

Bernard of Clairvaux provides us with yet another example. After he left home with his brother Andrew to enter the austere monastery of Citeaux, his sister Humberlina came, richly dressed, to visit them. Andrew greeted her, "Why so much solicitude to embellish a body destined for worms and rottenness, while the soul, that now animates it, is burning in everlasting flames?" Humberlina answered, "If my brother Bernard, who is the servant of God, despises my body, let him at least have pity on my soul. Let him come, let him command; and whatsoever he thinks proper to enjoin I am prepared to carry out."[108] Some time thereafter she entered a convent. In Bernard's famous sermons *On the Song of Songs,* in which some theorists say Bernard demonstrates the ultimate "liberation of the soul," sisterly virginity and the theme "my sister as my wife" (*soror mea sponsa*) are sexualized, in a manner familiar from other sibling saints.[109] Curtius remarks that it is not far from "the mystical love of the Madonna" of Bernard, who "spiritualized love into a divine love," to such cynical descriptions of erotic orgies at convents as the Latin poem *The Council of Love at Remiremont* (ca. 1150).[110]

Almost an exact contemporary of Elizabeth, Saint Teresa of Avila ran away from home with one of her nine brothers, Rodrigo, at the age of seven, and with another, Antonio, at the age of twenty, to a Carmelite convent (1534).[111] Originally Teresa had wanted Antonio to become a Brother and herself a Sister.[112] But in her *Life* she seems to forget Antonio as brother from the moment she enters the Sisterhood. In her *Meditations on the Song of Songs* (1566), Teresa seeks to replace the fraternal love she once had for Antonio by a "spiritual marriage" and rebirth into a Family where earthly kinship distinctions do not exist. Just such a transcendence of consanguinity was Teresa's essential goal for the Discalced Carmelite Order that she helped to found. Teresa writes, "For the love of the Lord refrain from making individual friendships, however holy, for even among brothers and sisters such things are apt to be poisonous."[113] Reminding her Sisters in her *Exclamations of Soul to God* to "think of the brotherhood which you share with this great God" as "children of this God," she exhorts them, and all Christians, "to make our actions conform to our words—in short, to be like children of such a Father and the brethren of such a Brother."[114] Saint John of the Cross, an ideological mainstay of Teresa's order, wrote in his *Precautions,* "You should have an equal love for or an equal forgetfulness of all persons, whether relatives or not, and withdraw your heart from relatives as much as from others."[115] In precisely this way, Teresa erased and raised herself above differences between family and nonfamily. The saintly Teresa, who verges in *The Book of her Life* on confessing to spiritual incest—a biographer, pressing too hard, might conclude that she made love with a certain Dominican Brother—came to accept the ordinary taboo on sexual intercourse with a brother only when she accepted the extraordinary taboo on sexual intercourse with any human being.[116]

The monachal attempt to transform sibling love into Sibling love was therefore widespread—as was the attempt to test that transformation. A few Catholic orders allowed close physical communication between siblings or Siblings in "double cloisters"—monasteries and nunneries standing side by side. Among the Faremoutiers, who developed double cloisters in the seventh century, Saint Cagnoald ruled the monks in one wing and his sister, Saint Burgundofara, ruled the nuns in the other.[117] The Order of Fontevrault encouraged nuns and monks to sleep together in the same bed,[118] for the "mortification of the flesh"—the basis of one of Marguerite of Navarre's tales.[119] The Brigittines adopted the same organization. The only major order founded in England, the Order of Saint Gilbert of Sempringham, was a double cloister of cohabiting Brothers and Sisters.[120]

In all these examples, we see that the attempt to flee from the family to the Sisterhood or Brotherhood, from mother and father to Mother Superior and Pope, relocates the problem of incest from the consanguineous family to the Christian Family. Consanguineous *siblings* (brothers or sisters) who become religious *Siblings* (Brothers or Sisters) avoid incest either because they thereby adopt chasteness or because the distinction between chastity and incest is thereby erased.

The New Way: *Siblings* Becoming *spouses*

We ourselves groan within ourselves, waiting for the adoption, to wit, the redemption of our body.

—Rom. 8:23

The orders and the sibling/Siblings considered thus far tried to transcend the desire for incest through chastity; they solved the problem of the desire for incest by getting over it, rising above earthly sex by lifting it from earth to heaven. But there were other orders, ultimately more influential in shaping our present world's emphasis on *fraternité* and *liberté*, that tried to solve the problem in the obverse way. They adapted to earth the universalist love that the traditional orders had reserved only for heaven. For them there is a corporeal redemption that does not so much spiritualize the body for the "perfect" as liberate it even for the "imperfect."

Just as at the origin of Catholicism there are siblings who become Siblings, that is, brothers and sisters who become Brothers and Sisters, so at the origin of Protestantism there are Catholic Siblings who become spouses, that is, Brothers and Sisters who become husbands and wives. In the context of the English Reformation, this incestuous marriage psychologically constitutes the decisive characteristic in the life of Elizabeth and sociologically constitutes the decisive moment in the history of Renaissance sexuality.

Among Brothers who became husbands were monks who directly influenced the education of Princess Elizabeth. One was Bernardino Ochino, whom Elizabeth knew and whose *Sermo de Christo* she translated for her half-brother Edward in 1547. (Ochino was driven from England during the Catholic Mary Tudor's accession to the throne in 1553.)[121] Another was John Bale who, having argued for the impossibility of absolute temperance, forsook his monastic habit and got married almost as an act of religion. (In Bale's play, *The Three Laws of Nature* [1538, 1562], the allegorical figure "Sodomy" appears dressed as a monk.)[122] In the material he attached to Elizabeth's *A Godly Medytacyon of the Christen Sowle,* Bale attacks the false doctrine of kinship that inheres in Catholicism; and in *The Image of Both Churches* he insists that neither popish orders nor gossipry (god-parenting) bring with them any diriment impediment to marriage: "No more shall that free state of living be bound under the yoke of thy damnable dreams, either for vows unadvised, nor for popish orders, nor yet for any gossipry, but be at full liberty."[123]

There are other precedents for Sibling marriage—for example, Leo Judae, a disciple of Zwingli who married a Beguine in 1523, and the Franciscan monk François Lambert, who proposed in 1528 that his monastery might become a school of marriage! (Consider also the remarkable papal legitimatization of Brother Rabelais' bastard children.[124]) None of these Sister-Brother marriages, however, is as important as Brother Martin Luther's 1525 marriage to the Cistercian Sister Catherine von Bora. Luther's doctrine of justification by faith instead of by acts and corresponding view of the relationship between intent and act sparked the Reformation. An Augustinian eremite who thought that his unfulfilled desires made him prey for the devil, Luther argued that few if any men were "perfect" enough to be celibate.[125] Thus he denounced both monastic vows and distinctive dress for the clergy.[126]

Any Catholic in England knew that such marriage as Luther's to Catherine was incestuous. Friars and monks—spiritual or "ghostly" fathers, Bale calls them—were regarded just as biological fathers. Thus we read in the thirteenth century, "Incest, thet is, bituhe sibbe, fleschliche oder gasteliche";[127] and in the fifteenth century, "Inceste . . . is bitwene sibbe fleshli or gasteli."[128] The *Treatise on the Ten Commandments* includes the rule that "incestus is he that delith with nonne, with kosyn, or with a maydon, the wich is called defloracio."[129] And in Lydgate's *Fall of Princes* it is written that "incestus is . . . trespassyng with kyn or with blood, Or froward medlyng with hir that is a nunne."[130] Similarly, it was considered incest for anyone to have sexual relations with a godparent. (Chaucer too had written, "For right so as he that engendreth a child is his fleshly fader, right so is his godfader his fader espirituel. For which a womman may in no lasse synne assemblen with hire godsib than with hire owene flesshly brother.")[131] In this tradition Sir Thomas More, author of the great *Utopia* (1516)—which exhibits a stoic cosmopolitanism and propertal communism that reflects aspects of the Benedictine order—accused

Martin Luther in 1528 of committing incest when he married Sister Catherine.[132] Concerning Luther he argues that any human being who becomes a nun or friar commits incest when she or he has sexual intercourse with anyone, whether with a consanguineous kinsperson or not. For Thomas More, therefore, clerical marriage "defileth the priest more than double or trebel whoredom";[133] and in his *Confutacyon with Tindale*, More accused Luther and his wife, "the frere and the nunne," of incest:

> Let not therefore Tyndall (good reder) wyth his gay gloryouse wordes carye you so fast & so far away, but that ye remembre to pull hym bakke by the sleue a lytle, and aske hym whyther his owne hyghe spirytuall doctour mayster Martyne Luther hym selfe, beynge specyally borne agayne & new created of the spyryte, whom god in many places of holy scrypture hath commaunded to kepe his vowe made of chastyte when he then so far contrarye there vnto toke out of relygyon a spouse of Cryste, wedded her hym selfe in reproche of wedloke, called her his wyfe, and made her his harlot, and in doble despyte of maryage and relygyon both, lyueth wyth her openly and lyeth wyth her nyghtly, in shamefull inceste and abominable lycherye.[134]

Incest of a kind was the charge not only against such secular notables as Anne Boleyn, Elizabeth's earthly mother, but also against such religious notables as Bernardino Ochino and John Bale, Elizabeth's spiritual fathers.

LIBERTINISM AND LIBERTY

> *Thou'rt my Mother from the Womb,*
> *Wife, Sister, Daughter, to the Tomb.*
>
> —Blake, "The Gates of Paradise"

If sex is what most men and women want, whether they know it or not, and if "All ye are brethren," as Jesus says they are, then all men and women want incest.[135] If, in this context, grace means doing what one wants guiltlessly, then Christian liberty allows for—and even mandates—a libertinism of the body. Liberty, or libertinism, amounts to universal incest enacted without guilt.

The idea that guiltless incest is a sign of grace appears even at the historical origin of Christianity. For example, the Corinthian sect's acts of incest, which Paul calls "such fornication as is not so much as named among the Gentiles, that one should have his father's wife,"[136] were not "deed[s] done secretly out of weakness but . . . ideological act[s] done openly with the approval of at least an influential sector of the community."[137] In fact, the sect had actually taken Paul's own words about freedom from the law—"All things are lawful for me"—to indicate, among other things, freedom from such Old Testament rules as those concerning incest. ("The nakedness of thy father's wife shalt thou not uncover.")[138] The sect of the Essenes offers potentially much the same endorsement of incest, as Rabelais' parody of it suggests.[139]

In the medieval period, the endorsement of guiltless incest was crucial to the lay order called the Brethren of the Free Spirit. (It had, throughout Europe, perhaps hundreds of thousands of adherents between the thirteenth and seventeenth centuries.)[140] For the Brethren, the spiritually incestuous relations of the Virgin Mary to God were to be reproduced in the Edenic or paradisiacal state of grace. Their motto was the Pauline rule *Ubi spiritus, ibi libertas.*[141] When the spirit of the Lord is in one, then the law is erased and one is raised above the law.

The similarity between the so-called heretical libertine sect of the Brethren and traditional Christian orders, for whom incest is anathema, should not be overlooked. Similar motivations and appeals to grace are involved (1) when a religious celibate in the traditional orders overcomes sexual desire and loves everyone equally as universal siblings and (2) when a religious libertine in the Brethren of the Free Spirit overcomes the restrictions of law or conscience and loves everyone equally, including siblings. Both the celibate of the traditional orders and the libertine of the Brethren hypothesize a universal siblinghood in which sleeping with a brother is no worse than sleeping with any other man. The religious celibate seeks liberty from physical desire; the libertine seeks liberty from rules that restrict physical intercourse. But for both, in the words of Saint Paul, "Where the Spirit of the Lord is, there is liberty."[142] (From this viewpoint, the old way up and the new way down—the "ascent" from sibling incest to Sibling chastity and the "descent" from Sibling chastity to sibling incest—are one and the same.)

The key spokesperson for the sect, Marguerite Porete, though she was burned at the stake by the official church, remains our most trustworthy source about the early Brethren: most all other sources are obviously hostile.[143] Her remarkable *Mirror of the Simple Souls* stands as an important document in the history of Christian thought, with links to the "libertine" Marguerite of Navarre's *Miroir de l'âme pécheresse*[144] (which Calvin attacked as antinomian in *Against the Fantastic Sect of the Libertines*)[145] and to the doctrine of "liberty, equality, and fraternity" of the French Revolution.[146]

The *Mirror* of Marguerite Porete was influential in England.[147] In the sixteenth century, several Middle English translations circulated, just when John Champneys was arguing that God condones for his chosen people such "bodily necessities" as "fornication, adultery . . . or any other sin."[148] The Free Spirit affected directly the doctrines of the Elizabethan "Family of Love" and their communal sexual practices. And it later influenced the seventeenth-century Ranters and Levelers, who attempted to transcend ordinary norms of "right" and "wrong," or chasteness and incestuousness,[149] eventually involving both the doctrine of liberty and fraternity of the French Revolution and the quasi-medieval ideologies of other nineteenth-century utopian projects, including John Noyes' incest-practicing Perfectionists in Oneida, New York, who looked to the Diggers and Levelers of Commonwealth times as well as to the Brethren as their precursors.[150] The American hippie movement of the sixties, especially the San Francisco Diggers, seems similarly influenced by the Free Spirit.[151]

England of the sixteenth century was rife with Antinomian, Anabaptist, and

Wycliffite Lollard trends.[152] Anabaptists in London asserted "that a man regenerate could not sin; that though the outward man sinned, the inward man sinned not"[153] and insisted, probably in response to juridical questions about incest in the Holy Family, that Jesus did not take flesh from his mother.[154] Bale takes up the cause of the martyred Antinomian Ann Askew in his "Conclusion" to the *Godly Medytacyon*. In booklets published in November 1546 and January 1547, Bale similarly defends Askew's heterodox views regarding transubstantiation. He published these, as well as Elizabeth's *Godly Medytacyon* (1548), in Reformation Germany, where the Anabaptists of Münster practiced a kind of polygamy that did not allow for the sort of distinction of "mother" from "aunt" on which traditional society, as well as typical Elizabethan works such as *Hamlet,* rely.[155] Queen Catherine Parr, author of *Lamentation of a Sinner* and Elizabeth's stepmother in 1544–45, was herself suspected of being a favorer of Ann Askew.[156] Joan Boucher, colleague of Ann Askew and a member of an English Anabaptist community in England, was martyred in 1550 for her heretical views concerning the Virgin Mary as spiritual and fleshly mother—and as the daughter, sister, and wife of God. The English Anabaptists, like their European counterparts, tried thus to erase and rise above the old distinctions between good and evil, even chastity and incest, and they asserted publicly, in the spirit of the Free Spirit, that "When the spirit of the Lord is in one, one can do no sin."

The doctrine of the Free Spirit is largely defined by this attempt to return to prelapsarian innocence and "perfect" liberty. The Second Clementine decree, promulgated at the Council of Vienne in 1311, announced of the Siblings that "those who have achieved this state of perfection and absolute freedom are no longer subject to obedience and law or obligated to follow ecclesiastical regulations, for where divine spirit rules, there is liberty."[157] For at least one Brother of the Free Spirit, John Hartmann of Achmansteten, spiritual liberty meant "the complete cessation of remorse and the attainment of a state of sinlessness."[158] This entailed total transcendence of the post-Edenic taboo against incest:

> The free man could do as he wished, including fornicate with his sister or his mother, and anywhere he wished, including at the altar. He added that it was more natural with his sister than any other woman because of their consanguinity. His justification lay in being perfectly free and bound by no law or ecclesiastical statutes and precepts; such a man was a Free Spirit in contradistinction to the gross ones who were subject to existing authority of the church. His sister, far from losing her chastity, increased it as a result of their intercourse.[159]

The same liberty to have intercourse with mother or sister appears in the testimony of Conrad Kannler, who said that he was "at liberty to have sexual intercourse with mother and sister, although he did not believe God would permit it for the imperfect."[160]

In her *Mirror of the Simple Souls,* Marguerite Porete writes, "Friends, love and do what you want."[161] In the state of spiritual liberty Marguerite experiences "that love [that] maketh of the innocents that thei don nothing . . . but if it please them."[162] In such a state, says Marguerite, "the soul taketh leeve of vertues";[163] "the soul . . . giveth to nature al that he askith withoute grucchynge of conscience."[164] It is as if one were restored to a state of Edenic or paradisiacal simplicity, a state where the ability to commit incest without feeling guilty is itself a sign of grace itself, as it was for certain English Ranters.[165] When the spirit of God is in me, "I belong to the liberty of nature, and all that my nature desires I satisfy. I am a natural man."[166] According to one adept, "the Spirit of Freedom or the Free Spirit is attained when one is wholly transformed into God."[167] "Then is God in us, and all we are in Him, and He in all men," wrote Princess Elizabeth.[168] A person's guiding principle must be "Do what you want"—the single rule of Rabelais' anti-abbey of Thélème.[169] Not to enact what one desires to enact would be in itself a sign of disunion with God.

This appeal to freedom was Marguerite Porete's radical heresy: that perfect and free souls are fourfold kin with God. Perfect souls, she writes, have transcended regular kinship and become "daughters, sisters, and spouses of kings," and as such they attain the state of pure liberty.[170] Marguerite of Navarre suggests much the same in the *Miroir*: "O what union is this since, since (through faith) I am sure of Thee. And now I may call Thee son, father, spouse, and brother. Father, brother, son, husband" (as Elizabeth puts it in her "Glass").[171] Marguerite Porete was heretical in other ways as well. She translated the bible into the vernacular,[172] insisted that becoming one in Christ raises men and women above gender differences,[173] and claimed that one could be saved by faith, without "good" works.[174]

The Brethren's liberty was often misinterpreted as empty libertinage or mocked as pagan sexual communism.[175] But this sort of characterization does not go to the heart of the issue. What characterizes a celibate fraternal order (such as the Franciscans) is its liberty *from* flesh, or its razing the desires of the flesh and raising them to heaven. What characterizes a libertine order such as the Brethren of the Free Spirit is its graceful liberty *of* flesh ("Now Libertines are named after the liberty of the flesh, which their doctrine seems to allow").[176] Although it might at first seem that between the two kinds of sexual freedom there is all the difference in the world, religious libertinism and religious celibacy are significantly linked, and both sides resonate with the larger antinomian and Manichaean debates of the sixteenth century.[177]

SIBLING LOVE VERSUS RESPECT FOR PARENTS

To endow parents with the authority of wisdom, it is first of all necessary to look upon them as nonsexual beings, i.e., as not-possible objects of sexual desire. The prohibition against incest embodies this reverence.

—Benardete, on Thomas Aquinas

Tous de goneis tima. [Honour thy parents.]

—Bale, "Conclusion," in Elizabeth, *Godly Medytacyon* (1548)

Teleologically, incest dissolves the *pater* (father) in the *liber* (son) and replaces the patriarchy with a radical egalitarian liberty.[178] Sexual liberty would restructure kinship relations by destroying the crucial distinction between generations. In 1544 Elyot and Chapman thus use the term *libertine* to mean "any man of bonde ancestour" and "an urban freeman."[179] The tension here between the tendency towards liberty in religion and parentarchy in politics is inevitable, because the practice of incest, whether of the ordinary sort (as by libertines) or the extraordinary sort (as by religious celibates as members of the Holy Family), which true liberty would seem to entail, shows a politically consequential disrespect to one's parents and hence to parentarchal political order. The Roman *respectus parentelae,* the reverence due to near kin, is the most frequent argument against incest advanced in Western culture.

In sixteenth-century England, the Family of Love, a radical libertine religious sect, claimed that all its members were "one Being" with their leader, who was "Godded man: and so bee all named Gods and Children of the most highest."[180] They assumed that all "are equal in degree among themselves; all Kings, and a kingdome of kings"[181] and announced a communist society where a new brother's "goodes shal be in common amongst the rest of his berth."[182] With this subversive aspect, the Family of Love was savagely parodied on the more politically conservative English stage.[183]

Like libertinism, traditional monachism had a subversive aspect. Originally or ideally monachism was an essentially revolutionary "sibling" movement against more conservative "parentarchal" authority.[184] Even before the formal establishment of eremitic communities in the early Church, ascetics joined together in single residences in familylike sibling relationships that excluded intergenerational hierarchies. The fourth-century pseudo-Clementian epistle "To Virgins," for example, refers to such ascetics as were living in "spiritual marriage" as brothers and sisters, and it emphasized Jesus' injunction (Matt. 23:19) to "call no man *father.*"[185] Patriarchal ecclesiastical critics like Eusebius of Emesa viewed such practices with alarm because they furthered "either radical asceticism or radical libertinism."[186] They feared not only a communal libertinism but a corresponding radical propertal communism and even a propertyless condition. Saint Jerome writes, "Since you have been consecrated to perpetual virginity, your possessions are not your possessions [*tua non tua sunt*], because they now belong to Christ," and Saint Gregory praised his sister, Sister Macrina, because she "found delight in temperance" and at the same time "thought it affluence to own nothing."[187] In his essay on virginity, Saint John Chrysostom says "Now is not the time for matrimony and possessions; rather it is the time for penury and for that unusual way of life that will be of value to us in the time to come."[188] In this spirit, certain Catholic orders do not merely endorse poverty; they raze all property relations just as they raze all sexual relations. Saint John Chrysostom succinctly expresses the politically threatening aspect of this rejection of possessions by calling virginity *isangelos politē,* an extreme homogenization at once communist and incestuous.

In its essential form, then, monachal fraternity militates against any and all propertal and sexual ownership. This attitude makes monachism in and of itself a cause of political conflicts. Many Elizabethans feared the political dissolution which is implicit in the rejection of ordinary kinship structure (the parentarchy) and ordinary economic structures (property).

During the Tudor Renaissance, as throughout the history of the Christian West, a political ruler was viewed generally as a parent and her/his people as her/his children.[189] For a child to beat his father was thus, according to the Elizabethan Nashe, tantamount to his upsetting the natural and political order: "It is no maruaille if euery Alehouse vaunt the table of the world vpside downe, since the child beateth his father."[190] And they understood the link between political dissolution and that sexual liberty which would destroy "respect for the parent."[191] Bishop Stephen Gardiner thus compares the radical libertine position "all is for the flesh, women and meat with liberty of hand and tongue" to the political "dissolution and dissipation of all estates."[192] Moreover, few persons in Elizabethan England would overlook the connection between the monastic orders and a more specific threat of political chaos—revolt. The orders had been prominent among the opponents of Henry VIII, and one of the reasons Henry VIII gave for his notorious Dissolution of the Monasteries was a fear that monks would incite the commons to rebel.[193] During Elizabeth's reign, about two hundred Catholic priests were executed; and monks had figured in anti-Tudor plots such as the Northern Rebellion and the Archpriest Controversy.[194]

By the early seventeenth century the challenge by proponents of Brotherhood and/or Sisterhood to religious and political parentarchy was growing into modern liberalism. In that century the proponents of absolute monarchy had such spokespersons as King James I of England, who in his *Trew Law* stressed the identity between the duty of a subject and that of a child, and Robert Filmer, who in his *Patriarcha* provided both sides of the debate (liberals and parentarchalists) with a major rallying point. John Locke, among others, came to be the political spokesperson for the proponents of liberalism. Locke criticized Filmer's conceptual reliance on a "strange kind of domineering phantom, called 'the fatherhood'"[195] and opposed to it his own idea of all men free and equal in the state of nature. For Locke, as Norman O. Brown puts it, "liberty . . . means equality among the brothers (sons)."[196] The *liberi*, or free sons, of the French Revolution—and to a lesser extent the English Revolution—sought "to turn the world upside down" by killing the king who heads the family.

Whether monachal resistance to the parentarchal English monarch in the sixteenth century can be explained as a speculative extension of monachal fraternalism, as well as by the specific historical circumstances surrounding the Dissolution, there is little doubt that in these times such rulers as Henry VIiI himself feared a tendency toward liberty in certain religious and lay movements. They believed (as who does not?) that the public needs to believe, or they wanted the public to believe, in the "strange kind of domineering phantom" of the father. The absolutist king and his apologist were right in one sense: the struggle between liberty and absolute monarchy did involve the breakdown of the old family order and the development of a new political one. Lawrence Stone contends, in *Crisis of the Aristocracy,* that in the sixteenth and seventeenth centuries "the most remarkable change inside the family was the shift away from paternal authority" and that "it was slowly recognized that limits should be set not merely to the powers of kings . . ., but also [to] those of parents and husbands."[197] And libertine religious groups, arguing that "where liberty is, there is the spirit of God," eventually provided the liberal revolutions of England and France with their most extreme ideologies. The Marquis de Sade was emphatic in arguing, in his proposals for the French Revolution, that France would become a real republic only when men could call their sisters Mother.[198]

In the Renaissance tug-of-war between spiritual liberty and political parentarchy, Princess Elizabeth in her "Glass" and Queen Elizabeth in her reign as Queen toed the line between spiritual libertinism's claim that consanguinity and the particular family do not count and parentarchy's claim that they do. There was, for example, the old debate about the real substance of nobility, and hence about the right to rule of a bastard child whose parents hardly deserved (full) respect. Here Bale helps Elizabeth to establish a notion of kinship that would serve her as Virgin Queen. It would enable her to develop an ideology of family and nation that would endure for centuries.

In his "Epistle Dedicatory" and "Conclusion" to Elizabeth's *Godly Medytacyon,* Bale introduces the following propositions:

1. Elizabeth is of noble blood, whether she is legitimate or illegitimate: "Nobility . . . she hath gotten of blood in the high degree, having a most victorious king to her father, and a most virtuous and learned king again to her brother."[199]

2. Noble blood does not confer true nobility. Nobility resides neither in "renowned birth or succession of blood" nor in "worthiness of progeny": it is by plain virtue that one becomes God's "child of adoption." "By that means becometh he the dear brother, sister and mother of Christ."[200]

3. Elizabeth's production and publication of the "Glass" itself proves that Elizabeth has in abundance true nobility as virtue. "Of this nobility have I no doubt (Lady most faithfully studious) but that you are, with many other noble women and maidens more in this blessed age." For evidence that Elizabeth is of noble kindred Bale points to her book as a "spiritual exercise of her inward soul with God" and to her allowing the book to be published: "By your godly fruits" you shall be known, he reminds us.[201]

4. Whether consanguinity counts (in which case Elizabeth has it) or doesn't count (in which case Elizabeth does not need it), Elizabeth does not hate, or even regard indifferently, her consanguineous parents. The liberal Jesus of the New Testament is sometimes interpreted as encouraging his fol-

lowers and all God's children to hate their parents: "If any man come to me, and hate not his father, and mother, and wife, and children, and brethren, and sisters . . . he cannot be my disciple."[202]

5. Elizabeth, on the contrary, respects her parents, according to the rules of the Roman *respectus parentelae* and the Hebraic Ten Commandments. Bale draws our attention to Elizabeth's Greek phrase *tous de goneis tima* to prove that Elizabeth respects parental authority. "Your . . . clause in the Greek inciteth us to the right worshipping of God in spirit and verity, Jn. 4, to honoring of our parents in the seemly offices of natural children, Eph. 6, and to the reverent using of our Christian equals in the due ministrations of love, 1 Pet. 2. Neither Benedict nor Bruno, Dominick nor Francis (which have of long years been boasted for the principal patrons of religion) ever gave to their superstitious brethren so pure precepts of sincere Christianity."[203]

In the old debate about whence nobility comes[204] bastardy and incest generally play a major role, as reflected in Shakespeare's *All's Well That Ends Well.*[205] But Bale and Elizabeth manage in the *Godly Medytacyon of the Christen Sowle* to walk the line between rejecting and reverencing consanguineous familial and national parenthood, between the spiritual mysticism where everyone is one in God and the political fact where someone lords it over the others.

THE ROMAN VESTAL AND THE BRITISH EMPRESS; OR, FROM "AVE MARIA!" TO "VIVAT ELIZA!"

The central motif of Elizabeth's "Glass"—the spiritually incestuous relationship to one's Lord—was important not only in the Christian devotional and monachal institutions of Roman Christianity. It was so, as well, in the political and religious institutions of the ancient Roman Empire, where Roman Christianity first took hold on the political life of Europe. Indeed, Roman Christendom seems to have incorporated both the political and the religious institutions of the Roman Empire concerning spiritual and physical "fourfold kinship." In the political sphere, for example, the Roman Empress Livia (wife of Augustus Caesar, who vied with Jesus for the title of Lord God), stood in multifold kinship relations to Caesar. And Agrippina (wife of that Claudius who first attempted to conquer Britain in Rome's name) stood precisely in a fourfold relationship to Rome's God. Tacitus, the historian, remarks of the incestuous Agrippina that her kind of distinction was "traditionally reserved for priests and sacred objects" since she was remarkable as "the daughter of a great commander and the sister, wife, and mother of emperors."[206] And in the religious sphere, the figure of fourfold kinship informed the Roman Empire's institutions of adrogation (adoption) and of the vestal virgins, of which institutions the nunneries of Christendom were the Roman Christian end.[207]

The incorporation and supposed transcendence of Roman imperial institutions (the empress and vestal) by Roman Christian ones involved the eventual transformation of the vestals' Palladia—the archaic idols symbolic of an older matriarchy, probably Trojan, to which the vestals were devoted—into the Christian Sisterhood's spiritually incestuous Virgin Mary. The transformation meant a certain power, or at least ideal autonomy, for the Sisters. Incest, which Empress Agrippina practiced on the physical plane and the Roman vestals raised to a spiritual plane, helped provide some women of imperial Rome and Roman Christendom with a certain independence. As a violation of propertal as well as sexual "norms," incest provided a potential refuge for those who refused to traffic—or to be trafficked in. Having rejected their consanguineous families, the vestals were exempt, for instance, from the *patria potestas:* no human had the *patria potestas* over the vestal—only the divine *pontifex maximus,* the vestal's "religious father" [*pater*]. (Thus the vestal could free any prisoner she happened to run across on the way to execution and, more important in the present context, she could dispose of her property at will.) The nunneries of Christendom, whose family ties inform texts written by reform-minded queens of almost imperial power, had much the same liberating and economic effect on its inmates. "A convent was . . . the only place where a girl could escape her father's absolute authority."[208] And its ideal Mother Superior, transformed to the political world, would make for a real empress.

In preparing the stage for the supposedly illegitimate and incestuously conceived Elizabeth to assume a new political place in the English nation, the Reformist John Bale sought to define in British terms the matriarchy underlying Roman institutions. Imperial Rome's original institutions, suggests Bale, were the Trojan matriarchal ones. Aeneas' ancestors ruled in Britain no less than in Rome, says Bale. (Bale relies here on the old British tradition, memorialized in Geoffrey of Monmouth and elsewhere, that the British and Trojan nations are one and the same.) And, in this same vein, Bale recalls the old view that in the reign of Belinus (among the first British kings) the Trojan and British succession was by primogeniture.[209]

That British monarchs should succeed in the supposedly Trojan fashion, by simple primogeniture regardless of gender, was no small point. And central to Bale's pro-British attack on Roman Catholicism is his myth of the British *domina*—an empress or *regina* who holds the throne in her own right of inheritance.[210] Women who inherited the throne were not common in Christendom.[211] Marguerite of Navarre, certainly, knew the traditional "Salic Law," which supposedly kept women from inheriting the throne.[212] The Salic Law, known to readers of Shakespeare's *Henry V,* insisted that a royal or aristocratic woman was a kind of property that could not herself inherit property;[213] though she might rule as coheir or regent—as did Catherine de Médicis, who became Queen of France in 1547[214]—she could not inherit the throne. Bale, understanding how the idealization of woman in Christian institutions had disempowered her in the political sphere, wanted to prepare the way for Elizabeth to inherit the kingdom in uncommon fashion.

Among early defenses of women (Bale lists Boccaccio and Plutarch), Bale's catalogs of famous women are unique in their praise of women for political and military as well as religious accomplishments[215] and for their direct attack on the prevailing derogatory view of women as expressed in the retrograde *Contra doctrices mulieres.*[216] By the same token, they are informed by anti-Roman and anti-Catholic sentiment, as when Bale endorses the struggle of Boadicea against the Roman emperor Claudius or the cause of the early Celtic church (and Wycliffe) against that of the Roman popes.[217]

In a nationalist and matriarchal spirit, then, Bale culls from British histories and legends the names of great queens who ruled by might or right: Boadicea who fought the Romans, Helena, who inherited the throne like a man, Gwendoline, Cordeilla, Marcia, and others.[218] Among British women who ruled by reason of inheritance *(iure hereditario)* were Sexburth, who succeeded in 672 to the royal throne of the Kingdom of Wessex, and Aethelflaed, the widow of King Aethelred and the daughter of King Alfred, who ruled as *domina* from 912 to 918. And there was the famous Empress Matilda, who ruled with her husband as a kind of "imperial jointress, to this warlike state"—like Gertrude in *Hamlet.*[219] Matilda's political status as *domina* was controversial (as is Gertrude's), since it was ambiguous "how far the [British] empress . . . was to reign independently of her second husband, married 1128, a few months after Henry had imposed her as his heir."[220] Edelradus in Saxo Grammaticus' twelfth-century *Historiae Danicae*—a source for *Hamlet*—may well have said "aliquanto speciosius mares quam feminas regni usum decere nouerat."[221] But British women ruled and, in Elizabethan times, ruled well.

The position that women should have a hereditary right to rule independently of their husband was, in the context of the British Reformation, a Protestant and nationalist turn, by which Bale was able to steer an ideologically effective course between native British and alien Roman custom. It was a course that would help transform the cultic figure of the Roman vestal virgin or Catholic Virgin Mary—which placed women on an ideal pedestal (already manipulated to Reformist purposes in Marguerite's *Miroir*)—not only into the image of a martyred Protestant saint (like the anti–Roman Catholic Ann Askew in the sixteenth century and the anti–Roman Imperialist Saint Blandina in the second century)[222] but also back into the supposedly original Trojan figure of propertied, matriarchal and victorious empress who plays, in the real sphere of politics, the role of Mother Superior in a national siblinghood.

NATIONAL SIBLINGHOOD

ENGLANDE: Thes vyle popych swyne hath clene exyled my hosband.
KING JOHAN: Who ys thy husbond? Tel me, good gentyll Yngland.
ENGLANDE: For soth, God hym-selfe, the spowse of every sort
That seke hym in fayth to ther sowlys helth and comfort.

—John Bale, *King Johan*

In the intellectual development of the orphan princess and in the ideological life of the nation during the period of the English Reformation, Elizabeth's "Glass" represents a complex and politically fruitful vacillation between physical and spiritual incest. It is a vacillation that suggests how a nation, reeling from the conjunction of English and French identities that began in 1066, would now be informed by a new ideology of siblinghood and reformed by a great queen.

Bale prepared Elizabeth and the English people for her monarchy. Even her gender, he suggests, is sufficient to make her "king."[223] Bale imagines that the princess will become in time a kind of Protestant spouse of Christ. "If such fruits come forward in childhood, what will follow and appear when discretion and years shall be more ripe and ancient?"[224] But Elizabeth actually became neither an ordinary monarch nor a Protestant nun—despite some public identification by herself with Saint Elizabeth and with the heavenly Mary as Virgin Queen.[225] She became instead the *sponsa Angliae*. She participated, as the mature leader of Britain, in making a nation state in ways Machiavelli did not foresee, and in providing that nation with liberty and sovereignty.

Elizabeth participated, for example, in bringing to political fruition an ideology of spousal political economy. Two thousand years earlier, Aristotle had written in his *Politics:*

> Since there are three parts of expertise in household management [*oikonomia*]—expertise in mastery, which was spoken of earlier, expertise in paternal [rule], and expertise in marital [rule]—[the latter two must now be taken up. These differ fundamentally from the former, since one ought] to rule a wife and children as free persons, though it is not the same mode of rule in each case, the wife being ruled in political, the children in kingly fashion.[226]

As Thomas Aquinas underscores, the difference that Aristotle remarks between a spouse as ruler and a parent as ruler marks one distinction between a mere "monarchy" and a genuine "polis."[227]

In the early medieval period a prevalent notion was that the ruler was a monarchal *pater* or papa to his people. When the metaphor of the Christian ruler marrying the body politic did occur, it was based mainly on the way a bishop at his ordination became the spouse of his church.[228] But by the sixteenth century, the jurist Lucas de Pennas's influential comparison of the ruler with a spouse had been published (1544). Charles de Grassaille now called King Francis I—Marguerite of Navarre's brother—the king *maritus reipublicae* (1538/45).[229] Other French political thinkers—including François Hotman, Pierre Grégoire, and Jean Bodin—similarly chronicled and encouraged a politically crucial transformation of a bishop's marriage with the *corpus mysticum* (the Church) into the monarch's marriage with the *corpus politicum* or with the *corps politique et mystique*—as the jurist Coquille put it.[230] In earlier times the ring that the prince received at his coronation had been interpreted as a *signaculum fidei*, not as a mark of marriage.[231] But now the coronation ring was understood as a "marque de ceste réciproque conjonction" between royal ruler and kingdom.[232] At the coronation of Henry II of France in 1547, the king thus married the realm itself—"le roy espousa solemnellement le royaume."[233] And throughout the century such theorists as René Choppin were declaring that "Rex curator *Reipubliciae* ac *mysticus* . . . ipsius *coniunx*" [the king is the mystical spouse of the *res publica*].[234]

In England, too, jurists were interpreting the "mystical body" of the realm as a royal spouse.[235] There, too, there was a gradual movement away from an ideology of the ruler as an intergenerational parent toward an ideology of the ruler as an intragenerational spouse in a new siblinghood.[236] At her accession in 1558, Elizabeth, educated in the wake of these events in France, thus showed the Commons her coronation ring and reminded them not only that the English were her "children and relations" but also that England was her "husband."[237]

But Queen Elizabeth went beyond insisting that the monarch was both a parent and spouse towards the end or transcendence of all kinship. Just as the young Elizabeth meditated in the "Glass" on the significance of the collapse of kinship distinctions for spiritual libertines (Marguerite of Navarre), for nuns or Sisters (those whose houses Henry VIII dissolved), and for physically incestuous persons (Anne Boleyn), so the mature Elizabeth institutionalized the collapse on a national plane that was at once secular and chaste. The "breaking of kindred," to which the young Elizabeth first refers in the "Glass" (1544), was the first step in establishing the new (kind of) state.[238] "You know a kingdom knows no kindred," Elizabeth wrote to Henry Sidney in 1565, a few years after her coronation.[239] "I love and yet am forced to seem to hate," Elizabeth wrote about her suitor.[240]

The second step in establishing the new kind of state was Elizabeth's institutionalizing the head of England—or its church—in multiple national roles both parentarchal and liberal. In her government speeches and private literary writings, Elizabeth thus emphasizes that all men are equal in terms of consanguinity. All are brothers and sisters, including Elizabeth herself. Spiritualizing and secularizing the idea of incest, moreover, Elizabeth as Queen established herself as the virginal Mother and Wife of the English people. In a speech before the Commons in 1558, when she had succeeded "Bloody Mary" to the throne and John Bale had returned from exile, Elizabeth put things this way:

> I have long since made choice of a husband, the kingdom of England. And here is the pledge and emblem of my marriage contract, which I wonder you should so soon have forgot. [She showed then the ring at the accession.] I beseech you, gentlemen, charge me not with the want of children, forasmuch as every one of you, and every Englishman besides, are my children and relations.[241]

In the ideology of the sixteenth century, it was not merely that Elizabeth became a kind of Virgin Mary transformed to Protestant ends,[242] as though, in John Dowland's phrase, an "Ave Maria!" could simply become a "Vivat Eliza!"[243] It was mainly that Elizabeth adjusted a *specific* ideological commonplace—that the Virgin

Mary is at once the parent and spouse of God—to the general political requirements of her monarchal maturity.

Perhaps this adjustment served also a psychological requirement as well. The fact that Elizabeth never married has puzzled her biographers for centuries. Adduced to explain the fact are domestic political situations, religious differences between herself and her suitors, desire to make use of courtships in international politics, fear of childbirth, recognition of infertility, unhappy love affairs, and so on. And truly, Elizabeth did have a real aversion to marriage; Salignac reports that Elizabeth said to the French ambassador, "When I think of marriage it is as though my heart were being dragged out of my vitals, so much am I opposed to marriage by nature."[244] But aversion or no, the psychological requirements of her girlhood—linked, no doubt, to her remarkable education in the ways of marriage and incest in the 1530s and 1540s—, together with the political requirements of a nation where Brothers and Sisters no longer played their old roles and the Roman Pope, or papa, was a mere father figure, demanded that Elizabeth explore interactions between family and politics anew. Both as traumatized princess and as brilliant statesperson, marriage for her as secular ruler, like marriage for the clergy, was almost completely out of the question. (Elizabeth strongly disapproved of clerical marriage, but did not make a legal issue of it.)[245] For Elizabeth, marriage—that "earthly paradise of happiness"—seemed possible neither as an earthly wife nor as a paradisiacal nun but only as the royal mother/wife of England.[246] The great queen of England was so successful at establishing herself as this spiritual Mama—in contradistinction to the Romish spiritual Papa—that Pope Sixtus V, in the late 1580s, "allowed his mind to dwell on the fantasy of a papal union with the English crown; what a wife she would make for him, he joked, what brilliant children they would have."[247]

In relation to the monarchs of Europe, then, Elizabeth portrayed herself in multiple kinship roles reminiscent of the *sponsa Christi*. In letters of the 1580s to her godson James I, King of Scotland, for example, not only does she employ the normal kinship terms used by European monarchs—thus frequently calling her royal Scottish godson "my dear Brother and Cousin" and saying that she "mean[s] to deal like an affectionate sister with [him]"[248]—, but she also extends that employment beyond the norm. In a letter of 1593, for example, Elizabeth writes ambiguously to James as if she were not only godmother, cousin, and sister but also a mother at once consanguineous and virginal: "You know, my dear Brother, that, since you first grieved, I regarded always to conserve it as my womb it had been you bear."[249]

In domestic politics Elizabeth became, in these terms, the "commere" of England, ruling her people not only as mother and wife but also as godmother. Thus Elizabeth, who would have no natural children of her own, became the controversial godmother, or spiritual mother and sponsor, of more than a hundred English subjects.[250] She often said that her subjects should never have "a more natural mother than I meant to be unto all."[251] And John Harington wrote of her as "oure deare Queene, my royale god-mother, and this state's natural mother.''[252] John Bale, in the essays he attached to Elizabeth's *Godly Medytacyon,* forcefully criticized the institution of "gossipry" as practiced by Roman Catholics, because it entailed a diriment impediment to marriage;[253] and Elizabeth adapted the institution to her own ends, replacing the role of the human godparent in the family with that of the divine monarch in the nation.

How Elizabeth replaced "biological" kinship with national gossipred is hinted at in Shakespeare's *Henry VIII.* Here we learn that the possibly illegitimate babe is in some ways less akin to her *pater* Henry VIII than to her paternal sponsor, or godfather, Thomas Cranmer. Cranmer, without consulting Henry, names the child "Elizabeth"—which means something like *sponsa Christi.*[254] In Shakespeare's *Richard III* the curse that Margaret (the widow of Henry VI) pronounces on Queen Elizabeth (the wife of Edward IV and one of Elizabeth Tudor's namesakes) depends for its effectiveness on denying to Margaret, in her relation to the rulers of England, what God had granted the Virgin Mary, in her relation to God: "Die neither mother, wife, nor England's Queen."[255] Queen Elizabeth Tudor, the wife of no man, was the *sponsa Angliae,* the country's regal Mother and Wife.[256]

Elizabeth was concerned to transform physical incest (of the sort that she had reason to fear in her childhood) and libertine spiritual incest (of the sort that she represented in the "Glass") into a kind of political incest based in the unity of the English people as a single family. If the fear of incest, or of a desire for incest, can help spur a woman to contemplate being truly *elishabet*—in Hebrew literally "consecrated to God," like a *sponsa Christi*—that fear and its concomitant political aspects also help to explain how Elizabeth defied nature's injunction to reproduce and became England's king of kings. Bale had hoped that Elizabeth would become an English Reform "nourish-Mother," like the Virgin Mary[257]—less a Catholic "Mother Superior" such as Bale feared in Mary Stuart than a Hebraic "nursing father" like Moses.[258] But Elizabeth became instead a secular version of the religious Mary she delineates in her "Glass." She became a national institution. In the iconography and visual representations of her reign, she is represented not so much as the Virgin Queen, which is how some modern day critics would have it, as the *Sponsa Angliae.*

Elizabeth's reign helped to transform the ideology of British monarchy.[259] In forging a national siblinghood, she helped to make England a "nation" state. Later rulers followed suit. The pattern of unchastity and (over)compensation in (public) virginity, or of tension between the libertinism one fears and the celibacy one believes to inhere in the "good" monachal or national institution, informs the reigns of later monarchs. A preeminent example would be her godson King James VI of Scotland, who succeeded Elizabeth as James I of England. James' concern about the fornication and bastardizing of his ancestors makes him her counterpart; and just as Princess Elizabeth, the supposedly illegitimate daughter of the adulterous and incestuous Anne Boleyn, made a claim to chastity, so did James, the grandson of the notoriously libertine and bastardizing James V and the son of the purportedly adulterous Mary (she had married her cousin, to whom she was supposedly unfaithful).[260] Like Elizabeth, James reacted against his family's past; despite a penchant for boys, he became as famous for chastity as Elizabeth. Richard Baker writes of James that "of all the Morrall vertues, he was eminent for chastity" and claims that James challenged comparison with Queen Elizabeth in this regard.[261] Beginning in 1603 with his first address to the English Parliament, moreover, James adopted Elizabeth's rhetoric of reincarnating the Holy Family on earth in secularist guise. Bale had called Elizabeth both a spouse and a "nourish-mother"; James claims similarly that he is both the spouse of "England"—"I am the Husband, all the whole Isles my lawful Wife"—and its "loving nourish-father."[262] Sovereigns would seem to represent in this view the middle term between a parent, who lords it over children, and a spouse, who is an equal in-law with them.

The main aspect of these ideological means of maintaining the Elizabethan regime were carried over for centuries, well beyond the purview of this chapter. Eventually, however, there exploded the contradiction between *liber* and *mater* that gives the Elizabethan slant on family and nation its full power. In France, where Marguerite Porete and Marguerite of Navarre lived, a more powerful and idealized form of reform took hold, at least for a while. Intellectual descendants of the libertines, if not of the English regicides,[263] French revolutionaries attempted brilliantly to bring *liberté, égalité,* and *fraternité* to a nation of free children. For a while it seemed that the libertine view of the Brethren of the Free Spirit—that *"Ubi spiritus, ibi libertas"* (Where the Spirit is, there is Liberty)—would win out. Yet something fell short in the transition from God to Earth. In America, with its Sons of Liberty and Daughters of the Revolution, a new American patriotism appropriated for its motto James Otis' sentence, "*Ubi libertas, ibi patria*" (Where Liberty is, there is Fatherland);[264] but liberty in romantic America in the nineteenth century—like *liberté* in France—really came to mean something like "my parents' place." We are warned in Genesis that young men and women "gotta get out of th[at] place if it's the last thing [they] ever do."[265] They must leave the heterogeneous family, whether nuclear or tribal, in order to become, as if we could, no more sons of a Fatherland or daughters in a Motherland but—like the sinful soul redeemed, ideally only, in Queen Elizabeth's "Glass"—siblings in the Promised Land of freedom.

5
HOODMAN-BLIND
OR
Hamlet and the End of Siblinghood

. . . What devil was't
That thus hath cozen'd you at hoodman-blind?

—*Hamlet,* 3.4.76–78

INTRODUCTION

The law will grant that brothers and sisters lie together if the lot falls out that way and the Pythia [at Delphi] concurs.

—Plato, *Republic* 461e

There is virtually no undeniable consanguineous kinship, as we have seen. Even where it is all but undeniable that one is the child of one's parent,[1] it is still, always, deniable. Posthumus' cry of rage in *Cymbeline* that "we are all bastards,"[2] or, in *The Winter's Tale,* Leontes' assertion, based on his fear of being a cuckold, that "many thousand on's / Have the disease and feel't not,"[3] may seem excessive or desperate, but concern about legitimacy, as well as about changelings and foundlings, is all but inevitable. Societies employ desperate measures in an effort to eliminate bastards and changelings (e.g., keeping women in harems or footprinting babies in modern hospitals). Here the dissonance of civilization sounds between the arguably political need to believe in the logical possibility of being able to know without doubt who's who—in order to avoid incest, say, or to determine succession—and the logical certainty of always being unable undeniably to know. *Nihil scitur* (nothing is known) wrote Francisco Sánchez in 1576.[4] The exceptions are the umbicularly conjoined mother and child and the conjoined (Siamese) twin—beings who both literalize the figure of consanguinity and suggest a need for legal and literary fictions and religious cults.

Some works of literature (e.g., *Oedipus the King* and the tale of Solomon's Judgment) assure us that we can know who are our kin. Yet without an all-knowing Delphic oracle or Solomonic God, how could consanguineous kinship be known? The possibility of being a bastard casts doubt on one's assigned father; the possibility of being a changeling casts doubt on both father and mother. It is this indeterminability of biological parenthood that ensures its always fictional or figurative character—even in a society like ours, which claims that it really knows the facts of life and that the real facts of life are biological. Montaigne, who was Sánchez's kinsperson, thus hints that kinship claims are essentially literary fictions.[5]

We need these fictions if, as Greek tragedians and latter-day theorists of the incest taboo aver, human beings are political animals and human politics requires belief in the possibility of knowledge of who's kin and who's nonkin. As we will see, even persons who believe themselves to be consanguineous parents will want standards of kinship other than consanguinity. For obeisance to the incest taboo requires not a genetrix but a *mater*, not a biological mother but a societal parent. It requires not a genitor but a *pater* of whom we may think "as of a father," in the words of Hamlet's uncle, Claudius.[6]

Many readers of *Hamlet* have assumed that the bond between members of the same nuclear family is the kinship structure at the play's center. This privileging of consanguineous as opposed to fictional kinship—of literal as opposed to figural—reflects a widespread and perhaps sociologically needful prejudice. Yet some social institutions do not privilege the consanguineous over the figural family: monachism and gossipred, for example, treat the *pater* as more important than the genitor even in determining diriment impediments to marriage;[7] and in some societies the milk-mother or wet nurse counts more than the blood-mother, as the always dubious Montaigne emphasizes.[8] More important in the context of a play like *Hamlet*, literature generally is largely about the very ambiguities in the distinction between literal and figural that any easy separation of literal from figural kinship ignores. *Hamlet*, as we shall see, offers precise access to these ambiguities.

Ignoring or misinterpreting the ambiguity of nonconsanguineous kinship relationships in *Hamlet* has often meant seeing in this tragedy mainly the destruction of the ordinary consanguineous nuclear or extended family through the consequences of incest and kin murder, as if *Hamlet* were a morality play that bolsters the incest taboo by threatening perpetrators with disaster. (That is precisely how *Oedipus the King* is often interpreted.) Yet *Hamlet* essentially involves the destruction not so much of a literal family as of a series of ever-extending figural families, a series that ends with the family of humankind. The tragedy of *Hamlet* resides in the heroic refusal to commit the kind of incest that any such idealized universalist siblinghood requires.

THE FIGURE OF SIBLINGHOOD

Therefore our sometime sister . . .
. . . .
Have we
. . . .
Taken to wife.

—*Hamlet,* 1.2.8–14

Most of the critical literature on *Hamlet* that considers its incestuous aspect focuses on intergenerational or mother-son incest. Yet the play is more obviously about intragenerational sibling incest, as in the case of Claudius' and Gertrude's liaison, termed incestuous several times both in the text and its "sources."[9]

The Claudius-Gertrude relationship is incestuous insofar as one's sibling-in-law *is* one's sibling. According to Saint Paul and the remarkable Catholic doctrine of carnal contagion, marriage transforms the flesh and blood of one's spouse into one's own ("man and wife / is one flesh"—4.3.54–55). And since, according to Catholic doctrine, a spouse is to all intents and purposes a lover fulfilled, extramarital sexual intercourse has the same effect of transforming one's lover's flesh and blood into one's own. It follows from this transformation of two into one that a man is no more or less akin to the brother of his wife than his wife is akin to her brother. Kinship is catching, and the agent of transmission is sexual intercourse.

The idea of kinship by contagion may seem like a mere "legal fiction." However, its seeming like a fiction does not essentially distinguish the idea of kinship by contagion from the idea of kinship by consanguinity. For the standard of consanguinity assumes that literal kinship resides in the blood, which may or may not be the case, and also that blood kinship is sometimes undeniably ascertainable and therefore "not a fiction," which is not the case.

Moreover, the way that the idea of carnal contagion conflates fictive with literal kinship informed the foundation of the Anglican Church during the decades of Henry VIII's several divorces. As we have seen, while the Jewish law of the Levirate states that one should marry one's deceased brother's widow (in which case she is hardly a sister), English jurists, during the years of the founding of the Anglican Church, took the opposite tack by adopting the doctrine of carnal contagion, according to which one should not marry one's deceased brother's widow because she is one's own sister. The case assumed importance in the first instance because, as we have seen, Henry VIII had married Catherine of Aragon. As the widow of his deceased brother Arthur, she was, to quote Claudius in *Hamlet,* his "sometime sister, now our queen" (1.2.8).[10] In 1526, Henry VIII and his counselors used the doctrine to argue that his marriage to Catherine was sibling incest. They used it in 1535 in much the same way when they sought to annul Henry VIII's marriage to Anne Boleyn—and hence to prove Princess Elizabeth a bastard. Anne's sister Mary, they said, had been Henry VIII's lover before Anne was. That made Anne Henry VIII's sister, and it made Henry VIII's marriage with Anne incestuous. For according to the doctrine of carnal contagion, as we have seen, there is no essential difference between a sister and either a sister-in-law (Catherine of Aragon) or a sister-of-one's-lover (Anne Boleyn).

The conflation of the terms *sister* and *sister-in-law* which the doctrine of carnal contagion requires lends *some* support to the usual interpretation of King Claudius' opening statement in *Hamlet*—he has his "sometime sister . . . taken to wife" (1.2.8–14)—to the effect that Claudius is using the term *sister* here as an abbreviation for *sister-in-law* and also as a synonym for it. According to the royal head of the Anglican Church, however, a sister-in-law really *is* a flesh-and-blood sister. That is why *Hamlet*'s author, making much of the difference between blood kin and kin-in-law in plays such as *All's Well That Ends Well* (where the distinction between daughter and daughter-in-law is all-important), makes nothing of this difference in *Hamlet.*[11] Gertrude, whether she is Claudius' consanguineous sister or not—which who can know?[12]—is to be thought of as sister in much the same way that Claudius would have Hamlet think of him "as of a father" (1.2.108). If Gertrude is not the literal sister of Claudius, she might as well be.

THE INDETERMINABILITY OF FATHERHOOD

Which of us has known his brother?

—Wolfe, *Look Homeward, Angel*

How can I know whether the physical or ghostly being standing before me is really my father? That is the question which some critics have said sums up *Hamlet.* Consider from this viewpoint the ambiguities surrounding the physical or spiritual (ghostly) paternities of the young men Hamlet and Laertes. These ambiguities confer on them the status of brothers—which is what Hamlet calls them (5.2.239, 249). And these ambiguities encourage both these sons to avenge the killing of the man each thinks of as his consanguineous father.[13] (Says Claudius to Hamlet: ". . . think of us / As of a father"—1.2.107–8.)

Who are Laertes' parents, really? Claudius, praising Polonius to Laertes, seems to imply that he, Claudius who holds the throne, is Laertes' father: "The head is not more native to the heart . . . than is the throne of Denmark to thy father" (1.2.47–49). (Compare Claudius' statement to Laertes that "I lov'd your father, and we love ourself"—4.7.34). Polonius himself seems unsure what things, if any, he may have fathered. Told by Claudius that Polonius has been ever "the father of good news," for example, Polonius responds, "Have I?"; and asked by Hamlet, "Have you a daughter?," Polonius has, as we shall see, only a fool's warrant to answer "I have" (2.2.42–43, 2.2.182–183).

But the main theme here is not who is Laertes's actual father (let us grant for the time being that it is Polonius), but Laertes' unallayable fear or hope about his paternity. For example, in the scene where Laertes returns from Paris and heads a popular insurrection ("'Choose we! Laertes shall be king'" shout the people, as if popular "election" were sufficient to secure royal succession), Laertes demands that Claudius tell him his father's whereabouts (4.5.106, 128): as if that father were a corpse in an unmarked grave or a ghost on castle ramparts or in bed chambers.[14] But Laertes' question "Where is my father?" also indicates fear and hope that he is not so much the legitimate consanguineous son of Polonius as the illegitimate one of Claudius. Laertes' fear/hope that he might be a bastard emerges when a "patriarchal" Claudius, threatened with a people's revolt that might "liberate" Denmark from his tyranny, asks Laertes to be calm. The *liber,* or son, Laertes responds:

> That drop of blood that's calm proclaims me bastard
> Cries cuckold to my father, brands the harlot
> Even here between the chaste unsmirched brow
> Of my true mother.
>
> (4.5.117–19; emphasis mine)

By the results of this test, Laertes *is* a bastard: Claudius does calm him down. Who then is the father that Laertes demands from Claudius? "Give me my father!" (4.5.128). Claudius might be able to give Laertes his father, or give his father back to Laertes, (1) if the dead Polonius were Laertes' consanguineous father and Claudius could arrange for Polonius to be resurrected as a ghost, much like the ghost of Old Hamlet, or (2) if Claudius were to state publicly that Laertes is his illegitimate consanguineous son, or (3) if Claudius were to adopt Laertes as a son in much the same way that he earlier adopted or adrogated Hamlet, in the fashion of Roman emperors (1.2.108–12). On the one hand Laertes wants to acknowledge Polonius as his consanguineous father, which would justify his continuing to revere his "true mother." On the other hand, Laertes wants to be—or to be recognized as—either the illegitimate or adoptive son of Claudius, which would give him some right more than mob election to the throne.[15]

Claudius plays Laertes' fears about what his mother was like against Laertes' hopes for who his father is. First Claudius questions Laertes' love for Polonius ("Laertes, was your father dear to you?"—4.7.106; compare the ghost's words to Hamlet: "If thou didst ever thy dear father love . . ."—1.5.23). Seeking to spur Laertes to kill Hamlet, who killed Polonius, Claudius exploits Laertes' uncertainty about his paternity and suggests that Laertes can best prove his legitimacy by killing Hamlet: "If you desire to know the certainty / Of your dear father . . ." (4.5.140–41). Claudius cleverly convinces Laertes that the way to prove himself to be the son of his "father" is "to show yourself in deed your father's son / More than in words" (4.7.123–24).[16] By this proof ("like father, like son"), Laertes' subsequent deeds tell us who his father is. Like Claudius, Laertes poisons Hamlet—the man who calls him "brother." (As Claudius' "brother's hand" killed Old Hamlet, so Laertes' "brothers' wager" kills Young Hamlet—1.5.74, 5.2.249.)

Hamlet's paternity is as ambiguous as Laertes'. Might not Hamlet's consanguineous father be Claudius? Did the "adulterate" Gertrude commit adultery with Claudius before Old Hamlet's death?[17] Might Hamlet be the offspring of that incestuous union? Gertrude says almost as much when she tells Hamlet "Do not for ever

. . . / Seek for thy noble father in the dust" (1.2.70–71). If Hamlet's real father is not in the dust, then where is he? Is Hamlet's father wandering the night as a ghost (the ghost of Old Hamlet)? Is Hamlet's father sitting on the throne (Claudius)? Claudius wants Hamlet to think of him "as of a father"—though Hamlet says that Hamlet is "too much in the sun"[18] and wants to be Claudius' son no more than he says later he wants to be Gertrude's son: "would it were not so, you are my mother" (3.4.15). Yet Claudius soon avows publicly an intent to name, or adrogate, Hamlet as his own son:

> . . . for let the world take note,
> You are the most immediate to our throne,
> And with no less nobility of love
> Than that which dearest father bears his son
> Do I impart toward you.
>
> (1.2.108–12)

What does this paternal adoption entail? In imperial Rome, which provides *Hamlet* with much of its historical backdrop, the Roman Emperor Augustus Caesar adopted the future emperor Tiberius, the son of his wife, Livia, when he married her. In the same way the Emperor Claudius of Rome, a namesake for Claudius in *Hamlet,* adopted the future emperor Nero, the son of his wife Agrippina. (This is the Nero Hamlet invokes in 3.2.385.) Similarly, the emperor Antoninus Pius adopted as his sons the future rival coemperors Marcus Aurelius (the Stoic writer) and Lucius Aurelius Verus. For these future rulers—Tiberius, Nero, Marcus Aurelius, and Lucius Aurelius Verus—their stepfathers' adoptions, or adrogations, were essential to their eventual successions (as we shall see in chapter 6).

Laertes and Hamlet fear and also desire the consanguineous and adoptive paternity of Claudius. Their rivalrous brotherhood is therefore one of love in hate, or vice versa. "I lov'd you ever" (5.1.285), says graveside Hamlet to Laertes—as Cain might to Abel, or Claudius to Old Hamlet. And the brotherly men eventually inflict mortal wounds on each other and die forgiving each other—much as Eteocles and Polyneices in the myth of Oedipus. Theirs is a fraternal embrace signifying a momentary reemergence of an old and gracious liberal society like the student friendships they left behind in Paris or Wittenberg.

SISTER AND *SISTER*

> *Children of the future Age,*
> *Reading this indignant page:*
> *Know that in a former time,*
> *Love! sweet Love! was thought a crime.*
>
> —Blake, "A Little Girl Lost"

What are siblings in love with each other to do? They may choose to commit sibling incest, as do Gertrude and Claudius, or they may choose not to commit incest. But what if all human beings are in an essential sense siblings—say insofar as we are equal children of God the Father? Or what if all creatures are doomed never to know for certain who are their real parents or siblings? ("How should I your true-love know / From another one?"—4.5.23–24.) In such a situation, as I have argued, any act of sexual intercourse is essentially incestuous or potentially incestuous. In such a situation not to commit incest means not to commit sexual intercourse with anyone. If Ophelia were Hamlet's sister, or if Hamlet were to think of her as a sister—say in the sense that Claudius asks him to think on Claudius "as of a father" (1.2.108)—what else could Hamlet do, if he were in love with her, but flee Ophelia's presence or demand that she become a nun, or Sister?

Many readers feel the essential or potential sibling quality in the affection between Ophelia and Hamlet—which is much like that between Orestes and his sister Electra in the court of their murderous stepfather Aegisthus and their mother Clytemnestra.[19] (Hints that a marriage between Ophelia and Hamlet would be a good idea are matched by hints that it would be incestuous. For example, Gertrude says to the dead Ophelia, "I hop'd thou shouldst have been my Hamlet's wife" [5.1.237], which means not so much that Hamlet and Ophelia can marry without commiting incest as that in Gertrude's view Ophelia was no more barred from marrying Hamlet than Gertrude was barred from marrying her "sometime" brother Claudius.)

But could we show that Ophelia is Hamlet's sister any more easily than that Laertes is or isn't Polonius' son? First, we might consider Hamlet's relation to Ophelia as a foster sister. In such sources of *Hamlet* as Saxo Grammaticus' *Historiae Danicae,* for example, Ophelia is defined as such. He and Ophelia share a common social education *(societas educationis)* and familiarity *(familiaritas):* "For both of them had been under the same fostering in their childhood; and this early rearing had brought Amleth and the girl into great intimacy."[20] Saxo's Hamlet has loved Ophelia "from her infancy"—which is as long as the Elizabethan Hamlet has loved Ophelia's brother Laertes ("I lov'd you ever"—5.1.285). And in Scandinavian literature as that elsewhere in Europe, collactaneous and foster kinship (that is, as *alumni)* generally pose a diriment impediment, both legal and moral, to marriage no less binding than that posed by consanguineous kinship or carnal contagion.[21]

Second, we might consider Hamlet's relation to Ophelia as an adoptive sister, as suggested by the Roman background to the plot of the play: Ophelia has a role like that of the adoptive sisters/wives of the first Roman emperors. Hamlet as would-be successor to the throne of Denmark finds himself in a relation to his uncle and adoptive father, King Claudius, like that of the would-be Roman imperial successors—Nero, Tiberius, and Marcus Aurelius—in relation to their adoptive fathers, emperors Claudius, Augustus, and Antoninus Pius. It was not enough for Nero, Tiberius, and Marcus Aurelius as would-be heirs to be adopted by the emperor; they also had to marry the emperor's consanguineous daughter. (Such marriage between adopted siblings in the Roman ruling family, generally defined as incest,[22] was regarded with something like the awe reserved for gods.) Thus Tiberius married Julia, Nero married Octavia, and Marcus Aurelius married Faustina. Who in *Hamlet,* besides Ophelia, might be the consanguineous daughter of Hamlet's adoptive father Claudius? Who else besides Ophelia should the would-be successor marry? In the old Hamlet story "the hero [Hamlet] weds the daughter of . . . his foster father."[23] (We will see later that not only the sororal Ophelia but also the maternal Gertrude is somehow marriageable.)

The motor of Shakespeare's plot, however, is not that Ophelia and Hamlet are somehow brother and sister, as hinted in literary sources like Saxo and historical situations like that of imperial Rome. It is the recognition and denial, tragic in Roman Catholic and universalist contexts, that any two people might be or are essentially siblings. After all, brothers and sisters who fall in love with each other sometimes deny they are kin. They may argue that particular kinship doesn't matter, in which case they become absolute libertines, like Claudius perhaps. Or they may seek to replace or transcend the particular kinship that makes them brother and sister with a universal kinship that makes them friar and nun, or Brother and Sister, in which case they become absolute celibates. The situation where siblings become Siblings out of love for each other is not unusual. Many of the saintly founders of the Catholic religious orders loved each other perhaps a little too amorously, and, in an attempt to transcend their particularized affections, endorsed that universal love of a Brother or Sister for all human beings, including a sister or brother. According to the terms of this love, physical kinship no longer counts for much, and sexual intercourse with a person previously regarded as a consanguineous sister or brother would be regarded as no worse (or better) than with any other person.[24]

Hamlet, I would speculate, involves just such a transition from brother-sister to Brother-Sister love. Having loved his collactaneous sister from infancy, Hamlet begins later in life to think of her not only as a buddy but also as a lover. All is well in young Hamlet's Edenic, paradisiacal, or innocent "garden of love," as Blake might call it,[25] until he becomes aware of the horror in sibling incest: until he experiences in *thought* the identity between chastity and incest—that is, between Hamlet's own "innocent" relationship to his collactaneous foster sister Ophelia and his uncle/father's "guilty" relationship to his sometime sister Gertrude. For Hamlet, the thought of marrying his beloved sister now appears just as taboo as Claudius' having married Gertrude. His new knowledge of chastity and incest, or absolute "good" and absolute "evil," changes his love into something like cruel indifference, on the one hand, and near-worship on the other. Cruel to be kind (3.4.180), Hamlet says to Ophelia, "I did love you once"; then he reverses himself, saying, "I loved you not" (3.1.115, 118–19). Because he fears the incestuous aspect of his brotherly love for Ophelia, Hamlet withdraws intimacy from his "sister" just as he hopes that Gertrude will withdraw intimacy from her "brother" Claudius.[26]

Toward the end of the play, Hamlet howls at brother Laertes—who embraces the dead Ophelia in her grave as Hamlet will soon embrace him—that

> I lov'd Ophelia. Forty thousand brothers
> Could not with all their quantity of love
> Make up my sum.
>
> (5.1.264–66)

What distinguishes Hamlet's relationship to Ophelia as brother from that of an ordinary consanguineous brother and "reckless libertine"—which is how Ophelia jokingly refers to Laertes (1.3.49)—is the desire to transform his ordinary sexual love into a love that transcends both physical sexuality and consanguineous kinship. Hamlet would love Ophelia not as a sister but as a universal Sister to all humankind.

In the nunnery scene (3.1) Hamlet wants Ophelia to transform such marriage vows as he had once made to her as lover ("almost all the holy vows of heaven"—Polonius had called them "mere implorators of unholy suits") into the nun's most holy and heavenly oath of marriage to God (1.3.114, 129). Five times, Hamlet demands that Ophelia "to a nunnery go" (3.1.121–51). Ophelia should go both to a nunnery, because she is whorelike, and to a brothel, because she is nunlike. Critics who, like J. Dover Wilson, insist that Hamlet's *nunnery* means primarily "whorehouse" ignore the vacillation in *Hamlet* between saintly Sisterhood, with its holy liberty from the flesh verging on spiritual incest with God, and profane sisterhood, with its commercial liberty from the flesh verging potentially on literal incest with human beings. Ophelia in *Hamlet,* much like Isabella in *Measure for Measure,* vacillates precisely between the prostitute's "beautification" and the nun's "beatification." In Shakespeare's *Pericles, Prince of Tyre,* the nunlike daughter Marina, a latter-day Virgin Mary, incestuously gives birth to her own father in a sort of reconstituted brothel: "Thou that beget'st him that did thee beget."[27]

The ceremony of becoming a Sister that Hamlet here requires Ophelia to undergo generally comprises three parts: the service for the dead, in which the novice Ophelia would die to the world and hence free herself of her consanguineous ties to such persons as ordinary parents, siblings, children, and spouses; the service for rebirth; and the marriage service by which the reborn and renamed Ophelia would become the *sponsa Christi* and God would become her only father, brother, child, and husband. None but a nun can have this incestuous, yet chaste, fourfold relationship to another being. The evil of incest can be undone only by a nun's relationship to God. ("Bad child, worse father, to entice his own/To evil should be done by none.")[28] Hamlet wants Ophelia, whom he thinks of as (of) a sister and as (of) a lover or wife, to become a Sister so that the oppressive taboo might be annulled.

That Hamlet wants the sisterly Ophelia to become a Sister—to give over her old paternity—helps to explain why his principal question to Ophelia in the nunnery scene is "Where's your father?" (3.1.130).[29] Ophelia answers that he is "at home." Perhaps she is purposefully lying about his whereabouts. (Doesn't Ophelia know that Polonius is "seeing unseen" this very scene?—3.1.33.) Or does she think that Polonius is *not* her father? ("Have you a daughter?"—2.2.182). But one way or another, Hamlet's question seems connected with the requirement that a nun break off her earthly kinships. Whether or not Polonius is Ophelia's genitor (which he thinks he is: "I have [a daughter]"—2.2.183), the real father of Ophelia as novice is that God who is, like Our Father in the Pater Noster, not "at home" but "in Heaven."

A MOTHER TONGUE

Is't not possible to understand in a mother tongue?
—*Hamlet,* 5.2.125 (Tschischwitz ed.; conjectured by Johnson)

For what should I be called, my son?
—Alexander Neville, *Oedipus* (1563)

Nuns, modeling their family relations on those of Mary, stand in a fourfold relationship to God. They are mothers of God as Mary was the mother of Jesus, daughters of God as Mary was the daughter of God the Father, wives of God as Mary was the wife of God, and sisters of God as Mary was the sister of Christ. (As we have seen, Queen Elizabeth provides a standard definition of this fourfold kinship in her version of Marguerite of Navarre's *Miroir de l'âme pécheresse.*)[30]

At first blush, it might seem that no being could be more different from the chaste nun that Hamlet hopes Ophelia will become than the incestuous mother that he fears Gertrude is. But when we compare the physical kinship relations Gertrude has to earthly rulers with the spiritual (ghostly) ones a nun has with God, the structural identity between Hamlet's mother and his near fiancée emerges. For Mother Gertrude stands in the same relationship to the rulers of Denmark as a Sister stands in relation to God. She is the wife of one king (Old Hamlet) and perhaps also another (Claudius), and the mother of a man who might properly be king (Young Hamlet); according to Belleforest, Geruthe = Gertrude is "daughter to a king."[31] Perhaps most important in this play where figural and literal kinship are often conflated, Gertrude is also a king's sister.[32]

From Gertrude's fourfold relationship to the king that was and is the question of these wars (1.1.114), it would seem that the royal succession in Denmark goes with the queen. *Hamlet* suggests a wide variety of other ways, of course, to execute succession. There is dueling as entered into by Old Hamlet and Old Fortinbras, warfare as entered into by Fortinbras and Poland, consanguineous kinship of the sort Hamlet claims,[33] and filial adoption by the king of the sort Hamlet gets and Laertes wants. Sometimes Denmark appears to be a fratriarchy where the oldest brother passes the kingdom on to the next; at other times it seems to be a fraternal diarchy where two brothers share a kingdom, as in Saxo's version of the story[34] and as in the Roman imperial history of the peaceful co-Emperor brothers Marcus Aurelius and Lucius Aurelius Verus. Moreover, we encounter hints of popular election of the sort that would choose Laertes and also of aristocratic election of the sort that would choose Fortinbras.[35] Yet throughout the play the queen is not only part of the kin of Denmark; like an earthly Mother Superior, she is also transcendent to the kin.[36] She is the quintessential Greek *guné.* In Denmark the polyandric rule of succession is that he who would be king must marry the queen.

Gertrude, as "the imperial jointress to this warlike state" (1.2.9), thus conjoins four different rulers:

1. As daughter she receives the kingdom—a jointure, or dowry for a husband—from Gertrude = Geruthe's father Rodericke.[37]

2. As wife she was involved with the gift of the kingdom to her first husband Old Hamlet = Horvendile—perhaps as a dowry settlement. Upon his death she got it back. (Coke-upon-Littleton writes that a jointure is "a competent livelyhood of freehold for the wife, of lands and tenements; to take effect, in profit and possession, presently after the death of the husband; for the life of the wife at least.")[38] And then she was involved with the gift of the kingdom to her second husband Claudius. In *Hamlet* Claudius tells the filial Laertes that the "jointress" Queen is "conjunctive" to Claudius (4.7.14). It is as if Claudius became king only through his wife—as when in 1567 James Bothwell, sometimes said to be the illegitimate father of King James, had the title of King conferred upon him thanks to his marriage with Mary Queen of Scots.

3. As sister Gertrude gave the kingdom to her sometime brother Claudius, helping him to consolidate power through marrying him.[39]

4. As mother whom the son must marry in order to be king, Gertrude is the stuff of the play. In Saxo's account, indeed, it is explicit that the son must marry the queen mother as a precondition for coming to possess the kingdom. For Amleth = Hamlet, as for Claudius and Old Hamlet, "the sceptre and her hand went together."[40] During his adventures in Britain, Amleth, who is already married to the daughter of the King of Britain, has a sexual affair with the queen whom his royal father-in-law sent him to woo on the King's behalf; Amleth's eventual bigamous marriage to his queenly mother-to-have-been gains him the kingdom, according to the tradition that "whomsoever she thought worthy of her bed is at once king, and she yielded her kingdom with herself."[41] In *Hamlet* the filial Laertes is similarly told that the throne is to be conjoined somehow *with* the queen ("we will our kingdom give, / Our crown, our life, and all that we call ours"—4.5.205–6; compare Claudius' claim that Gertrude is "conjunctive to my life and soul"—4.7.14). In 1581 Mary Queen of Scots, imprisoned by Queen Elizabeth since 1568, desperately stated that she would allow her son James title to reign as king of Scotland conjointly with herself.[42]

Gertrude in *Hamlet,* like Clytemnestra in the *Oresteia* and Jocasta in *Oedipus the King,* might be viewed as "an instrument for conveying the property to another male." By virtue of relations with father, husband, brother, and son—by virtue also of her becoming "one flesh with her husband[43]—she is the unionizing joint of a matrilineal and vestigially patriarchal state.[44]

If, as Stanley Cavell writes, "Gertrude's power over the events in Denmark has not been fully measured,"[45] it is because, like any secularized *sponsa Christi,* she lacks "character."[46] But she makes up for this lack in her polyandric structural position in family and state: a position that allows her to join and separate rivalrous fraternalistic powers. It is a role that arbitrates war and peace. For what keeps kings at warlike peace is "a commere 'tween their amities . . . (5.2.42)."[47] "Commere" means something like "godmother," but it also has a salacious aspect involving *commatres*, or Sisters, who commit incest with Brothers.[48] In many polyandrous societies, where fathers are called "uncles," a fraternal peace reigns among husbands who call one another "brother"; but there is no such peace in Denmark—and most especially not for Hamlet, who calls Claudius "uncle-father" (2.2.372) and is unseemingly called upon to marry his "aunt-mother."

INCEST AND MATRICIDE IN ROME

Soft, now to my mother.
O heart, lose not thy nature; let not ever
The soul of Nero enter this firm bosom;
Let me be cruel, not unnatural.

—*Hamlet*, 3.2.383–86

The history of imperial Rome pervades the story of *Hamlet*'s Denmark. There are the figures of Aeneas, the Trojan founder of Rome in the wake of the Trojan War; Marcus Junius Brutus who killed Julius Caesar—Brutus' natural father, perhaps, and his mother's lover; Lucius Junius Brutus, whose name means "dim-witted" and is connected by Belleforest with the Danish *Amleth;* Seneca, philosopher and author of tragedies; and Roscius, the well-known actor.[49] Among other links between Hamlet's Denmark and ancient Rome is the analogy between Hamlet and Nero, nephew-son of the Roman Emperor Claudius. Claudius was the first Roman Emperor to set foot in Britain. (The nurse in the Senecan play *Octavia* has it that Claudius was "the man who first made British necks to bow,"[50] which is one theme of *Cymbeline*.) And together with the Emperor Claudius "Gothicus"—who was called both "uncle" and "father" by Emperor Constantine the Great (the probably illegitimate son who "Christianized" Rome in the fourth century and thus founded Roman Catholicism)—Claudius helped to provide *Hamlet* with its villain's name.[51]

Certainly when it comes to their actual or potential relations with their mothers, Hamlet's situation is like that of Claudius' eventual successor, Nero. Consider, for example, the role of Agrippina, Nero's mother, as imperial jointress of the state during a period when imperial Roman mothers played a key role in choosing or making successors.[52] The powerful Agrippina was in Roman history what Gertrude is in *Hamlet*'s Denmark and what the Virgin Mary is in Roman Catholicism: "the daughter, sister, wife and mother of an Emperor," as Tacitus says.[53] (In Racine's *Britannicus,* Agrippina arrogantly and aptly calls herself "the daughter, wife, sister, and mother of [the Romans'] masters.")[54] Agrippina was the daughter of Germanicus, the sister of Caligula, the wife and niece of Claudius, and the mother of Nero. In order to become ruler and become a free man (*liber*) in his own right, Nero had to unite carnally with this matrilineal jointress—just like the hero of the *Oedipus* of

Seneca (Nero's actual tutor) when he marries his mother Jocasta. Tacitus and Suetonius, both influential writers in the Renaissance, stressed this incest,[55] and English Renaissance writers refer to Nero as one who "ripp[ed] up the womb / Of [his] dear mother" and gained thus advancement into the womb of his mother just as he gained advancement to the throne of the Empire.[56]

After this union with his mother Agrippina, Nero had to separate entirely from her by such means as filial matricide or maternal suicide—as perhaps does Oedipus from Jocasta.[57] Tacitus draws attention to the suicidal aspect of Agrippina's death when he tells how Agrippina had been warned by astrologers that a son of hers would become Emperor if he were to kill his mother. The ambitious Agrippina thus has to choose between killing her infant son now or being killed by him eventually. With suicidal courage and with eyes on the imperial prize, Agrippina—unlike Laius and Jocasta when confronted with a similar dilemma—chose her own death in order that her son might rule. "Let him kill me," Tacitus reports that she responded to the oracle, "provided he becomes Emperor."[58]

Gertrude in *Hamlet* plays a similar role in this context to that of Agrippina in imperial Rome. For so long as Gertrude lives, Hamlet can come to life politically only through incest with her or through her death. But Hamlet wants—or he strives to want—neither to marry nor to kill Gertrude. He wants merely to outlive her in order to succeed to the throne. Hamlet indirectly expresses his hope for Gertrude's death in his delight at the story of Pyrrhus the matricide, which is told from the Trojan/Roman perspective to the Carthaginian widow Dido. The name "Pyrrhus" should be understood here not only as Pyrrhus the father, that is, the Greek Achilles, who killed Priam as Hecuba watched. (That is the usual interpretation.) It should also be understood as Pyrrhus the son, that is, Neoptolemus, who was the son of Achilles and who killed his mother, or stepmother, Polyxena at the bidding of Achilles' ghost.[59] Pyrrhus's matricidal murder of Polyxena is like the murder that Hamlet the son fears he wants to commit and that Hamlet the father warns his son against committing: "Taint not thy mind nor let thy soul contrive / Against thy mother aught" (1.5.85–86).

Gertrude in *Hamlet* may understand that her continued existence bars Hamlet from succeeding to the throne in chaste and nonmurderous fashion, that is, that only her death would free Hamlet to succeed to the throne without incest.[60] (Agrippina and Jocasta, each in her own way, has this understanding.) Does Gertrude, as loving mother, have an inkling that the cup that Claudius gives to Hamlet with its awful "union" (5.2.269, 331) is poisoned? Does she overhear perhaps the conspiratorial conversation between Laertes and Claudius? (She interrupts them just as Claudius describes how he will supply Hamlet with a poison drink—4.7.161.) Apparently testing the drink on behalf of Hamlet and thus saving Hamlet for the throne, Gertrude commits a suicide that allows Hamlet his momentous "advancement" from womb to throne (cf. 3.2.331). Hamlet thus manages, in fact, to outdo the rule that he who would be ruler must marry or kill his mother. He outlives Gertrude for a few moments. His political triumph, a Pyrrhic victory, resides in the few moments that he enjoys as dying king apparent.

THE PLAY-WITHIN-THE-PLAY

So you mis-take your husbands
—*Hamlet,* 3.2.246

Hamlet hesitates to kill Claudius not only because this would mean killing a kinsperson but also because he fears that a newly widowed Gertrude would want to marry him according to the imperative of the Player Queen: "None wed the second but who kill'd the first" (3.2.175). By this imperative the Player Queen, like widow Dido, seems to condemn outright all remarriages.[61] That a widow should wed *none,* or no-one, and live like a *nun,* or Sister, is presumably what Hamlet wants for Gertrude and what he will try to prepare her for in the bedroom scene. (Tertullian's *De monogamia* calls all second marriages adultery.)[62] But what Baptista actually says is that a widow should marry her husband's killer; and marrying her husband's killer is what the Player Queen in the dumbshow seems to do and what Gertrude herself has done. According to the terms of dramatic irony, moreover, the Player Queen will do what she says: "O, but she'll keep her word," says Hamlet (3.2.226).

Hamlet, were he to kill Claudius, might have more to fear than marriage to his mother: The "adulterate" Gertrude may have conspired in the murder of her first husband. (Recall that Queen Mary was charged with helping to poison her first husband Darnely, the *pater* if not the genitor of King James.) Who could say whether Gertrude might not kill a second time? Hamlet indirectly hints that Gertrude is a husband-killer when he says of his own killing of Polonius that it was "Almost as bad, good mother, / As kill a king and marry with his brother" (3.4.28–29). In earlier versions of the Hamlet story, the Danish people believe that Geruthe murdered Old Hamlet in order that she might live in adulterous incest without restraint. In *Hamlet* there is no clear-cut evidence that Gertrude is a regicide like Clytemnestra or the husband-poisoner Agrippina. But the figures of the wife-killing King Henry VIII, father of Queen Elizabeth, and the husband-killing Mary Stuart, Queen of Scots, mother of King James, hover uncertainly about this play.[63] And there is in *Hamlet* the implication that Gertrude is figuratively a multiple murderer of her husband.[64] For according to the Player Queen, "A second time I kill my husband dead, / When second husband kisses me in bed" (3.2.179–80). Each time Gertrude sleeps with Claudius, she is accessory to the murder of Old Hamlet: she is guilty of killing a king.

Hamlet identifies with his father and his uncle. So he fears that if he were to kill Claudius, Gertrude would marry and then kill him. Remarkably, the counterpart to Hamlet in both Saxo's and Belleforest's versions of the story is killed by his wife Hermetrude = Gertrude, who conspired with Hamlet's uncle Wigelunde in the murder of Hamlet and eventually marries Wigelunde.

Just as the Player Queen, when she says "I kill my husband dead" (3.2.179), virtually counsels kin murder, or fratricide, so the Player King, when he encourages the Queen to find after his death "one as kind / For husband" (3.2.171–72), effectively counsels kin marriage, or incest. Men, if they should not be their brothers' keepers ("kindly"), should be at least their brothers' brothers ("kin"); but in *Hamlet* men are their brothers' killers—as Cain was Abel's and Claudius Old Hamlet's. And, if men should not be their sisters' keepers, they should be at least their sisters' brothers; but in *Hamlet* they are their sisters' sexual partners—as Cain was Calmana's and Claudius Gertrude's. In *Hamlet,* things are "a little more than kin, and less than kind" (1.2.65), and kinship makes only for fratricide and incest taken together ("the primal eldest curse"—3.3.37). In this context what can one do but be cruel in the name of kindness? "I must be cruel only to be kind" (3.4.180).[65]

The queen is really the thing of the play. Though Hamlet affects an interest in catching "the conscience of the King" (2.2.601), he seems really to want to catch that of the queen ("If she should break it now"—3.2.219).[66] Thus, too, the ghost of Old Hamlet, though it calls for revenge against Claudius, actually changes Hamlet's mind less in relation to Claudius, whom Hamlet already despised, than in relation to Gertrude, against whom the Ghost tells Hamlet not to "taint" his mind (1.5.85–86) and of whom Hamlet comes to think in terms of what Gifford calls "filthy lusts."[67]

During the play-within-the-play, for example, Hamlet surveys Gertrude for a sign that she recognizes that she is an incestuous husband-murderer, and he hopes that Gertrude will now make a commitment never to marry again. (He hopes for a less ambiguous promise than the Player Queen's: "Both here and hence pursue me lasting strife, / If, once a widow, ever I be wife"—3.2.217–18.) But he has grounds to fear that Gertrude will not "break" off either her marriage ("The lady doth protest too much, methinks"—3.2.225) or her custom of marrying the murderous kinsman of a former husband. For soon after the performance of "The Mousetrap," Gertrude invites Hamlet, the would-be murderer of her husband Claudius, into her "closet ere [he] go to bed" (3.2.322–23).[68]

Hamlet receives Gertrude's invitation to the closet with trepidation. He is afraid of being driven, like Nero, to commit incest or matricide or both. Although Hamlet decides to accommodate Gertrude ("We shall obey, were she ten times our mother"—3.2.324), he wishes that Gertrude were not his mother ("would it were not so, you are my mother"—3.4.15). If Gertrude were not, then sexual commerce with her would of course not be maternal incest, and killing her would not be matricide. But since she is, Hamlet's decision to visit her in the bedroom is an act of moral courage—he knows that he will have to face down the challenges of matricide and maternal incest as well as his own death.

Belleforest remarks that "the thing that spoyled this vertuous prince was the over great trust and confidence hee had in . . . Hermetrude."[69] Gertrude's bedroom is no less dangerous to Hamlet, who "lacks advancement" (3.2.331), than the lair of the Sphinx was dangerous for Theban boys, strangled by her sphincter muscles: she caught her victims in the vagina or birth canal during intercourse, killing them on their advance towards a womb, or *delphos,* almost as if they had never advanced from one.[70] For a time, Nero was trapped in just this way in the incestuous grip of his mother, Agrippina. Hermetrude, a Sphinx-like "Amazon without love . . . who despised marriage with all men . . . in such a manner that . . . there never came any man to desire her love but she caused him to loose his life," seduced Hamlet and then conspired with his uncle Wigelunde to kill him.[71] Hamlet's visiting this mother is the moral turning-point of the play.

HYPOCRISY AND ACTING

I will speak daggers to her, but use none.
My tongue and soul in this be hypocrites.

—*Hamlet,* 3.2.387–88

Freud argues that in *Hamlet,* "the child's wishful phantasy of doing away with his father and taking his father's place with his mother is repressed."[72] Yet Hamlet does not hide the fact that he has wishes, even needs, along lines patricidal and incestuous. He brings them into the open in the dumbshow, for example, and in the many comparisons of himself with Claudius, who killed Hamlet's father and married Hamlet's mother, and with Nero, who committed incest with his mother, Agrippina, and killed both his mother and his adoptive brother, Britannicus. And *Hamlet* brings the desire to commit kin murder and incest into the open by other means, as in the references to the two men named "Pyrrhus," who between them murdered a father and killed a stepmother (upon the urging of the ghost of a dead father), and to "Cain," who murdered his brother and committed incest with his sister. What is really unique about Hamlet is not his unconscious wish to be patricidal and incestuous, but rather his conscious refusal to actually become patricidal and incestuous. Hamlet is quite clear about this in his prayer: "Let not ever / The soul of Nero enter this firm bosom" (3.2.384–85). Hamlet would be neither Nero nor Neronian.

Like Hamlet, Nero was a playactor, playwright, and spectator. While Rome burned, he dressed for the stage and sang the "Sack of Ilium," the tale that Hamlet asks the Player to recite (2.2). While watching tragedies, Nero would weep real tears over such fearful and pitiable sufferings on stage as he himself inflicted on others in real life: "What's Hecuba to him, or he to her, / That he should weep for her?" (2.2.553–54).[73] Nero's favorite plays included *Orestes the Matricide* and *Oedipus Blinded,* articulating themes which have drawn the attention of Shakespearian scholars.[74] A third play, equally relevant to *Hamlet* and to Nero's stagecraft but overlooked by the critics—they generally focus only on problems of intergenerational incest—is *Canace parturiens* [Canace Giving Birth]. In the Renaissance version of this play, by the Italian Speroni (1546), a sister gives birth on stage to the child of her incestuous union with her twin brother Macareus, on the twin siblings' common birthday. *Canace parturiens* thus argues against generational difference and succession in a more radical manner than do the other two plays. It levels all differentiation along lines of both kinship (where homogenization means incest) and gender (where it means androgyny).[75] It is surmised by his biographers that Nero delighted to act in such plays as the *Orestes, Oedipus,* and *Canace* because he committed incest with his mother, killed her, and committed both crimes against his adoptive and step- siblings.

Hamlet, confronted with situations similar to those facing his Greek and Roman counterparts, denies himself the way of Oedipus—that is, to unknowingly commit maternal incest and patricide—the way of Nero—that is, to knowingly commit maternal incest and matricide—and the way of Orestes—that is, to knowingly commit matricide. Equally important, Hamlet denies the method of Macareus—that is, to knowingly commit sibling incest. Hence not wanting to commit the acts that Nero performed both on stage and in life, Hamlet must play the scene in his mother's bedroom as though Gertrude both were and were not his mother ("would it were not so . . ."). He becomes an actor in life—literally, in Greek, a *hypocrite* ("My tongue and soul in this be hypocrites"). In the first scene at the Danish court, Hamlet excoriated hypocrisy as a vice ("I know not 'seems' "—1.2.76). But now, fearing that he sincerely wants to commit matricide, Hamlet wants to be other than he is. He wants to keep his acts inadequate to his wants.

Instead of a choosing to become a hypocrite, Hamlet might choose to be nothing, as in a suicidal wish "not to be" (3.1.56), or to do nothing, as in a wish to be "neutral to his will and matter" (2.2.477). He might want to become like Pyrrhus the son, pausing—perhaps forever—before killing the archetypal parent:

> So, as a painted tyrant, Pyrrhus stood,
> And like a neutral to his will and matter,
> Did nothing.
>
> (2.2.476–78)[76]

Doing nothing, instead of becoming hypocritical or murderous/incestuous, is the basis for melancholic delay in *Hamlet.*

Pretending to be virtuous is, for Hamlet, the first step toward becoming really virtuous. Thus he demands of his mother, "Assume a virtue if you have it not" (3.4.162). Gertrude, whom Hamlet believes to have married Claudius for "base respects of thrift, but *none* of love" (3.2.178), should now hypocritically "put on" a "frock or livery" like the habit of a *nun* (3.4.167, 166). The livery is artful, of course, but the "habit" of wearing it will eventually transform Gertrude into a chaste woman.[77] The medieval saying, *Cucullus non facit monachum* [The cowl, or hood, does not make the monk], which the Fool in *Twelfth Night* uses against "Madonna" Olivia, is here reversed. In *Hamlet* the actor's habit really does make the nun, or so it seems.[78]

"Such an act / That blurs the grace and blush of modesty, / Calls virtue hypocrite . . ." (3.4.40–42). Hamlet resolves to "speak daggers to [Gertrude], but use none" (3.2.387). Though his "tongue and soul in this be hypocrites," Hamlet will show Gertrude "the inmost part" of herself (3.4.19). This part is her soul, which Hamlet hopes to reveal and cure by speaking verbal daggers into her ear

(3.2.387).[79] Or it is her gut, which Hamlet might spill with the same dagger he used to kill Polonius. Or perhaps that part is her womb, toward which a man advances with his penis—or "dagger," in the Elizabethan idiom. Gertrude fears that Hamlet plans to use his dagger to spill her gut: "Thou wilt not murder me?" (3.4.20). Her fear is not obtuse.[80] But when mother and son are left "all alone," "nature makes [mother and son] partial" (3.1.184, 3.3.32). Without such external restraint on the will of mother and son as a living Polonius might have been or represented, the way is open for Hamlet to advance to that inmost part of Gertrude which is her womb.

The famous kiss in the Hamlet story is at once chaste, of the kind that mothers are ordinarily expected to give sons, and unchaste, of the kind that mothers are ordinarily expected not to give sons. Belleforest, anticipating a debate about the quality of the kiss, protests that Gertrude kisses Hamlet in the bedroom in the chaste manner, "with the like love that a virtuous mother may or can use to kisse and entertaine her owne childe."[81] But the relationship between Hamlet and Gertrude, if it is not incestuous in the particularist sense, becomes so in the universalist sense. *Hamlet* gradually transcends the difference between chastity and incest in a higher condition of general liberty that threatens the political and familial order as a whole: "His liberty is full of threats to all" (4.1.14).

SHAMELESSNESS

. . . proclaim no shame . . .

—*Hamlet,* 3.4.85

The conversation between Hamlet and Gertrude in the closet indicates, we shall see, a moral condition beyond the ordinary distinction between chastity and incest (literally, unchastity), or good and evil. Tacitus assigns to Nero's mother, Agrippina, the incestuous shamelessness that Hamlet fears in Gertrude: "Agrippina through a burning desire of continuing her authorite and greatness grew to that *shamelesnes* that in the midst of the day, when Nero was well tippled and full of good cheer, she offered herselfe to him drunk as he was, trimly decked and readie to commit incest."[82] In Shakespeare's play Hamlet imagines the same shamelessness:

> . . . O shame, where is thy blush?
> Rebellious hell,
> If thou canst mutine in a matron's bones,
> To flaming youth let virtue be as wax
> And melt in her own fire; proclaim no shame . . .
> (3.4.81–85)

If it is shame or moral cowardice that alone keeps most men good ("conscience does make cowards of us all"—3.1.83), then Hamlet's qualified proclamation of shamelessness announces a liberty to murder and make love with whomsoever he chooses, including mother and father. If his "matron" (3.4.83) is thought of as Gertrude and the "flaming youth" is thought of as Hamlet himself, then his proclamation of shamelessness amounts to a call for father-daughter and mother-son incest.[83] Some readers might say that Hamlet should "blush for shame" ("You Neroes, . . . blush for shame!") in speaking words that can be so interpreted. (By the same token they might say that Nero should have blushed when he contemplated matricide and incest—and that Laertes should have blushed when he contemplated cutting Hamlet's throat "'i th' church" [4.7.125].)[84] However, Hamlet pitches his appeal to shamelessness to a level beyond that of the ordinary "sinner," who generally commits acts he himself believes to be wrong in the eyes of others or of God: Hamlet's pitch is to the level of an extraordinary "saint," from whom all external compulsion to be good, including shame and moral cowardice, has been removed.

In *Hamlet,* transcending the difference between "good" and "bad" ("there is nothing either good or bad but thinking makes it so"—2.2.249–50) means transcending that between kin and nonkin and hence the ordinary taboo on incest. The condition of "no shame" thus implies a withdrawal from particularized kinship with one's family—whether as a religious celibate (saint) or as a secular libertine (sinner). Hamlet is thus like the Roman Coriolanus, who wants to divorce his mother Volumnia,[85] or the Christian Jesus, who tells Mary that she is not His parent. Hamlet wants Gertrude to be not his kinsperson but a stranger, not a mother but an other: "Would it were not so, you are my mother." If Gertrude were no longer his mother, then killing or sleeping with her would be no better or worse than killing or sleeping with any other woman. Hamlet's wish to alienate Gertrude from himself as son thus expresses a cosmopolitan and universalist—Christian—love of liberty like that of the Stoics.[86]

Ordinary rules of kinship entail regulations concerning kindness to particular kinspersons ("Be the keeper of your brother") and concerning sexual activity with kinspersons ("Do not have sexual intercourse with your mother"). So divorcing particular kinspersons, or transcending altogether the standard of particularized kinship, is often the precondition for those ordinary people—that is, for those people who are neither saints who celebrate freedom from the flesh nor natural men who celebrate freedom of the flesh—who want to commit acts of revenge or incest in a psychologically bearable fashion. In Belleforest's story, when the king of England learns that his son-in-law Amleth = Hamlet has killed the king's old friend Feng = Claudius, the king is torn between the obligation to avenge the latter's death, upholding sworn faith to a friend—the king "had determined of old, by a mutual [fraternalistic] compact, that one of them should act as avenger of the other"—and the obligation to protect the life of a kinsperson, upholding marital affinity. The king chooses the former obligation.[87]

In the same manner, the familial bond that kinship presupposes can be replaced by a fellows' obligation to revenge. In Belleforest's equivalent of the closet scene, the revenge-seeking Amleth attempts to divorce his uncle-father—and, significantly, his aunt-mother:

> I, for my part, will never account Feng for my kinsman nor once knowe him for mine uncle, nor you my deer mother, for not having respect to the blud that ought to have united us so straightly together, and who neither with your honor nor without suspicion of consent to the death of your husband could ever have agreed to have maryed with his cruell enemie.[88]

The effect is to neutralize such terms as kin-murder and incest: so long as Claudius and Gertrude are not kinspersons, killing one and sleeping with the other is merely adultery and murder.

The transcendence of particularized kinship to the point where one person can play four family roles at once—a point figured by Gertrude as jointress and Ophelia as nun—sometimes requires not so much divorce of particular kinspersons as recognition that we do not know who our particular kinspersons are insofar as sexual generation is always subject to surreptitious substitutions in the bedroom and cradle. In this context, the whole world, though it is neither the seat of the Roman imperial family nor the nunnery of Roman Catholicism, is a masquerade peopled with hooded Sisters and "mobbled queen[s]" (2.2.498). At this masquerade, the reveler is like a bedtrick dupe who does not really know whom he kisses. The whole world is a bedtrick-ridden play where the lover does not really know whom he sleeps with.[89]

Denmark is this endless game of hoodman-blind or blind man's buff. Here blindfolds remain over the eyes forever and cousins pass for noncousins as easily as two coins of the same denomination. Hamlet accuses Gertrude of being a dupe in such terms: ". . . What devil was't / That thus hath cozen'd you at hoodman-blind?" (3.4.76–77). Gertrude has been cozened, or tricked, into taking one cousin, or kinsperson, for another ("This was your husband . . . / Here is your husband"—3.4.63–64). She has taken Old Hamlet for Claudius, Hyperion for a satyr, or vice versa.[90]

Where liberty requires incest or where the inascertainability of parenthood makes incest inevitable, human beings must choose to be either saintly friars or sinful libertines, to be either chastely hooded monks or incestuous players at hoodman-blind. A friar, universally akin to all men, transcends the incest taboo by becoming a Pauline "eunuch for God" through a "castration of the heart."[91] A libertine, in contrast, delights unashamedly in "sullies" and he acts out the Parisian "liberty" of which Polonius asks Reynaldo to accuse Laertes,[92] whom his sister Ophelia taunts with the term "libertine" (1.3.49; cf. 2.1.32). For a libertine, with his liberty of the flesh, as for a friar, with his liberty from the flesh, sexual intercourse with any person, including a sister or mother, is as chaste or unchaste as with any other person.

In *Hamlet,* then, it is not only "revenge [which] should have no bounds"

(4.7.127) but also love: "No place indeed should murder sanctuarize" (4.7.126).[93] In the moment between Hamlet's outcry—"Proclaim no shame!"—and Old Hamlet's appearance in the bedroom, there is "no place" that is holy for Young Hamlet. Or, to put it otherwise, every place is holy for him. No place should incest sanctuarize unless, as in the Christian church—with its adoration of the spiritually incestuous Holy Family and credo of universal siblinghood—it is every place.

CHRISTIANITY AND STOICISM

We are all sprung from heavenly seed. All alike have the same father, from whom all-nourishing mother [alma mater] *receives the showering drops of moisture.*

—Lucretius, *On the Nature of the Universe*

As Hamlet lies dead, the English ambassadors bring the news that "Rosencrantz and Guildenstern are dead" (5.2.376). What are Rosencrantz and Guildenstern to us that we, at this moment of high tragedy, should care for them? Rosencrantz and Guildenstern, I would suggest, do as "the indifferent children of the earth" (2.2.227). The societal end of the political institution of friendship that they represent is the same as the aesthetic telos of *Hamlet* itself.

Hamlet's friendly relationship to his Wittenberg buddies or alumni has an aspect of kinship. "Friendship," writes the anthropologist Shmuel Eisenstadt, "always involves an aspect of submerged kinship."[94] Hamlet and his university pals share the same alma mater, or nourishing mother—a Lucretian term suggesting common descent and nurture that the Renaissance humanists applied to the university.[95]

Equally important, Horatio's counterpart in *Hamlet*'s sources is Hamlet's actual nurse-sibling; and Horatio values his collactaneous tie to Hamlet more than his political obligation to the king: "Among these [men appointed by the King and Queen to attend Hamlet] chanced to be a foster-brother of Amleth, who had not ceased to have regard to their common nurture; and who esteemed his present orders [from the King and Queen] less than the memory of their past fellowship."[96] Not only Horatio (and, as we have seen, Ophelia) but also the "schoolfellows" Rosencrantz and Guildenstern were "of . . . young days brought up with [Hamlet], / And . . . neighboured to his youth and haviour" (3.4.204, 2.2.11–12)—as in the medieval institution of foster brotherhood in which young men growing up together became intimate friends.[97]

What were the characteristics of this familylike fellowship of alumni? One was equality. Hamlet's first words to Horatio—"Horatio, or I do forget myself" (1.2.161)—already suggest that he and Horatio are equal or even interchangeable beings. Hamlet would be "even" with Horatio: "Sir, my good friend, I'll change that name with you" (1.2.163; cf. 254). In the same vein, Hamlet begs Rosencrantz and Guildenstern to "be even and direct with [him]" (2.2.287). The indifference or interchangeability of Rosencrantz and Guildenstern, their twinlike coequality, is their memorable aspect (2.2.33–34). They are "Dead for a ducat, dead" (3.4.22)—like the bits of wergeld—literally man-money—for which the lives of their counterparts in the sources of *Hamlet* are exchanged.[98] The relationship between the two halves of this duo thus expresses a monstrous perversion of the fraternal or liberal principle of equality.

A siblinghood potentially open to all human beings not only collapses distinctions between families, including the distinction between Ophelia's family and that of her precisely ambiguous foster brother Hamlet, but also between classes. Thus Hamlet, playing the princely pauper, jokes liberally about how "the toe of the peasant comes so near the heel of the courtier he galls his kibe" (5.1.136–38); Polonius considers that the love between Hamlet and Ophelia may be appropriate after all (2.1.114–17); and Horatio, although he is no landed aristocrat like Osric and although he lacks "revenue," has for his fellow the apparent heir to Denmark (3.2.56–58). Horatio thus manifests in himself that Stoic spirit of leveling political distinctions and of indifference to the goods of this world (*adiaphora*)[99] which in *Hamlet* extends even to the physical world, where the night becomes "joint-labourer with the day" (1.1.81).

A fellowship in which all human beings are equal or indifferent siblings leads to one of several conditions. It may lead to academic celibacy of the kind that the scholarly brotherhood in Navarre in *Love's Labour's Lost* would practice. Or it may lead to religious celibacy of the kind that Brother Martin Luther had practiced in pre-Reformation Wittenberg. But as I have argued, such Brotherhoods as Luther's old order always involve problems of incest, Thomas More thus following the great Catholic tradition when he insists, in his attack on Luther's marriage to Sister Catherine, that membership in a Catholic Brotherhood makes any sexual intercourse incestuous.

One way for a society to skirt the problem of incest that the hypothesis of universal siblinghood entails is to encourage all human beings to remain at a presexual or supersexual stage of physical or spiritual development—like children or Blackfriars. "Consonancy of youth" and "ever-preserved love" (2.2.284–86) are the qualities that Hamlet praises in the Wittenberg fellowship. Joining a troupe of child-players with low statures and high voices—like the cry of child-players that performed at the Blackfriars in the 1590s[100]—would appeal to Hamlet if it might halt the process whereby one generation succeeds and is distinguished from the next—hence arrest "this fell sergeant, Death" (5.2.341). In the ordinary course of events, however, adulthood succeeds childhood, boyish voices crack (2.2.423–25), and child-players "exclaim against their own succession" (2.2.349).[101] Child-players become adult-players such as the ones with whom Hamlet—who play-acts at madness and produces a "Murder of Gonzago"—fraternizes and might have joked about getting "a fellowship" (3.2.271–72).

The society of "indifferent children of the earth" in *Hamlet* suggests a utopian fellowship of free sisters and brothers (*liberi*), like the fellowship we might profess to admire in *As You Like It*'s Forest of Arden. Here people of different generations and different birth orders can be or pretend to be "co-mates and brothers in exile."[102] But unless Hamlet and his fellow siblings somehow manage to avoid or transcend sexual maturity, they will become incestuous libertines, committing sibling incest, much as do Claudius and his Gertrude. The only real question is whether their incest will be earthly or Edenic/paradisiacal—sinful or guiltless.[103]

The Stoic basis for the Christian conjunction of saintliness and shamelessness is worth pursuing here because *Hamlet* is as much a Stoic tragedy, with its emphasis on "cosmic" kinship and "shamelessness," as it is a Sophoclean or Aeschylean classical tragedy. The Stoic thinkers invoked the Pythagorean notion of a "union of friends" (*haetery*) to hypothesize a republic transcending consanguinity, where people would share all things sexual and propertal and be fully equal. The Stoic "union of friends" was thus both political and familial—or it was neither.[104] According to the "cosmopolitan" theorist Epictetus, man in this transcendentally fraternal association is essentially neither Athenian nor Corinthian; man is *kosmos* and *huios theou.*[105]

For Plato, a fraternal fellowship approaching national autochthony is the noble lie of the ideal polity,[106] and Plato understands that fellowship to require universalist incest. Plato only retreats from stating this ideal requirement outright because he had too much shame (*aidṓs*) or because he judged it politic to feign shame.[107] The Stoics, on the other hand, were not ashamed and did not feign shame. For them, such distinctions as those between family members (e.g., between Hamlet and Ophelia) and between social classes (e.g., between Hamlet and Horatio) are transcended.[108] The Stoic Zeno, in his *Republic,* and Diogenes of Sinope, in his lost work of the same title, thus consider how fraternity in the *politeia* transcends *all* kin ties, as in a fully politicized version of the Greek fraternalistic ceremony of the *apatouria,* in which the barrier between children and parents (e.g., between Hamlet and Gertrude) is dissolved.[109]

This Stoical transcendence required a new shamelessness (*anaideia*). Thus Chryssipus asserted that virtue is merely conformity to nature and that vice is merely deviation from nature. Besides this, for the Stoics and Cynics, there is, in *Hamlet*'s terms, "nothing either good or bad but thinking makes it so" (2.2.249–50). In his *Republic* and in his *On Things for their own Sake Not Desirable,* Diogenes permitted marriage with mothers, daughters, and sons.[110] Diogenes of Sinope, a prototype of the Cynics and author of tragedies based on Stoic principles, says that incest is natural and that "what is natural cannot be dishonorable or indecent and . . . should be done in public."[111] The Stoic philosopher Zeno, who did not distinguish between the nature of men and that of dogs—*cynic* means "doglike"—says that incest is not wrong in itself.[112]

In some respects traditional Christianity does not differ from Stoicism in regard to shamelessness. (As I have suggested in chapter 1, sixteenth-century thinkers such as Montaigne were much influenced by Stoicist ideas.) In Christianity there is an innocent shamelessness, characteristic of many universalist and chiliastic sects, which affects the wholly natural and also the wholly saintly human being.[113] For such people, the sign of Pauline grace and liberty is doing what you want, including committing incest, without feeling guilty: "Where the spirit of the Lord is, there is liberty," said Saint Paul, and all things are allowed;[114] Rabelais' "Abbey of Thélème," which demonstrates a universalist kinship structure, has for its motto, "Do what you want."[115] According to some Christian groups, as we have seen, not to commit incest out of a fear of "disgrace" was itself sinful. Incest was actually required by some Christian sects, including the ancient society in Corinth and the Elizabethan Family of Love.[116] They, too, would "proclaim no shame."

FROM NUN TO NONE

Boy, What sign is it when a man of great spirit grows melancholy?

—*Love's Labour's Lost,* 1.2.2

The Door of Death I open found
And the Worm Weaving in the Ground
Thou'rt my Mother from the Womb
Wife, Sister, Daughter to the Tomb
Weaving to Dreams the Sexual Strife
And weeping over the Web of Life

—Blake, *For Children* (1793) [see fig. 7]

Caetera silere memineris. [On the rest see thou keep silent.]

—Amleth to his mother (Saxo, *Historiae Danicae*)

The tragedy of *Hamlet* resides in the revelation that marriage, monachist celibacy, secular libertinism, and imperial joinery—institutions which at first blush seem quite different from each other—are one and the same. Such a nun, or Sister, as Hamlet enjoins Ophelia to become in her relation to God and such a jointress or mother superior as Gertrude is in her relationship to the imperial ruler are structurally identical. Everything in *Hamlet* expresses this uniformity. In this sense Denmark is no different from Wittenberg, England, and Paris. They compose a prison nightmare with no exit. Like Sophocles' *Antigone, Hamlet* can end only in incest or death or both together.

In comedies about kinship, the dilemma that the incest taboo poses is often dissolved through the sublation of incest in some kind of marriage. In *Hamlet* this comedic dissolution is figured in the essentially brother-sister love between Hamlet and Ophelia. It is a love that might have ended in marriage—as Gertrude, who married her brother, had hoped it would end. (Ruskin calls Ophelia Hamlet's "true lost wife.")[117] But from the moment that Hamlet is called upon to abhor his mother's remarriage—or to avenge the death of his father—the possibility of such a marriage is denied.

Besides marriage, as with a sister (Ophelia or Gertrude), another comedic way

Fig. 7. Illustration from William Blake's For Children *(1793), with caption, "I have said to the Worm: / Thou art my mother & my sister." (Beinecke Rare Books and Manuscripts Library, Yale University)*

to attain resolution is figured in *Hamlet:* the way of universal fellowship in national *socii*. Such fellowship would dissolve the tragic problem that incest and its taboo poses by denying the possibility of incest in the ordinary sense. The destruction in *Hamlet* of such fraternities as the Wittenberg fellowship (Hamlet, Horatio, Rosencrantz, and Guildenstern), the cry of children in Old Hamlet's court (Hamlet, Horatio, Ophelia, and Laertes), and the association of child-players is of the greatest consequence; that is why the deaths of sister Ophelia and brethren Rosencrantz and Guildenstern, sometimes dismissed by critics as incidental to the main movement of the plot, figure prominently in its overall structure. The announcement by the English ambassador near the end of the play that "Rosencrantz and Guildenstern are dead" marks the close of an era.

"You do surely bar the door upon your own liberty," says Hamlet's old school chum, without a fraternalistic friendship (that is, "if you deny your griefs to your friend"—3.2.329–30). Rosencrantz threatens Hamlet with prison, or loss of liberty, should Hamlet refuse to share griefs with him; equal brothers should share everything. But friendship or siblinghood universalized as an association of *liberi* necessarily entails incest, as in the case of the foster siblings Hamlet/Ophelia and of the siblings Gertrude/Claudius, who are related both "in-law" and by carnal contagion.

Hamlet finally gets the familial and political "advancement" that he said he lacked and that we have hoped for him. He gets it in the regal moment following his mother's death, between the death of his uncle-father Claudius and that of Hamlet himself. And thanks to his liberal brother Horatio he gets more, or rather we do. From Horatio—who like his counterparts in the sources never puts obligation to the *polis* above that to friendship—Hamlet has expected no advancement (3.2.57). Yet it is Horatio who absents himself from felicity awhile to tell the story of Hamlet as Hamlet's body is brought "high on a stage" (5.2.383). The tale that this cosmopolitan and scholarly Stoic tries to tell is *Hamlet* once removed: instead of getting incest in the play we get the play itself.

Some people have said that *Hamlet* is a domestic tragedy of incest like Sophocles' *Oedipus the King*. But I am here arguing that *Hamlet* is a tragedy of incest denied, in which the liberal, or fraternal/sororal, community—the only community that offers hope for comedic liberty—is destroyed by an inherent contradiction within itself. The tragedy does not lie principally in the destruction of a regular nuclear family, or even of a larger "extended" family or nation, through the consequences of incest and kin murder—a destruction which is the theme of *Oedipus the King*. Rather, it lies in the destruction of siblinghood by the incest taboo that the nuclear family, or the larger extended family, presupposes and perhaps requires. In *Hamlet,* fratricide (such as that of Claudius) is punished only as siblinghood (such as Horatio's and Ophelia's) is destroyed forever.

From nun to none. From the start of his first soliloquy, Hamlet puts before us two kinds of death: an extraordinary or religious carnal deliquescence and an ordinary or secular suicide.

> O that this too too sullied flesh would melt,
> Thaw and resolve itself into a dew,
> Or that the Everlasting had not fix'd
> His canon 'gainst self-slaughter.
> (1.2.129–32)

Hamlet wishes for a "self-slaughter" that is at once in apposition to the "death" of Christian celibates, since they "die to the world," and in opposition to the Christian canon 'gainst self-slaughter. "Literal" suicide was endorsed, however, by such Stoic counterparts to Hamlet as the anti-Christian Emperor Marcus Aurelius and the moral philosopher Seneca (Nero's tutor). In this Stoic tradition, Montesquieu, in his *Considerations on the Causes of the Greatness of the Romans and of Their Decline,* admires suicide since it "gives every one the *liberty* of finishing his part on the stage of the world in which scene he pleases."[118] The ability to choose "not to be" (3.1.56), as did many a man like Horatio ("more an antique Roman than a Dane"—5.2.346), frees and equalizes everyone. And for Hamlet, the end is to become not a comedic nun (or friar) but none of living humankind. There is throughout the play a pervasive vacillation between ascent into absolute chastity (the nunnery) and absolute incest or death (universal homogeneity), but finally it is not the religious celibate's "death to the world" that obviates physical incest but rather Death. Death homogenizes everything in *Hamlet*;[119] the worm of death—"your fat king and your lean beggar is but variable service" (4.3.23–24)—is the true "Diet of Worms." It is the night of *la mort*—the centaurlike knight named "Lamord" (4.7.91)—that Hamlet, like the Stoic hero Hercules, has to face down.[120]

We are "indifferent children of the earth" because we are, like Adam, "kin" to dust (4.2.5). Our indifference to such emperors as Alexander and such kings as Old Hamlet resides principally in the dust of which we and imperious Caesar are made (5.1.196–209). "Do not for ever . . . / Seek for thy noble father in the dust" (1.2.70–71.)

Consider in this context of autochthony the Roman historian Livy's story about the brutish (i.e., "mad") Lucius Junius Brutus—the Roman political hero who was Saxo's and Belleforest's model for the Danish hero Hamlet.[121] Brutus, son of Tarquina, was the indigenous liberator of Rome from his uncle, the tyrannical Tarquin. Tarquin had murdered Brutus' father just as Claudius in *Hamlet* has murdered Hamlet's father. Brutus succeeded in liberating Rome from patriarchal and avuncular tyranny principally by acknowledging a universal siblinghood with all human beings and downplaying so-called blood kinship. Thus Livy writes that Brutus went with a group of aristocratic young men, including two sons of Tarquin, to the oracle at Delphi—literally, to the womb of the earth. The oracle spoke these words to them: "Which of you (O yong men) shal first kisse your mother, he shal beare chiefe and soveraigne rule in Rome." All but one of the young Roman bluebloods raced home to kiss his consanguineous mother—all but Brutus. He "touched the ground with his mouth and kissed the earth, thinking this with himself, that she was the common mother of all mortal men."[122] By thus denying particularized consanguineous kinship and embracing national autochthonous siblinghood Brutus liberated himself and politicized Rome.[123] He turned Roman society from one governed by a pseudopatriarchal tyranny into a society of free brothers, or "liberi"—a democratic republic of "liberty."

Livy's tale is pleasant, but it is also utopian. It does not resolve the old conflict between family and state, for example. The consequence of Brutus' denying particularized kinship and embracing a more general political siblinghood comes to the fore, for example, when he inflicts capital punishment on his own sons, who were political traitors to Rome.[124] Liberal democracy, even as the Renaissance came to know it, requires the sacrifice of generational heterogeneity. An alliance with the *liberi*, or *socii*, replaces an alliance with the divine or regal *pater*. For the sake of the national society a son might now kill his father and a father his son.[125] Already in the fourteenth century, Coluccio Salutati, surveying the prospects for a new nationalism, writes in this spirit:

> Thou knowest not how sweet is the *amor patriae* [love of fatherland]: if such would be expedient for the fatherland's protection or enlargement, it would seem neither burdensome and difficult nor a crime to thrust the axe into one's father's head, to crush one's brothers [*fratres*], to deliver from the womb of one's wife the premature child with the sword.[126]

"Justice it is to defend the brothers [*socii*] even at the expense of the consanguineous brother [*frater*]," insisted Lucas de Penna.[127]

The notion of a common social parenthood in *Hamlet*, while it may well serve to underpin a national ideology, is not without its awesome social contradictions. For Shakespeare's play does not allow for the real appearance of the relatively happy liberation that Livy hypothesizes for Brutus and Rome. And Roman historiography, like our own, is filled with tales of false "liberations" from vestigially patriarchal tyrants. In the *Pumpkinification of Claudius*, for example, the author (sometimes thought to be Seneca) reports that on the day of Emperor Claudius' funeral "the Roman people walked like free men;"[128] but Nero, "put on" the throne by means of matricide and incest, was soon to become monster and master of cruelty. Does Fortinbras, son of Old Hamlet's sparring partner, really know what kind of king Hamlet would have made had he been "put on" (5.2.402)? And what of Fortinbras, who cuts off Horatio's tale about Hamlet—the tale that is our *Hamlet*—as if he knows already that the tragedy of siblinghood is too much for the Danish political order to tolerate? What kind of king will Fortinbras be?

The union with which *Hamlet* ends is the tragic and genocidal conjunction of kinspersons all in death ("Is thy union here?"—5.2.331). We might hope that the union in *Hamlet* is also a conjunction in genuine burial in the earth and reunion with kin, of the sort that Old Hamlet sought and that the childless Antigone, the last of her line, sought underground with her mother, father, and beloved brother. But the old brotherhood or sisterhood, with all its beauty and horror, is gone.

6

CHILDREN OF THE NATION

OR

France, Orphanhood, and Jean Racine

Arise, lift up the lad, and hold him fast by the hand; for I will make him a great nation.

—Gen. 21:18

INCEST AND CLAUSTROPHOBIA

Sometimes I feel like a motherless child.

—Richie Havens, "Freedom"

Jean Racine's consanguineous mother and father died before he was three. In 1643 he went to live with his grandmother and grandfather, whom he called "mother" and "father."[1] In 1649 he went to live with Fathers and Mothers, or Brothers and Sisters, at the cloister [*claustrum*] at Port-Royal. Here the spiritual Christian Family, whose kinship rules the Port-Royalists strictly enforced, more or less replaced the orphan's consanguineous family.

Angélique Arnauld, the Port-Royal abbey's principal Sister-Mother, had distinguished her Cistercian order by its unusually rigorous principles of spiritual Orphanhood that made all people equally kin and nonkin as members of an "original family."[2] Rejecting her own consanguineous family and refusing her father entrance to the house, Angélique argued that all religious human beings likewise should divorce their consanguineous kin and be spiritually reborn as "children of adoption."[3] Racine, in his *History of Port-Royal* (1697)—"a sacred chronicle . . . of holy history rather than of history"[4]—describes Angélique's praise for her spiritual Sister-Daughters' diremption, or divorce, of their consanguineous kin; he pays special attention to how "God shears from them fathers, sisters, and children"[5] in the solemn rite of "profession," where a novice becomes both Spouse to Jesus and Sister to all (other) men. And Racine notes Angélique's emphasis on schools for such foundlings or orphans as himself, who passed most of his childhood at the Port-Royalist School of the Granges.[6]

Spiritual Orphanhood at Port-Royal erased the difference between foundlings or orphans (those who had lost their consanguineous kin) and nonorphans (those who still had them) by rising above the ordinarily crucial difference between kin and nonkin. Both kin and nonkin became Kin: On the one hand, In Port-Royal, Racine called his consanguineous uncle Antoine Sconin, Bishop of Beauvais, "Father";[7] he called his consanguineous sister Marie and his grandmother's sister "Sister,"[8] and his consanguineous aunt Agnès, or Mère de Sainte-Thècle—later to become Mother Superior—both "Mother" and "Sister." This "aunt-mother"—we will recall, Hamlet's term for Gertrude—generally spoke with Racine only through the confessional grill, and according to some biographers, served Racine as something like a "mother."[9] On the other hand, Racine called the physician Jean Hamon "Father," as he did the *solitaire* Antoine Le Maître (one of Racine's schoolmasters at the Port-Royalist *collège* at Beauvais), in the same way that Le Maître called him "Son."[10] Thus was Jean Racine's consanguineous orphanhood universalized in the spiritual Orphanhood at Port-Royal.

In the first section of this chapter, I will discuss the Port-Royalist universalization of consanguineous orphanhood and spiritual Orphanhood. The second section concerns Racine's *Britannicus* and the actual relationship of the Roman Catholic ideology of adoption to the Roman imperial ideology of adrogation. The third section considers what this relationship might suggest for understanding anew the romantic national ideology that wants—in places as diverse as revolutionary France and contemporary China—to make everyone an equal "child of the nation."

Port-Royal

Orphanhood and adoption are generally accepted, informing elements of Christian "holy history" (hagiography), but they are usually dismissed by secular anthropology as "figural" and by secular psychoanalysis as "hysterical." For the last century Racine's biographers, accepting these secularist views of the family, have helped to bolster an ideology of literal and normative family life. They have treated Sisterly hysteria as a subject only for prurient or pornographic interest. (They follow here in the footsteps of Mirabeau's *Erotica biblion*.)[11] Or they have attacked the Catholic notion of Universal Siblinghood as an occasion for mere parental disobedience. (Mauron's psychoanalytic critique of Angélique Arnauld is an example.)[12] Psychoanalytic biographers, especially, treat such Siblinghood as that at Racine's Port-Royal as a sociologically neurotic "substitute" for or imitative "extension" of the nuclear family. Perhaps as psychoanalysts they must. Psychoanalysis, after all, actually grounds its epoch-making considerations of hysteria in critical analyses of religious celibates.[13] (Freud says that "the asylum of the cloister is nothing other than the social counterpart to the insulation of illness of individual neurotics."[14]) An exceptional text here is the early biography of the seventeenth-century French "hysteric" Sister Jeanne des Anges, a quasi-orphan who had been seduced by a Catholic Brother—he had written a best-selling book on celibacy and spiritual sexuality—and whose visions had Jesus saying, "I will love you both as my daughter and my spouse."[15] (Jean Charcot was the biography's editor.)

In contrast, Sainte-Beuve's biography of Racine has at least the virtue of not ruling out the question of what a conjunction of Racine's orphanhood with Port-Royal's theology of orphanhood might imply. "Jean Racine," writes Sainte-Beuve, "was an orphan from infancy, *if* one can call orphan a person in the midst of so numerous and saintly a family. . . . Port-Royal developed in Racine all the sentiments of a family; Racine was never an orphan."[16] The primary issue, though, for understanding Racine—and for understanding siblinghood generally—is not whether Racine was ever an orphan but whether human beings are really ever anything other than orphans. This issue informs the Port-Royalist and Jansenist idea of rebirth into a universal siblinghood. In the same way, we shall see that the question of whether human beings are to be adopted into a national family or *fraternité* informs the French Revolution's doctrine of *régénération* in a national *Fraternité* (as promulgated by such Jansenist *philosophes* as the foundling d'Alembert, son of an ex-nun, and such Jansenist revolutionaries as the Abbé Grégoire) and also that doctrine's concern with the question of whether a national fraternity requires either celibacy or incest (as discussed by the revolutionary Marquis de Sade: "And what's the point of a revolution without general copulation?").[17]

Even as he sojourned with the "original family" at Port-Royal, suspended between religious and secular orphanhood, Racine's thinking was much concerned with the sect of the Essenes, understood as a proto-Christian familist sect with which sojourned the essentially parentless Jesus, *filius nullius* or "son of no one." The young Racine, himself "son of no one alive," reports from Josephus' *Jewish War* that this sect, whose family organization the Port-Royalists regarded as near ideal, "takes into their group a few outside children . . . and regarding them as their own blood, forms and educates them."[18] And Racine reports from Josephus' *De vita contemplativa* that the Essenes admitted to membership in their family "only those young people . . . with a perfect love for the most sublime virtue"[19] and that these adopted children "served the Essene mothers and fathers in the way that well-born children serve their mothers and fathers . . . regarding [their new parents] just as communal fathers [and mothers], and having for them the same tenderness as those who share the same blood."[20]

Where most all human beings are siblings—as in Port-Royal or in the world perfectible according to Port-Royalist principles—it is not possible to obey the Old Testament obligation to be fruitful and multiply without violating the Old Testament laws against incest. And so the Port-Royalists, like other orders, called "incest" any act of sexual intercourse among Siblings.[21] The Port-Royalists explained the Essenes' aversion to marriage in such terms—as well as in terms of the supposed sexual incontinence of women, which the Essenes emphasized and which could only lead to cuckoldry, bastardy, and hence more incest.[22] Port-Royal itself could hardly allow for exogamy or endogamy in the ordinary senses: Where all human beings are members or potential members of one group, marrying inside the group and marrying outside it are equally chaste and unchaste. Port-Royal was a familial cloister where claustrophobia meant incest. If one were to seek an escape from this prison-house without also rejecting its doctrine of essential universal kinship, one could be only a libertine or a practitioner of incest.

Racine left Port-Royal in the early 1660s.[23] Libertinism and incest then informed his life and work. Marie Champmeslé and Thérèse Du Parc were among his mistresses, and Mademoiselle de Beauchâteau and he became both spiritual parents (as godfather and godmother to a child of their friends)[24] and consanguineous parents (to an illegitimate child whom Racine, himself an orphan, put out for adoption).[25] More important in the present context, libertinism and incest informed Racine's dramatic work where, as Jean Giraudoux claims, "the stage is nothing other than the sanctuary of the family, or the central jail." "All the theater of Racine," writes Giraudoux, "is a theater of incest."[26] The plays Racine wrote during his productive period outside Port-Royal (*La Thébaïde,* the first, was written in 1663 and *Phèdre,* the last, in 1676) are rife with incest. In his dramatic work Racine staged both the ordinary incest taboo, where it is incestuous to have sexual intercourse with some people and not incestuous to have intercourse with others, and the extraordinary incest taboo, where there is no difference between chastity and incest since "All ye are brethren." It is incestuous for Phaedra to love her stepson Hippolytus, but it is chaste for her to love her ordinary husband Theseus. *Alexandre, Andromaque, Les Plaideurs, Britannicus, Bérénice,* and *Iphigénie* involve a similar, ordinary illicit love. (Other plays track that love even from birth, as in the brother-sister love in *Bajazet,* or from the womb, as in the rivalry between the two brothers in *La Thébaïde* and *Mithridate.*[27] In *La Thébaïde* Mademoiselle de Beauchâteau, Racine's co-godparent, played the role of "Antigone," Oedipus' nunlike sister-daughter, whose name means something like "Against generation" or "End of the bloodline.") *Britannicus,* we shall see later, delineates at one and the same time the ordinary and extraordinary structures of incest in Racine's drama.

After his "secular" years outside Port-Royal, Racine wanted to become a religious celibate or solitary. But "he who does not marry cuts off a large part of life, which is succession," as Racine's Josephus reports the Essenic doctrines. And "if everyone followed [his] example, all the race of men would die out."[28] And so, not surprisingly in view of Racine's previous life, his Port-Royalist counselors encouraged him to marry and raise a family. In 1677 he married a woman from far outside his own intellectual group (Catherine de Romanet had never heard a line of his plays). In 1679 he returned to Port-Royal to join his Mother Superior and aunt-mother Agnes ("It is she that God used to bring me back from my wanderings and miseries of fifteen years").[29] And he ceased writing for the secular stage, the "seat of scorners," which his Port-Royalist master Pierre Nicole had already attacked in 1660. (At that time the young Racine, on the verge of leaving Port-Royal, responded to Nicole's attack in a public letter mocking Mother Angélique, Le Maître, Nicole, and other members of his family by adoption.) In his last years Racine ordered his son Louis never to set foot in the theater. And emphasizing the privilege of Christian spirit over family blood, he insisted that he be buried not beside his consanguineous father but at the tomb of Jean Hamon, the solitary whom he used to call "father."

The Siblings Pascal

Once there was a little girl named Jacqueline. . . . She had a mama who was named Mrs. Jacqueline. Her papa was named Mr. Jacqueline. The little Jacqueline had two sisters who were both named Jacqueline, and two boy cousins who were named Jacqueline, and two girl cousins who were named Jacqueline, and an aunt and uncle who were named Jacqueline.

—Ionesco, *Story Number 1*

Ecclesiastical history and literature is rife with brothers and sisters who, barred by the distinction between kin and nonkin from consummating their love, become loving Brothers and Sisters.[30] They replace the terms of ordinary consanguineous kinship with the terms of extraordinary spiritual kinship.

Racine was interested in this Sibling love of siblings. His translation of Saint Basil's *De institutionibus monasticis* attests as much.[31] (Basil's sister helped to preside as Sister over a community of nuns that followed an adaptation of her brother's rules for monks, much as did the sisters of Pachomius and Benedict, who were also founders of cenobitical orders.) And as Racine knew, the community at Port-Royal had its fill of sister/Sister associates. Many Port-Royalist siblings followed Saint Bernard of Clairvaux, the founder of the Cistercian order, which provided the early Port-Royalists with basic theological tenets. As mentioned earlier, Bernard loved his sister Scholastica, converting her to Sister, and his sermons on the *Song of Songs* conflate "wife" with "sister" (*soror mea sponsa*). In the Port-Royal family, several members of the Arnauld and Le Maître families had been consanguineous kin before their "conversion" to Kin, for example. And Racine reports of the sister-loving Claude Lancelot that he "was not touched with grace until the taking of the veil by his sister, when, in a torrent of tears, he saw her advancing, feet naked, carrying an enormous black cross."[32]

The most famous sibling relationship at Port-Royal was that of Jacqueline and Blaise Pascal. In 1655, while Racine was a student at Port-Royal,[33] Pascal—mathematician, philosopher, and author of the pro-Port-Royalist *Lettres provinciales* (published that year)—followed his sister Jacqueline when she became a Port-Royal Sister. Blaise's "knowledge of religion," says Racine in his *History of Port-Royal,* "was enabled by means of the mademoiselle, his sister, Sister in this nunnery."[34]

Pascal followed his sister-turned-Sister to the Siblinghood mindful of the love they had felt for each other since their childhoods. Jacqueline had written:

> For though I knew his love who followed,
> Yet was I sore adread
> Lest, having him, I must have
> Nought beside.[35]

Gilberte Pascal, a sister to Blaise and Jacqueline who had married their cousin Florin Perier, said of the sibling couple that "their hearts were but one heart."[36] Perhaps fearing a forbidden consummation to their love, Jacqueline sought to transcend consanguineous kinship by joining a Sisterhood,[37] where her love for Blaise would be no more or less taboo than for any man. Jacqueline thus contrasts the unequal love of a consanguineous brother for his particular sister with the universal love of God for every human being. She tells Blaise, "You know well enough that it is from God alone that all love comes."[38] At first Blaise opposed Jacqueline's intentions to "die to the world," citing propertal considerations. But Jacqueline argued successfully that he should join her as Brother. "If you do not have the strength to follow me at least do not hold me back."[39] Finally Blaise gave his sister his blessing for her plan to become his Sister, and renounced intercourse with the outside world. Gilberte said that "not only did Blaise not want to have attachments to other people, but he did not want others to have attachments for him."[40] Denouncing marriage ("Il est injuste qu'on s'attache"),[41] he declared it to be "homicide, and almost deicide." Absolute aloneness or at-oneness, said this solitary, was "the only proper regimen for a Christian."[42]

Sainte-Beuve, when he considers the conversion of Lancelot thanks to his Port-Royalist sister, evokes Chateaubriand's romantic story, in *René*, of René's love for his sister/Sister, Amelia, which centers on a dramatic Sistering like Jacqueline Pascal's. In Chateaubriand's novel, Amelia, when she decides to enter a Sisterhood, expresses the hopes that her brother will find a wife in whom "you would feel that you had found a sister again" and, since "many times we used to sleep together . . . we might one day be together again in the same tomb."[43] René, who realizes neither that he is in love with his sister nor that she is in love with him, suspects that Amelia wants to become a Sister in order to escape "a passion for a man that she dare not avow." He cannot bring himself to acknowledge that he is himself the man Amelia loves—until a moment of peculiarly Christian tragedy. This occurs in the intersection, as it were, between the sister Amelia's death to the world (her Pascalian "detachment" from it), which is the first part of the ceremony for becoming a nun, and her rebirth as a Sister by spiritual marriage to her Brother Jesus, which is the second part. At that instant—the intersection between repressed consanguineous sibling love and expressed universalist Sibling love, spiritual death and spiritual marriage—the brother René overhears his sister Amelia begging God to forgive her "criminal passion."[44] René then expresses his love for his sister: "The horrible truth suddenly grew clear, and I lost control of my senses. Falling across the death sheet I pressed my sister in my arms and cried out: 'Chaste spouse of Jesus Christ, receive this last embrace through the chill of death and the depths of eternity which have already parted you from your brother.'" "For the most violent passion," writes Chateaubriand, "[religion] substitutes a kind of burning chastity in which lover and virgin are one."[45]

The Tragedy of Catholic Profession

René de Chateaubriand, who loved his sister Lucile and contemplated becoming claustrated as a Brother in the Catholic orders,[46] comments in *René* on the dramatic spectacle of profession, where a kinsperson becomes a Kinsperson: "nothing can ever again be tragic to a man who has witnessed such a spectacle, nor can anything ever again be painful for one who has lived through it."[47] In the religious celibate's profession, the Christian aesthetic more than matches Greek tragedy.

Profession outdoes classical tragedy when it comes to kinship. For a nun's arrival at a fourfold kinship relation to a single person (God is her Spouse, Parent, Child and Sibling) transcends the situation in Greek tragedy where one person plays more than one kinship role. In the *Iliad,* for example, tragic Andromache speaks thus to Hector, her doomed parent-sibling-spouse: "You are my father and my lady mother, you are my brother and you are my husband."[48] And in Aeschylus' *Oresteia,* Electra speaks similarly to Orestes, her long-lost parent-sibling:

> To call you father is constraint of fact, and all the love I could have borne my mother turns your way, while she is loathed as she deserves; my love for a pitilessly slaughtered sister turns to you. And now you were my steadfast brother after all.[49]

The Christian culmination of this tradition of manifold kinship relations erases the Greek standard of consanguinity (as in Shakespeare's Roman plays)[50] and raises that standard to a spiritual level (as in Sophocles' *Oedipus at Colonus*). In Dante's *Divine Comedy,* the Jansenist hero Saint Bernard thus calls the Virgin Mary "the daughter of her own son"—a holy mystery, since the incest involved is beyond the distinction between chastity and its opposite.[51] And Nietzsche, in a christological passage of his *Birth of Tragedy,* writes that "Sophocles with his [variously incestuous] Oedipus strikes up the prelude to a transcendent victory hymn for the saint."[52]

The ceremony of Catholic profession marks the transcendent end of ordinary kinship with human beings (where spouse, child, parent, and sibling are different persons) and the beginning of extraordinary kinship with God (where Spouse, Child, Parent, and Sibling are the same Person). In the "profession," a religious novice enters a coffin, hears the extraordinary "service for the dead" recited over her dead body, is reborn, and marries Jesus in the course of an extraordinary "marriage service."

Profession as such was Racine's favorite theater, and his favorite protagonist was the profession's sacrifice or *victime* —a term meaning "religious novice on the point of being 'sacrificed' as a nun in the rite of profession" as well as "tragic heroine." (It also means "oblate," or child offered up to the Church, discussed in a later section.) Racine wept hard at ordinary funerals, as if he were again losing his consanguineous parents,[53] but it was the figural "service for the dead" in the rite of profession that really attracted him.[54] Louis Racine, the playwright's son, writes of his father that "he never attended such ceremonies without crying, however much he was *indifferent* to the victim"; and, in reference to an upcoming profession, he remarks that "Mme. [Françoise d'Aubigné, marquise] de Maintenon knows that 'Racine who wants to weep, will come.'"[55]

Racine wept all the more at the ritual professions of people with whom he was not wholly *indifferent* in terms of consanguinity,[56] including especially the professions of his daughters Anne, Elisabeth, and Marie-Catherine.[57] Here the fear- and pity-provoking aspect of the profession elicited an especially powerful response from Racine. (Aristotle argues that men, in seeing their own families threatened with sacrifice or victimization, are able to fear only for themselves, only for those who are homogeneous with them in danger; and in this case fear drives out pity.[58] In specifically Christian religious "victimization," however, the repulsion from death is balanced by the attraction of rebirth.) Concerning the effect on him of his daughter Anne becoming his Sister by the rite of profession, Racine thus wrote to Louis: "Without flattering your sister, you know she is an angel; your mother and your eldest sister cried a great deal, and as for me, I did not cease weeping, and I think that it even contributed to deranging my feeble health."[59]

The "Father" Pasquier Quesnel, who presided at the "sacrificial" quasi-profession of Marie-Catherine, suggests that Racine—who had sent his illegitimate child by Thérèse Du Parc to the Foundling Hospital—educated his legitimate children precisely to become victims of this adoption by God into universal siblinghood:

> The good education that Racine gave her and the sentiments for religion with which he inspired her have led her to the altar of sacrifice. She has believed what he said: that within every human being there are two beings:
>
> L'un tout esprit et tout céleste
> Veut qu'au ciel sans cesse attache
> Et des biens éternels touche
> Je compte pour rien tout le *reste*.
>
> Please assure Jean Racine that I have offered his victim on the altar.[60]

ROME AND CHRISTENDOM; OR, *BRITANNICUS*

Roman Adrogation and Christian Adoption

Arrogation they saie is, when he which is his own man, and at libertie, is receiued in steede of a sonne. But Adoption is, when hee which is receiued, is vnder an other man's power.

—Marbeck, *Book of Notes*

Mindful of the twice-orphaned Racine's interest in adoption and the claustrophobic aspect of universal kinship, and with a view toward interpreting his *Britannicus* as the pivot in an *oeuvre* about both consanguineous and spiritual incest, I shall here begin to explore the ways that the kinship structure of "secular" Rome's imperial family—the apparent focus of *Britannicus*—is in certain respects identical to the kinship system of "spiritual" Rome's Holy Christian family—the subject of all Racine's work. The kinship structure of Rome—which we will consider here in terms of Roman adrogation and Christian adoption, Roman empress and Christian Virgin Mother, Roman vestals and Christian nuns, and Roman sublation and Christian Oblation—expresses the political incarnation of Christianity's Holy Family (God and Mary) and hence Racine's Port-Royalist "original family."

In imperial Rome succession to the throne often depended upon adrogation by the divine emperor as God, just as in Roman Catholicism success meant spiritual adoption by the divine Jesus. Nonconsanguineous adrogated sons like Tiberius and Nero were heirs to the throne, and even the consanguineous son of the Emperor had to be adopted by him if he were to succeed. Emperor Augustus—whom Seneca, in *The Pumpkinification of Claudius,* presents as the patriarchal archetype of Rome[61]—adopted Tiberius and willed him the throne, after rejecting as heir his consanguineous grandson Agrippa Postumus. Similarly, Emperor Claudius did exactly the same for Nero after rejecting Britannicus—his consanguineous offspring by his third wife Messalina.[62] The emperors thus replaced consanguinity—or at least patrilineal consanguinity—as the standard for inheritance with such legal affiliations as adrogation and marriage.

Men like Tiberius and Nero, who were the sons of the emperor both by adrogation and by marriage (as they were the consanguineous sons of the emperor's wife), generally could lay claim to the throne only if the ruler had no consanguineous sons.[63] Ambitious adrogated sons or their mothers therefore chose either to kill the emperor's consanguineous sons (as Nero killed Britannicus)[64] or become the sons-in-law to their imperial stepfathers, thus circumventing the need to kill. Tiberius thus married his stepsister Junia (the namesake for the character in *Britannicus*) and Nero married his stepsister Octavia. Since Roman law generally did not allow for sexual relations between adopted and consanguineous children, these liaisons of Tiberius and Nero were "incestuous" in much the same way that the famous liaison between Caligula and his consanguineous sisters was incestuous (see fig. 8).[65] Marriages like those linking Tiberius with Junia and Nero with Octavia were therefore regarded as sacred. (The term *taboo* means both "sacred, consecrated" and "dangerous, forbidden.")[66] And these marriages—including that of Nero and Octavia, described in *Britannicus*—were frequently sexless and barren.[67]

In this way the imperial family of Rome, called divine by the Romans, like the Holy Family of Roman Christianity, transcended the difference between consanguineous kin and nonkin. That is, the Emperors Augustus and Claudius partly erased the standard of consanguinity by disinheriting their sons and, as divine Emperors, they partly raised themselves above that standard by arranging for the empire to pass to adrogated sons who had married the emperors' daughters. The adrogated sons, Tiberius and Nero, upon becoming emperors, continued this dia-

Fig. 8. Bronze Roman coin. Reign of Emperor Caligula (A.D. 12–A.D. 41). The coin depicts Caligula and his three sisters. (American Numismatic Society, New York)

lectic of kinship. They divorced their sister-wives and killed their adoptive brothers. Tiberius divorced his stepsister Julia when he came to power and arranged for the death of Agrippa Postumus;[68] Nero divorced Octavia, and raped—according to Roman historians—and killed Britannicus, the once official heir and fiancé to Junia.

Racine times *Britannicus,* his study of Nero's birth as political monster, with the period of Nero's divorce from his stepsibling Britannicus. Then Nero might have completely transcended the distinction between consanguineous and nonconsanguineous kinship. (Christians believed this transcendence had been accomplished by Jesus of Nazareth, who was crucified during the reign of Tiberius.) To Junia, betrothed to Nero's brother-by-adoption Britannicus, Nero says, "The sister touches you less than the brother."[69]

Roman Empress and Virgin Mother

> *Even animals devoid of reason as they are and accused by us of cruel ferocity spare their own kind: wild beasts respect their own likeness. But the fury of tyrants does not stop short of their own relations: they treat friends and strangers alike.*
>
> —Seneca, *De Clementia* (a letter to Nero at the age of twenty-two)

Agrippina's place in *Britannicus*' imperial Rome involves the power that some women, though they could not rule on their own, had as people who ruled, or were feared to rule, through their husbands or sons.[70] By analogy, Livia, the first Roman empress, came to rule through her son, Tiberius. Livia divorced her husband (Tiberius Claudius Nero), married the divine Emperor Augustus,[71] convinced Augustus both to reject as imperial heir Augustus' own consanguineous grandson (Agrippa Posthumus) and to adopt her own consanguineous son by her earlier marriage (Tiberius), and eventually ruled through the son. As did Livia, so did Agrippina. As

Racine tells the story, Agrippina divorced her husband (Domitius Ahenobarbus), incestuously married her uncle, the divine Emperor Claudius,[72] convinced Claudius to adopt her consanguineous son by her earlier marriage (Nero)—and when Claudius was poisoned (perhaps by Agrippina), seemed to rule through the son. Her voice was imperial: "Ma voix / Ait fait un Empereur."[73]

Concerning Agrippina's power, the Roman political historian and theorist Tacitus remarks: "That a woman should sit before Roman standards [as Agrippina did] was an unprecedented novelty. She was asserting her *partnership* in the empire her ancestors had won."[74] Tacitus explains the sacred (taboo) quality of Agrippina's role as "imperial jointress of the state" (as *Hamlet's* Claudius calls it) in the mythic kinship structure of Rome: "Agrippina enhanced her status [by entering] the Capitol in a carriage. This distinction, traditionally reserved for *priests* and *sacred* objects, increased the *reverence* felt for a woman who to this day remains *unique* as the *daughter* of a great commander and the *sister, wife,* and *mother* of emperors."[75] Tacitus means that Agrippina is the daughter of the hero Germanicus Caesar, the wife of Emperor Claudius, the sister of Emperor Caligula, and the mother of Emperor Nero. Racine's Agrippina thus claims she is at once "fille, femme, soeur, et mère de vos maîtres."[76]

Behind Agrippina's claim in *Britannicus* that she is the "daughter, wife, sister, and mother of your masters," says one reader, is "the patrician arrogance of . . . seventeenth-century France."[77] I would argue, however, that the claim should be understood no more in terms of France than in terms of ancient Rome—the seat both of the "secular" institution (the Empire) and of the "religious" one (Christianity). No one in France besides nuns—who are to some extent, as I will argue, Roman vestal virgins converted to Christian ends—could make a claim of fourfold connection to God like the Christian claim for Jesus' mother Mary and Tacitus' claim for Nero's mother Agrippina.

Rome's Agrippina in relation to the divine Emperor is like the Virgin Mary in relation to the divine Jesus. She is his/His spouse, daughter, wife, and sister. Chateaubriand saw in Racine's Iphigenia the "Christian daughter," in Racine's Andromache the "Christian mother," and in Racine's Phaedra "the Christian spouse."[78] But Chateaubriand failed to mark the role of Racine's Agrippina as a figure with an incestuous fourfold relationship to emperor or god—a role like that of the Virgin Mary in relationship to God. And so, despite Chateaubriand's focus on religious profession, he missed the full dimensions of Racinian tragedy.

Thanks to his mother Agrippina, Nero is caught in the confusion (even conflation) of parent with child. It is Agrippina's power to withhold political birth from Nero, as well as her role as quadrifold kin to the divine Roman Emperor(s), that makes Agrippina the Volumnia-like imperial linchpin of *Britannicus*. That power explains why Racine says in his second preface to the play that he was concerned with the faithful portrayal of her above all else. Having *delivered* Nero to the world on his first birthday, Agrippina refuses to fully *liberate* him politically. Nero thus lacks political "advancement," like Hamlet, and he is driven, like Orestes, to commit a matricide that might gain him what the liberated [*affranchi*] Narcissus calls *liberty* ("Vous serez libre")[79] or free filial status. So long as the mother lives, the son lives only in her. Racine's Agrippina is thus a "symbol of seizure [*agrippement*]"[80]—like, we have seen, the Theban Sphinx, strangling or *gripping* her victims in vaginal sphincter muscles. For upon learning that Nero had formed an attachment to Junia—a woman other than his stepsister Octavia—Agrippina used to her advantage the tension between Nero as her consanguineous son and as *Pater Patriae*, seducing him and keeping him under wraps.[81]

Writing in imperial Rome, Nero's tutor Seneca—unhappily absent from the city, in *Britannicus*—argued in *The Pumpkinification of Claudius* that the ruler is the father of his people.[82] In *Britannicus*, though, Nero has to vacillate between filial status (liberty) and parentarchy. At first he seems more interested in liberation from the tyranny of his uncle-father Claudius than in himself becoming a parental authority. As Suetonius says, he refused the title of imperial *Pater Patriae* or *Parens Patriae*—a term meaning "the ruling father of one's tribe or people."[83] The position of *Pater Patriae* (which Nero eventually accepted) has incestuous implications, since the father of everyone entails being the father of one's mother. Only after Nero was really "put on" (as Fortinbras says of Hamlet)[84] did Nero kill *liberi* like Narcissus and require that all Romans call him "Father."[85]

According to the Roman historians to whom Racine refers in his preface, Agrippina knows that she must die in order for Nero truly to rule. Tacitus tells how an astrologer announced to Agrippina that her as-yet-unborn son will become emperor, provided that he kill her. Agrippina, more loving of her son Nero than was Laius of his son Oedipus, responds, "Let him kill me provided he becomes Emperor."[86] For Agrippina, it was enough to rule from the grave.

Roman Vestal and Christian Nun

What is there to say . . . about Junia's seeking refuge at the Vestals and being placed under the protection of the people, as if the people protected someone under Nero?

—Sainte-Beuve, *Portraits littéraires*

Junia in *Britannicus* is, like Agrippina, an imperial jointress of the state: she is the last blood ancestor of the man-god emperor Augustus Caesar, and Claudius (Nero's uncle-father) had once promised her the empire as dowry if she married Britannicus (Claudius' consanguineous son).[87] But Nero, adopted son of Claudius, has killed Britannicus, the consanguineous son, and wants to marry Junia. A quintessential tragic hero in the oeuvre of Racine, she must now choose between becoming empress through marriage to Nero and becoming vestal through "death to the world" and spiritual marriage to the man-god.

Like such Racinian tragic heroes and heroines as Titus,[88] Berenice, and Phaedra, Junia dies, but she dies, like a Catholic victim, or sacrifice, in the rite of profession, only to the world, on an altar both Roman Catholic and Roman imperial. (Critics of *Britannicus,* without taking into account the role of the vestal virgins in the overall *mythos* of the piece, have complained since 1670 that Junia's joining the virgins is irrelevant to the plot as a whole and belies historical fact.)[89] Junia's flight to the vestals ends the "liberal" first part of Nero's reign (which had begun well, in much the same way that his adoptive father Claudius' reign had begun well) and marks the start of the second part, when he becomes a Western archetype for monstrous unkindness to kin and nonkin alike (for which his consanguineous father Domitius Ahenobarbus had been notorious). Nero begins to go mad with cruelty after Junia, who might have been the woman to match his lover-mother as imperial jointress of the state, becomes a vestal virgin.[90] Junia's profession fixes the sadistic Nero's fate and reputation.

Junia's death to the world also expresses the tension between the implicit Roman Catholic and the explicit Roman imperial aspects of the familial and political quandaries informing *Britannicus.* Junia's sublation, through the vestal Sisterhood in religious Rome, of Agrippina's part in the Holy Family of imperial Rome—and, incidentally, of the Amazon Sister Antiope's nunlike role[91]—thus helps to locate the point of vacillation, in Racine's claustrophobic oeuvre, between a turning toward the spiritual incest figured in the *claustrum* at Port-Royal and a turning away from it.[92]

Racine's linkage of Roman vestals—and mythic Amazons—with Port-Royalist Sisters has a historical basis. For the Christian nunnery is the Roman community of vestal virgins transformed to Christian ends. During the Roman Emperor Constantine's reign in the fourth century, when the official state religion became Christianity, the Pauline Christian ideology of adoption began to replace the imperial doctrine of adrogation, and the Christian Sisterhood devoted to the adoration of Mary began to replace the cult of the vestals devoted to safeguarding the Palladia—archaic idols symbolic of an older matriarchy that figured in Roman imperial politics in the persons of the linchpin empresses Livia and Agrippina. After all, Livia and Agrippina, are—like the Virgin Mary—daughters, sisters, mothers, and spouses of the ruler/Ruler.[93]

Christian Sisters thus have many of the same rituals and powers as vestal virgins.[94] In the Roman ceremony in which a novice becomes a vestal, for example, the priest shears her hair and the *pontifex,* or head priest, calls her *Amata* ("Loved One").[95] Moreover, a Catholic nun has familial and political privileges like those of a vestal. Nobody human had the *patria potestas* (a paterfamilias' absolute power) over the vestal; only the *pontifex maximus,* the vestal's "religious father," could lord it over her. The virgins are adrogated, not merely adopted, "children." (In Rome adrogation meant the person adopted was under no *potestas* and was "her own person.")[96] The vestal was above the ordinary law not only insofar as she could free any prisoner she came across on his way to execution, as suggested in *Britannicus,* but also insofar as she could dispose of her property at will. The nunnery in seventeenth-century France had a similar liberating effect, as suggested by the case of

Jacqueline Pascal.[97] "A convent," writes one historian, "was . . . the only place a young woman could escape her father's absolute authority."[98]

Junia is a Roman imperial or Roman Catholic version of Antigone—a woman at the end of the bloodline[99] reeling from the deaths of brothers[100]—and toward the end of *Britannicus* she is reported to kneel before an idol statue of the divine Emperor Augustus Caesar, who is both her spiritual *pontifex maximus*, insofar as she is vestal, and her consanguineous great grandfather, insofar as she is not. In *Britannicus* it thus becomes ambiguous whether God is the Caesar, to whom Junia bends the knee, or the Christian God, whom the Port-Royalist worshiped. The term *pontifex maximus*, after all, names not only Caesars—as on the inscription PONTIF.MAXIM in the denarius that the scribes and Pharisees presented to Jesus in their attempt to trip him up on the question of taxation—but also the Roman Catholic pontiffs who successfully appropriated Roman imperial terminology and power to Roman Christian ends. "Render unto Caesar what is Caesar's and unto God what is God's" said Jesus. But who is God? The Romans held that Caesar was, as suggested by the inscription on the other side of the same denarius: CAESAR. DIVI.AUG.F.AUGUSTUS [Tiberius Caesar Augustus, son of the divine Augustus].[101] If the Caesar is God, then everything should be rendered to the man-God Tiberius. Just so, if Tiberius Caesar's explicit archrival Jesus is Christ the Lord, as believed by Roman Christians, then everything goes to the man-God Jesus Christ—and nothing goes to Nero Caesar.

The Vestal Daughters of Jesus' Childhood

What, are they children? . . . Will they not say afterwards, if they should grow themselves to common players—as it is most like, if their means are no better—their writers do them wrong to make them exclaim against their own succession?

—*Hamlet*, 2.2.343–49

The crowd that saves Junia from Nero's wrath in *Britannicus*, says Lucien Goldmann, represents the people of God—a people that until *Britannicus* was both silent and invisible in Racine's work.[102] ("'Indeed, you are a hidden God.'")[103] *Britannicus* in this view represents the critical transition in Racine's works between the ambiguously secular tragedies up to and including *Phèdre*, which were written before his return to Port-Royal in 1679, and the later quasi-religious tragedies. The community of vestal virgins, whose members are (like Junia in *Britannicus*) bereft of all consanguineous kin, remains unseen in *Britannicus*, but it does appear in *Esther* (1689) and *Athalie* (1691). The orphan Racine wrote these plays about orphans at the request of the influential Mme. de Maintenon—a woman who had been orphaned at an early age (like Racine) and had founded the famous school at Saint-Cyr for foundling and orphan girls, as well as for impoverished aristocratic girls who had thus been "offered up" to the Church.

The orphans, foundlings, parentless and "suppositious children" depicted in

Racine's quasi-secular plays—the changeling Astyanax in *Andromaque,* who is substituted for another child without the wily Ulysses' ever noting the exchange, for example;[104] the apparently "sacrificed" Iphigenia in *Ihpigénie en Aulide,* who lives on elsewhere; and the seemingly massacred Joas in *Athalie,* who lives on as his duplicate, Eliacin (the grandson of Athalie, raised, like Samuel, in the Temple of adoption)—cannot but imply the inascertainability of kinship relations that makes human beings essentially orphans or foundlings.[105] The known existence of such children signals that any assignation of consanguineous kinship is deniable.

Roman Catholic ideology takes cognizance of how the consanguineous genitor is sometimes denied in favor of the political *pater*—as in the English legal proverb "Who that bulleth my cow, the calf is mine"[106] (according to which an illegitimate child is called the "son of the feudal master" even when the genitor is known to be someone else) and as in the German civil code (according to which "an illegitimate child and its father are not deemed to be related").[107] But Roman Christianity goes beyond this. Sometimes it maintains that a child whose father is not his genitor or who has, like Angélique Arnauld, denied her genitor has no consanguineous father at all. In Christendom an illegitimate child is thus sometimes called "the son of no one"—*filius nullius.*[108] Jesus of Nazareth, who can be understood as both the ordinary illegitimate son of Joseph and the extraordinary legitimate, or extralegal, son of God, is such a *filius nullius,* especially so if he is deemed consanguineously related either to no one human or to no one human except the Virgin Mary. *Filii nullius* are all like.

The orphan girls whom Racine chose as his troupe players were fifteen years old—"an age when childhood is at the height of its flower but youth has not yet bloomed."[109] This troupe, whom Racine called "Filles de l'Enfance de Notre Seigneur Jésus-Christ" [Daughters of Jesus' Childhood], portray the chorus of homeless girls in *Esther* and *Athalie.* Recalling (for us) *Hamlet's* child-players in their renovated "Blackfriars," they performed their plays just at the time that the Port-Royalist schools were under attack from the outside world: The distress of the young orphan Jewesses in *Esther* is thus similar to that both of the "Filles de l'Enfance" and of the Port-Royalist pensioners at the Toulouse convent of Jansenist converts, whose familial house had been closed by royal decree in 1686.[110]

Racine's *Esther* concerns orphanhood and Orphanhood. In the biblical Book of Esther (2:7), Esther, "orphan of father and mother," becomes "the daughter of her uncle," namely, Mordecai, who eventually comes to stand at the royal antechamber—the *port royal* (2:21). Similarly, in Racine's *Esther,* Mordecai, who "at the death of [Esther's] parents . . . took her with him as if she were his daughter," regards his niece as "the daughter of his brother," treats her as his own child, and plays for her the role of "father and mother."[111] It is crucial to Racine, however, that Esther's consanguineous lineage is never fully revealed.[112] True enough, Esther has a Jewish father. (The Book of Esther [2:15] says she is the daughter of Abihayl, and in Racine's play she tells King Assuérus that she "had a Jewish man for her father," Assuérus then repeating that she is "the daughter of a Jew man.")[113] But

Esther has no mother to speak of: she is apparently partly *filia nullius*. And since Judaism is matrilineal, as Racine knew from Josephus and others, descent from Abihayl would not entail that she was Jewish. Moreover, Esther's patrilineal link through Abihayl with the tribe of the Benjaminites, whose king, Saul, had conquered and decimated the Amalekite ancestors of Haman, cannot fully constitute her supposed offense against him.[114] Like the Marranos who made Esther their "patron saint," Esther was among those about whom one could never know for sure who they were.[115] The Chorus of Jewish orphans in Racine's play are "Sisters," as Esther calls them, with a monachal detachment from the world—"un détachement du monde au milieu du monde même"[116]—and Haman thus seeks "the blood of the orphan," just as these "daughters of Zion" seek a divine king to become "the father of the orphan."[117] (Queen Elizabeth I, whose mother was beheaded when Elizabeth was three years old, was called England's Queen Esther.)[118]

Racine's last play, *Athalie*, similarly involves figural Family and literal family—including perhaps Racine's own family. Joad, the adoptive father of Joas, is probably modeled on the spiritual "Father" Antoine Sconin, bishop of Beauvais, the consanguineous uncle whom Racine used to call "Father."[119] The last line of *Athalie*, which speaks of giving to "the orphan a father," is the last word of Racine's extant theater.

CHILDREN OF THE NATION

Roman Sublation and Christian Oblation

Roman Christianity, in its proselytizing accommodation of imperial Rome, not only internalized the Roman institution of vestal virgins, as part of the Christian ideology of Siblinghood and the institution of monachism, as we saw in our analysis of Racine's Port-Royal. It also internalized Roman practices involving child abandonment and adrogation, as part of the Christian ideology of abandonment and adoption and the institution of oblation. In this manner, Roman Christians allied themselves with the kinship institutions of pagan Rome, where child abandonment was common, against those of Jewish Jerusalem, where child abandonment was both illegal and rare.[120] That alliance, we shall see, is significant for understanding the development of the ideology of nationalism in modern secular Christendom.

The founding mythology of Christianity, like that of Rome, involves a god's abandoning the child that he generates by a virgin: God and Mary parented Jesus, who was raised by another father (Joseph) just as Mars and the Vestal parented Romulus and Remus, who were raised by another mother (the she-wolf). Jesus was interpreted consubstantially not only as a miracle child, born to the extraordinarily virginal Mary, but also as an ordinary bastard abandoned by his true father, God. Jesus' last words are: "Eli, Eli, lama sabachthani?" [My Lord, My Lord, Why have you abandoned me?][121]

The Church christologically interpreted Jewish stories condemning child abandonment and sacrifice in such a way as to support the Christian ideal of universal siblinghood. Drawing an analogy between Isaac, who was nearly sacrificed by his father, and Ishmael, who was nearly abandoned by his mother, church fathers applied the Roman term *sublation* to the salvation of both sons—sometimes to the point of insisting that Isaac and Ishmael were really the same foundling, or *asufi*.[122] According to the story, Abraham expels Hagar, mother of Ishmael, and Hagar despairingly "sets down" Ishmael to die in the desert. But God orders Hagar to "pick up" her abandoned son: "Arise, pick up [*si-i*] the lad, and hold him fast by the hand; for I will make [*a-si-me-nu*] him a great nation."[123]

The Christian Vulgate uses the word *sublatum*—from the verb *tollo, tollere*—to translate the Hebrew term *si-i*.[124] *Sublation,* a technical term from Roman philosophical and legal discourse, indicates either a parent's recognition of a child's consanguinity (as when a father picks up a newborn child from the ground in the process of the formal Roman rite of acknowledging consanguinity) or his agreement to adopt a child regardless of consanguinity (as when anyone, including a consanguineous parent, picks up a child as the sign of his intention to adrogate, or adopt, the child).[125] Luther translates the term as *Aufhebung,* which word in Hegelian Christian dialectics involves arguing by the *modus tollens* and suggests both the erasure of the standard of consanguinity and the raising of that standard to a spiritual level. In his mass presented at the dedication of the Orphanage Church, twelve-year-old Mozart included an especially magnificent chorus for the liturgical "Lamb of God, Who Sublates [*qui tollis*] the Sins of the World.")[126]

In imperial Rome, where perhaps two-fifths of children were foundlings or bastards, sublation was crucial to the kinship structure. It was similarly crucial in Roman Christianity, where all orphans through conversion become "children of adoption" to God.[127] Many Christian parents in the medieval era, mindful of this glorification of abandonment—and encouraged doubtless by their poverty—followed the lead of Hannah in the Book of Samuel when she left her son Samuel, miraculously born like Isaac and Jesus, to the temple.[128] And these Christian parents, instead of abandoning their children in the Roman fashion, deposited them at the church doorsteps as "oblates."[129]

By oblation—from the Latin *oblatum*[130]—Christian parents repaid God the Father for the gift, or "offering," of God's Son Jesus, or for God's help in providing them with (other) offspring of their own. Jews repaid this gift of human life by donating tax money, meant to represent the life, to the temple,[131] but Christians literalize the law of talion and pay life for life. Thus the young oblate is generally defined as a "person sacrificed" or "victim,"[132] like the protagonists in Catholic profession, when novices become nuns. Much as kinship in Christ took precedence over consanguineous kinship, parents who had given up their children as oblates lost all right to them forever: The first Christian emperor, Constantine, endorsed the primacy of oblative kinship when he "denied the previously supreme right of parents and owners to reclaim, and granted to foster parents or new owners absolute rights over [*alumni* and] children picked up";[133] and it was with reference to a quarrel with the Cistercian Saint Bernard, a model for the Port-Royalists, that twelfth-century monks made a landmark decision upholding the irrevocability of parental oblation over even monastic profession.[134]

Even more than monks and nuns, then, oblates were bound to serve out their lives in the monasteries and nunneries to which they had been offered or sacrificed without their own consent as infants; and they were compelled to lead lives as celibate as those of the Sisters and Brothers who had voluntarily denied their human parentage. To allow people to marry who did not know who their parents were, it was said, would lead to incest.

Christian doctrine can here be compared with Islamic and Jewish traditions regarding both abandonment and incest. Moslems and Jews interpret the stories of Abraham and Ishmael to mean that people ought not to abandon their children. (They should not, except in direst circumstances, even become foster parents to children abandoned by others, since that would encourage others to abandon their children.)[135] Nevertheless, Judaism raises on the ideal level the problem of the inascertainability of consanguineous kinship in the case of the *asufi*—the foundling about whom we cannot know whose child it is or whether it is a bastard. (Isaac, if he were a foundling, would be such an *asufi*.) By some Jewish accounts an *asufi*, if a male, cannot marry a legitimate Jewish woman, since he may be a *mamzer* or bastard, and he cannot marry a female *mamzer*, because he may be legitimate.[136] It is in this vein that Maimonides raises the issue of whether an *asufi* should marry at all, since he cannot know who his kin are, and then discusses incest as both the result and the cause of child abandonment.[137]

Maimonides' linkage of foundlings with incest remains at the ideal or theoretical level for most Jews (child abandonment being both outlawed and rare). But for Christendom, where the number of foundlings and known bastards sometimes exceeded the number of legitimate children raised by their consanguineous parents, Maimonides' observation that perhaps foundlings should not marry was applicable on the practical level. Indeed, fear of incest, occasionally reaching pathological proportions, was sometimes the only Christian argument made against child abandonment![138] (The Catholic Church's argument against "orphan trains"—the nineteenth-century American practice of loading tens of thousands of orphans and poor children onto trains and shipping them to the continent's "heartland" to be "placed out" with families—sometimes amounted to the position that, since the children's names were often changed, "brothers and sisters might meet and perhaps marry [incestuously].")[139]

Christianity tried to obviate the problem of incest, found ready-to-hand in the demographics of child rearing, by outlawing sexual intercourse both for actual foundlings whose parents had abandoned them as oblates and also for those spiritual foundlings who voluntarily had given up their consanguineous families and become Sisters and Brothers in Christ. Roman Christianity thus transformed Roman institutions like abandonment and adoption into Christian rites such as profession and oblation. The oblates and religious celibates of Western Christendom are the alumni of imperial Rome "transformed."

In the relatively secular Renaissance, state-supported foundling hospitals began to subsume the business of nurturing the abandoned and orphaned children of Christendom from the various institutions of the Catholic Church.[140] Orphans, as we shall see, once housed in institutions with names like "Children of God," were now placed in institutions such as the French "Children of the Nation." In an eighteenth-century France influenced by Port-Royalist and Jansenist ideas about universal siblinghood, aware of the ever increasing numbers of abandoned children and bastards—and of the potential for incest (as suggested, say, in Melville's Franco-American novel *Pierre*)—people turned to the republican purpose of homogenizing or legitimating everyone as equal in a spiritually conatal national *fraternité*.[141]

Rousseau's Bastards

> *Plato wanted all children raised in the republic. Let each child remain unknown to his father and let all be children of the State.*
>
> —Rousseau, letter to Madame Fancueil (April 20, 1751)

It was not a simple moral weakness that Rousseau, major ideologist of French national republicanism, abandoned his five illegitimate children to the foundling hospital, or Enfans Trouvés. Nor is it insignificant that Racine similarly abandoned his child by Thérèse Du Parc. Rousseau turned his children into orphan foundlings so that he would never be able to recognize them grown up. His children and nonchildren would be to him all alike.[142]

If as an adult Rousseau (or Racine) were to have made love to a young girl—so goes the literary topos of the period—she might well have turned out to be his own daughter. Thus Rousseau's republican project to foster orphanhood (like the Port-Royalist's to foster Orphanhood) is crucial to the Marquis de Sade's similarly republican project to foster universal incest.[143] Children of unknown consanguinity are more easily reborn as children of the French nation. Republicanism, thought Rousseau—in an egalitarian French tradition that includes Saint-Cyran and Angélique Arnauld—, requires a transcendence of the ordinary family unit. ("In leaving my children to public education . . . ," wrote Rousseau in his *Confessions,* "I was acting as a citizen and father, and looked upon myself as a member of Plato's Republic.")[144] For the revolutionary republican it is the nation that is the real parent.

Among attempts in eighteenth-century France to create a national siblinghood was, as we have seen, the outlawing of the distinction between legitimate and illegitimate children. There was also cofostering in foundling hospitals, like that where Mme. Claudine Alexandrine Guérin de Tencin, an ex-nun, deposited her son d'Alembert, eventually to be educated at the exclusive Jansenist Collège de Mazarin and to become the famous philosophe.[145] There was also the myth of autochthony, according to which—as in the Platonic tale and the French national hymn "La Marseillaise"—people of the same nation think of themselves as generated or regenerated from common ground. And finally, there was the rite of common lactation, linked with the historic foundation of the ancient Roman nation in lupine nurse mothering. During the decade of the 1790s—when medical ideologists debated whether children who drank milk from extrafamilial nurse-mothers thereby became essentially bastards,[146] and whether children who drank the milk of extraspecies animals thereby became essentially animals[147]—the idea of national regeneration through common lactation, already a theme in American politics,[148] was literalized at national milk-drinking rituals like the "Festival of the Unity and Indivisibility of the Republic" (an elaborate ceremony of August 10, 1793, orchestrated by Robespierre's associate, the painter Jacques-Louis David). A commemorative medal struck for the festival, entitled "Régénération française" (French Regeneration), depicts milk or water spilling from the breasts of a statuesque alma mater, raised on the ruins of the Bastille and inscribed "Ce sont tous mes enfants" (They are all my children). The *mater* provides consubstantial nourishment to a people regenerated as the new *liberi* of the French nation. Similarly "La Fontaine de la Régeneration," a drawing based on a work by Charles Monnet, shows the crowds who came to drink at the fountain (see fig. 9).[149]

The revolution also realized a French national siblinghood by means of the Terror. A latter day rite of profession in the secular sphere, the Terror transformed hundreds of children into orphans by executing their parents and then ritualisti-

Fig. 9. La Fontaine de la Régénération française, *by Isidore-Stanislas Helman, after a design by Charles Monnet, showing the fountain erected on the rubble of the Bastille. (Bibliothèque nationale, Paris)*

cally adopting them as members of the new nation of Frenchmen. Michelet reports that in 1793 the radical Pierre Gaspard Chaumette recommended that all Frenchmen adopt the children of freshly executed parents: "Happy example of the virtues of the Republic! . . . This child is an orphan under the law [her parents were executed under the Terror], and now she will receive, by virtue of your paternal embraces, an adoption by the Nation." (In his *History of the French Revolution* Michelet argued that, while the revolution and Christianity "agree in the sentiment of human fraternity," the latter "violently contradicts the spiritual notion of justice" insofar as it grounds fraternity on "a filiation which transmits, with our blood, the participation of crime from father to son.") The National Convention founded the asylum for orphans named "Children of the Nation" (see fig. 10). Similarly in funerals the human body was covered by the French tricolor, "the sacred flag of human regeneration," in order that "the dead person might depart this world clothed in mother France and enveloped by the Nation."[150]

Revolutionaries hoped by these myths, rites, and laws of universal kinship to "liberate" the children of the *ancien régime* from tyranny. Some associated that tyranny with political "parentarchy" and others with spiritual repression of the Cath-

Fig. 10. Daniel Urrabietal Vierge. Enfants de la Nation. *In Michelet,* Histoire de la Révolution, *8: 167 (Widener Library, Cambridge, Mass.)*

olic kind.[151] Certain leaders of the revolution, however, admitted that Catholic Christianity was not irreconcilable with modern conceptions of political liberty and equality. Racine's Port-Royalist and Jansenist ideology, they knew, required an Orphanhood (death to the world) and Siblinghood (adoption by God) not far removed from the tragic regicide and nationalist regeneration, or resurrection, in revolutionary France in the 1790s. Among these was the Jansenist Abbé Henri Grégoire, who wrote a prize-winning essay on the physical, moral, and political *régénération* of the Jews into French nationhood (1788). As a Jansenist in a Port-Royalist and Gallican tradition, Grégoire was an ardent republican. "Kings are in the moral order what monsters are in the natural," he said alluding to Nero in a speech at the first session of the National Convention in September 1792. In November of the same year, Grégoire was elected president of the Convention, an assembly over which he presided in episcopal dress even during the Terror that helped found modern France.

The Outlawing of Siblinghood; or, On China and the Romantics

As primogeniture consisteth in prelation, so unigeniture in exclusion.

—Bishop Pearson, *Exposition of the Creed*

I finished the first draft of *Children of the Earth* just as the ruling group in the Chinese government ordered its troops to turn out the lights over Tiananmen Square in Beijing and open fire on young citizen students demonstrating for "civil" liberties. Their blood soaked the earth as soldiers toppled a copy of the Franco-American Statue of Liberty, in an inversion of the crowd scenes in Sergei Eisenstein's *October*. The government soldiers burned the bodies of the dead so that they could not be counted by historians. Will these deaths count in history?

On Canadian television a bereaved Chinese mother reminded viewers that, according to a current Chinese law aimed at restricting population growth but with other consequences, married couples are allowed to have only one child. The loss of one's child at Tiananmen Square now meant the loss of family lineage. I thought of Antigone, whose name means something like "End of the Line." Creon buried Antigone alive because she tried to bury her brother properly. In a land where siblinghood is outlawed, what sibling will weep for the unburied slain?

What will it mean for the institution of politics in China when everyone is an "only child" without siblings? When all citizens are only-begotten sons or daughters? Forbidding a family to have more than one child has straightforward justifications and difficulties: A country like China needs to restrict population growth, for example, but where ancient traditions of male primogeniture predominate, unigeniture encourages female infanticide. But beyond these demographic, economic, and moral considerations there are other ideological and political consequences of unigeniture relevant to the development of a liberal state, not only as a conglomeration of acquisitive individuals, but also as a polis of *liberi*—free and equal brothers and sisters.

In Tiananmen Square the Chinese students had erected a makeshift copy of the Statue of Liberty, whose political sense, I suppose, involves that equality in fraternity/siblinghood which qualified the romanticist French Revolution's *liberté*. From the viewpoint of this *liberté,* what might be the political and natural implications of forbidding siblinghood?

If one precondition for the development, in the individual person's psyche, of the ideology of liberal siblinghood is the extension of consanguineous siblinghood to include everyone in the country, then one implication of communist Chinese unigeniture—which would prohibit people from ever experiencing consanguineous siblinghood—would be that communal or liberal siblinghood could now not develop. But that same extension of siblinghood to include everyone in the country would have to result in the practice of either celibacy or incest—each of which would result in the death of the body politic: celibacy through the eventual extinction of the members of the body politic, and incest through the destruction of the polis, or at least of the polis propped up by the incest taboo, which most social anthropologists and psychoanalysts tell us every polis must be. Does not the ideology of siblinghood thus require, along with the extension of consanguineous kinship, also a cancellation or transcendence of consanguineous kinship that is not only Rousseauist but also platonic and Christian?

Romantic political theorists, we have seen, have argued that sibling love is the basis for the republican affiliation which Plato suggests might be gained through the noble lie of autochthony and Jesus suggests might be through adoption by God. Romantic theorists say that sibling love should be extended to the point where one loves most all fellow "nationals" as if they were siblings. Coleridge, for example, observes that such love initiates all genuine society. From "pure" sibling love, he says, there develops first conjugal love and then love of most all other human beings. (Thomas De Quincey, in the same Wordsworthian vein, thanks providence "that my infant feelings were moulded by the gentlest of sisters.")[152] "By the long habitual practice of the sisterly affection preceding the conjugal," writes Coleridge, "this latter is thereby rendered more pure, more even, and of greater constancy. To all this is to be added . . . the beautiful Graduation of attachment, from Sister, Wife, Child, Uncle, Cousin, one of our blood, and so on to mere Neighbour—to Townsman—to our Countrymen. . . . " Coleridge says that he has observed the bad effects of a want of variety of "attachments" among the familist Quakers (whom he criticizes as being at once too hot and too cold in relation to other people) and among families in Italy (the setting, he points out, of Shakespeare's *Romeo and Juliet*). In Italy, "the young are kept secluded, not only from their neighbours, but from their own families—all closely imprisoned until the hour when they are let out of their cages without having had the opportunity of learning to fly—without experience, restrained by no *kindly* feeling."[153] (Similarly, in his novel *Contarini Fleming,* some decades before he became Prime Minister of England in the 1870s, the Marrano Benjamin Disraeli, son of Isaac, writes that "had I found . . . a sister, all might have been changed.")[154]

Coleridge retreated from the idea of an equal and universal siblinghood in which all people—including consanguineous sisters and natives of distant lands—are alternately neither siblings nor aliens or both at once. This idea was a logical precondition for pantisocracy, much as the substitution of autochthony for consanguinity was a precondition for Plato's *politeia,* but Coleridge withdrew from supporting it because he understood that the literal and figural realization of universal siblinghood requires either incest or celibacy.[155] Other political activists in the romantic period made the leap: Sade called the practice of incest essential to republican liberation, as we have seen, and Lord Byron, losing his love for everyone generally in a universalist love of his sister Augusta, wrote to her that "I have never ceased nor can cease to feel that perfect and boundless attachment which bounds and binds me to you—which renders me utterly incapable of a love for any other human being—for what could they be to me after you?"[156]

If the present policy in China continues enforced, soon the Chinese will have no brothers or sisters. There will be no gradual extension from sibling to citizen of the Earth for them, of the sort that Coleridge praises—or at least pretends to praise. Might this mean that they will have no political or psychological basis on (or against) which to build liberty, as a republican affiliation of free sons and daughters (*liberi*)?[157]

Without seeking to entirely replace the fiction of familial generation with one of abiogenesis or autochthony, might the Chinese now look to institutions of adoption or communalism to replace those of fictive consanguinity among siblings? One such institution is that of "minor marriage"—comparable, as we have seen, to the sibling relationships between Gertrude and Claudius and between Ophelia and Hamlet, and also to the relationship between such spouse/siblings as Sarah and Abraham in the pre-Abramic Hurrian fratriarchy.[158] In the Chinese practice, a (future) daughter-in-law is adopted into the family at an early age and raised by her (future) mother-in-law as a sibling to her (future) spouse. Such adoption/marriage preserves exogamy while at the same time minimizing the threat to the extended family posed by a disruptive spouse brought in from the outside world.[159] Presenting marriage as simultaneously spousehood and siblinghood ("My sister [is] my wife"), it both allows and forbids the latter, thus encouraging a unigeniture that opposes the ideal of liberty.

7

THE FAMILY PET

OR

The Human and the Animal

INTRODUCTION

Pets are everywhere. In our homes there are millions of pet dogs, cats, gerbils, birds, fish, rabbits, snakes, and monkeys. Our political economy includes a multibillion-dollar service industry that provides veterinary medical care, food, breeding, and assorted paraphernalia.[1]

Why do we have such an institution? Pet owners and pet lovers join the pet industry in detailing benefits that pet ownership confers: Pets, they say, provide pleasure, companionship, and protection, or the feeling of being secure.[2] Owning pets decreases blood pressure and increases life expectancy for coronary and other patients.[3] Pets provide an excuse for exercise and a stimulus to meet people.[4] They help children to learn gentleness and responsibility, they help young couples to prepare for parenthood, and they give their owners some of the pleasures of having children without some of the responsibility. (From pet dogs and cats, writes Gomperz in *Moral Inquiries* [1824], "mankind may learn maternal, filial, conjugal, and in some cases paternal affections.")[5] Pets help people to deal with the death of a friend or relative.[6] Not least of all, pets are useful in many kinds of psychotherapy and family therapy.[7]

What is it about pets that makes them beneficial in these ways? In this chapter I will explore one possible answer to this question. I will suggest that pethood derives its power from its ability to let pet owners experience a relationship ever-present in political ideology—that between the distinction of those beings who are our (familial) kin from those who are not, on the one hand, and the distinction of those beings who are our (human) kind from those who are not, on the other. Pethood, it would seem, lets us experience and enjoy that crucial distinction in a harmless and even comforting way. And, indeed, we generally think of pethood as one of consumer society's innocuous and even trivial institutions.[8] We will see here, however, that the particular idealized articulation of kinship with kind that the contemporary institution of pethood helps to maintain conceals from would-be kindly human beings a brutally inhumane political reality.

A pet "is good to think on, if a man would express himself neatly," writes Christopher Smart in his poem "Of Jeoffry, His Cat."[9] That the individual pet is in some fashion the expressive mirror of its owner is a longstanding commonplace; Barbara Woodhouse, the dog trainer, goes so far as to claim that "we get the animals we deserve."[10] In this essay, however, I will be concerned not so much with the neat expression of an individual pet owner by his individual pet as with the general expression of Western familial and national structure in its unique *institution* of pethood. It is a generally accepted doctrine nowadays that "the human/pet relationship, while biologically derived and universal, may also serve a particularized psychopathologic purpose";[11] I here want to discuss the sexual, familial, and finally social role that the institution of pethood plays in contemporary politics and ideology.

THE KIND AND KIN OF PETS

A little less than kin and more than kind.

—*Hamlet,* 1.2.65

The *Oxford English Dictionary* defines "pet" as "any animal domesticated or tamed kept as a favorite or treated with indulgence."[12] This is a reasonable preliminary definition. And since it passes over, even obscures, certain potentially discomforting ramifications of what it may mean to domesticate animals and to indulge them, it is also a socially useful definition.

The Kind, or Species

The ordinary definition of the family pet as an animal tends to obscure the essential demarcation between human beings and other animals since it implies that any animal, including a human being, can be a pet. To put the matter this way is, however, to assume that there is an essential interspecies demarcation between human and animal beings, which pet lovers might deny. Pet lovers, after all, "find it difficult to separate people and animals," as Betty White confesses in her book *Pet-Love;*[13] they would have it that we humans can sometimes have a special, or super-special, kinship with the particular living being who is a pet of ours.[14]

In America today, our thinking of pets as human and our treating them as human has many aspects. We feed our pets human food, for example, and celebrate their birthdays.[15] More than half of American pet owners look upon their pets as "almost human," nine-tenths talk to their pets as though they were human,[16] and six hundred pet cemeteries in the United States imitate the burial or cremation service for human beings or bury animals alongside their human owners.[17]

For pet lovers this interspecies transformation of the particular animal into a kind of human being is the familiar rule. (It is the rule also in the legend of "Beauty and the Beast," where a friendly monster is metamorphosed into a family man, and in the Homeric tale of Circe, where men are metamorphosed into domestic animals.[18]) Likewise, it is the rule expressed in the typical English pet lover's practice of giving his animal a human name—a practice which suggests that the pet lover regards his pet as though it were human. Indeed, "to pet" can mean "to treat a human being as an animal."[19]

The tendency to erase—and, if you wish, also to rise above—the ordinary distinction between human and animal beings suggests the first potentially disturbing question raised, not only by the ordinary definition of *pet,* but also by the institution of pethood itself: "What kind of animal is a pet?," or "As what kind of animal is a pet *thought* of?" Another way to put the same question is, "What is (a) human being?"

The Kin, or Family

Ordinary definitions of *pet* obscure not only what man and animal are but also what the place of the pet in the family structure is. For pet lovers, as for Betty White, "animals have always been a part of [the] family."[20] The "cade lamb," which is the archetypal pet in the Scottish and English traditions, is a being raised by hand *in* the family; it is a being in the household as well as the house.[21]

For many pet lovers, their animals are thus not only surrogate family members that function as children, grandchildren, spouses, or parents,[22] or that are considered to be as important as family members.[23] Rather, pets *are* family.

But how can an animal be in my family, or be thought of as being in my family? What is my pet's kinship relation to me, or its kind of kinship relation to me?

Bestiality and/or Incest

Somehow the family pet is, or is thought of, as familiar enough to be both in the special family, or in humankind, and in the particular consanguineous family.[24] It is worth noting that although the French language has no single word to indicate the kind of being that we mean by *pet*—few, if any, languages do—French does bring out the relevant ambiguity of most pethood in its term *animal familier,* which is the closest equivalent in the language to the English word *pet. Animal familier* means "familiar animal" and "family animal." That is, the French term for *pet* indicates an animal that is at once part of the family's kinship structure and also, like an *animal domestique,* part of its property. If my pet animal is somehow human, or is thought of as being somehow human, and if my pet is also somehow in the family, or is thought of as being in the family, then might I not wonder whether I can love or marry my humanoid pet without somehow violating a basic taboo, or somehow thinking of violating one?

For all its outlandishness, the preceding question suggests how, at some level, pet love traduces (or transcends) two practices we ordinarily think of as being taboo. One of these practices is bestiality, or interspecies lovemaking, which is an effect of traducing the ordinary distinctions between human and nonhuman beings, or between kind and nonkind. The other practice is incest, or intrafamilial lovemaking, which is an effect of traducing the ordinary distinction between kin and nonkin.[25]

Pet love thus toes the line between chaste, or socially sanctioned, attraction (between a human being and a being from inside his species and outside his family) and either bestial attraction (between a human being and an "animal" being from outside humankind) or incestuous attraction (between a human being and a being from inside the particular kinship family). Or, as I am suggesting, whether we look at it from the viewpoint of the individual or of society, the institution of pethood allows us to toe the line between chaste attraction and both bestiality and incest, taken together.

In psychoanalytic and anthropological terms: Ontogenetically the pet is a transitional object,[26] and phylogenetically it is a totem.[27]

Puppy Love and Petting

Connections between kind, kin, and sexuality of the sort we are describing are hinted at throughout ordinary language. Consider, for example, the popular American terms *puppy love* and *petting*. On account of their humorous aspect, these symptomatic terms are able, each in its own way, to both conceal and reveal the bestial and incestuous aspect of pethood.

Puppy Love. One tendency of the institution of pethood is to make distinctions such as that between sexual and nonsexual feelings seem clear and uncontroversial. Thus some pet lovers may object to my discussion of the sexual significance of loving pets on the grounds that, although pethood possibly does blur the distinctions between kind and nonkind and between kin and nonkin, it does not blur the distinction between sexual and nonsexual love. (That is, so long as amatory relations with borderline creatures are nonsexual, there does not have to be ambiguity about whether an amatory relationship with a pet is essentially bestiality or incest.) This objection assumes that there *is* such a thing as essentially nonsexual love for a being who both is and is not kind and kin. Yet even ordinary language belies the assumption of essential difference between such sorts of love. Consider first the term *puppy love*.

Puppy love between human beings, we ordinarily say, is, like calf love, a sentimental and transitory affection between a young boy and girl;[28] we say it is, for all practical purposes, asexual. (It is the presumably asexual aspect of puppy love that helps to explain why the term is usually one of mild contempt.)[29] Puppy love is supposed to be as sexually innocuous as loving a "puppy" in the traditional sense of "a small dog used as a lady's pet or plaything, a *toy* dog";[30] the beloved being in puppy love is much like a *poupée* (or doll, the French term being the etymological source of *puppy*), and also much like a "puppet."[31]

We assume that puppy love is or should be just as sexually innocuous as loving close human kin is or should be. Put otherwise, we assume that it is no more or less bestial for a human being to love a puppy dog—a being from outside his species—than it is incestuous for a human to love a "puppy lover" from inside his consanguineous family.[32] It follows that if one wishes to avoid or sublimate both literal bestiality and literal incest—as who does not?—one way to do so would be to seek out a "snugglepup."

The term *snugglepup* indicates a pet puppy with whom one snuggles, in the sense that a child snuggles with its transitional object, or that the one-half of all the pets in the United States who sleep in the same bed with a member of the family snuggle or are snuggled by their owners.[33] Or *snugglepup* indicates a young man with whom one attends petting parties.[34] (Sometimes such a man is called a "pet.") Or, as I am suggesting, *snugglepup* may indicate both the beloved animal and the human lover taken together.

The idea of snugglepuppy love, or pet love, is a great commercial success. It is sold, in its feminine form, as the *Penthouse* "pet of the month" and as the *Playboy* "bunny."[35] (*Playboy*'s human bunny is a doll-like creature if ever there were one, as unlike a rabbit as a *poupée* is unlike a dog.) Snugglepup love is the commercial ideal of a relationship between living beings: for all its apparent sexuality, it is a relationship that is infertile and unthreatening. In the social and sexual institutions represented for us by the pet and the bunny, we grown-up human beings dress other human beings to look like animals (or we brand them with the insignia of animals).

Petting. We may "doubt if there's [really] such a thing as puppy love."[36] Freudians, after all, believe there probably is no such thing as entirely asexual love—or even essentially nonincestuous affection—in a human kinship family. Put another way, we may wonder at the simultaneously asexual and sexual significance of petting pets. Consider here the verb "to pet."

Petting means not only mere patting, "fondling or hugging,"[37] but also "sexual embracing" or "petting below the waist."[38] Our petting an animal that we say we love—a being whose kind we distinguish in a commonsense way from our own kind—is thus a kind of bestiality.[39] Our petting the child, sibling, or parent whom we love—a being whose kinship we identify in a commonsense way with our own kinship—is, by the same definition, a sort of incest.[40] (We give the nickname "pet"—and sometimes also "beast"—to the human beings with whom we are intimate.) And, by the same definition, our "petting" the family pet—a being who is at once neither our kind nor kin and both our kind and kin—is bestiality and incest taken together.

Some students of the various physiological benefits to pet owners of hugging and patting their pets assume, as we might expect, a distinction between "engaged" and "idle" petting. According to them, "idle" petting resembles the "absent-minded fondling of a child while attention is focused elsewhere . . . ," and idle petting "can provide reverie and relaxation."[41] One explanation of why this is so is that it allows us to mark and transcend an otherwise absolute and oppressive distinction between kin and nonkin, and between kind and nonkind, while at the same time allowing us to briefly blur without shame the distinction between sexual and nonsexual demonstrations of affection.

From this perspective on the kind and kin of pets, the way to rightly determine their familial and sexual role must go beyond analysis of psychotic or neurotic human-animal relationships—analyses of the kind that we encounter in studies such as Sigmund Freud's "Little Hans" and "Wolf Man," Helen Deutsch's "Chicken Phobia," and Sandor Ferenczi's "Little Chanticleer."[42] Such studies ignore the institution of pethood, except to make it a latter-day totemism. Freud, in *Totem and Taboo,* argues that in zoophobia, or fear of animals, the animal preserves the barrier against incest;[43] I should argue that especially nowadays and in America, it is through zoophilia/zooerasty, or animal love—i.e., in particularized pethood—that the effort to preserve this barrier is more typically made.

BEAUTY AND THE BEAST

The institution of pethood depends upon the individual pet owner having a different relationship to his animal than he has to other animals, or on his distinguishing between his family pet and unfamiliar animals in general. This dependence means that pethood militates against the idea of *general* interspecies kinship and may even exclude it. In pethood only family pets are familial kin, only they are human kind.

However, pet love *is* in some circumstances extensible to a brotherly (or, if you will, sisterly) love of all animals universally—to a *Kinship With All Life.*[45] Thus Chaucer's Prioress weeps not only when someone beats her familiar dog but also when an apparently unfamiliar mouse is caught in a trap.[46] Pet love also seems extensible to universal interspecies love in the case of Christopher Smart, a lover of his cat who writes that animals and birds are, together with himself, "fellow subjects of the eternal King."[47] For both Chaucer's Prioress and Smart, all humans are essentially children in one family under God the Parent. And all family pets, or all animals able to become pets (or able to be converted to the status of pethood), are part of a superhuman family.

"All ye are brethren." One consequence of hypothesizing a universal kinship among human beings is that such kinship makes any act of human sexual intercourse incestuous. That is one reason why religious celibates shun all intercourse as "spiritual incest": Religious celibates in the Catholic tradition have rejected their kinship ties with their consanguineous human families and, as "children of adoption" by God, claim that, in their new family, all men and women are equally their brothers and sisters and hence equally taboo.

If one wants to avoid incest and celibacy one must, of course, look for a sexual partner outside the family. But if, as I have argued, one maintains the traditional Christian and pagan belief in universal human kinship, then all human beings are from inside the family. The ultimate outcome of such an identity of species with family would be a search for an extraspecies and extrafamilial creature. While sexual relations with actual animals are not uncommon (marriage with them is an ancient legal problem),[48] the more frequent quest is for the symbolic Beast. (Human deities like Jesus Christ and Augustus Caesar might do just as well since, in varying degrees, they are also extraspecies and extrafamilial.)

Maybe this quest for the Beast—or this flight from incest—is not as bad as it sounds. Is not bestiality better than incest? Or "spiritual bestiality" better than "spiritual incest"? The social anthropologists call the taboo on incest, not that on bestiality, the law of laws. In the well-known folktale "Beauty and the Beast," one of several whose publication and popularity in France tend to mark the Romantic and post-Romantic eras of "liberty, equality, and fraternity,"[49] the heroine leaves her loving kinsman and kisses the beast.

The tale is of a young maiden whose agreement to marry—or at least to kiss—a fearsome animal corresponds to its physical transformation into a handsome man. Some interpreters explain away the tale's sexual and bestial aspect by arguing that it is a philosophical allegory of the rational soul's journey toward intellectual or spiritual love: The sensitive Beauty's insight into the spiritual beauty of the physically ugly Beast precipitates its transformation into a human, hence beautiful, being.[50] This reading is reassuring and enlightening, but it does not take into account that it is Beauty's embrace of the Beast—and in some versions of the story, her intercourse with it—and not her insight into its character that is the agent of its transformation from animal to man.

This sexual aspect of the tale has made it an attractive text for psychoanalytic criticism, which generally interprets it as expressing Beauty's fear of and eventual accommodation to human sexuality, but explains away the "bestial" theme. Thus some analysts say that the tale represents a young girl's reaction to a man who requests that she sleep with him, a request that at first she can only understand as beastly, but that she comes to understand as simply human.[51] According to a psychoanalytic elaboration of the tale, the beastly man behind the Beast is Beauty's own father; her father's picking the red rose in Beast's garden is a symbol of his desire, conscious or unconscious, to "deflower" Beauty.[52] But in their occasional symbolic sophistication, such readings do violence to the literality of "Beauty and the Beast": Beast is not only a human being in animal guise—the usual figure in animal fables and cartoon talkies—but also an actual animal. (In Jean Cocteau's brilliant cinematic version of the tale, Beast says that he is not really an animal while Beauty insists that he is—see fig. 11.) The tale, ultimately about bestiality and the human family, marks the definitive limits of the literary form called the "animal groom story."[53]

Who else is there for Beauty to marry but an animal? Almost all the male human beings in her world are her close kin—usually her father and three brothers,[54] so

Fig. 11. Beast watches Beauty at supper. Still photograph from Cocteau's Belle et Bête. *(Courtesy of Museum of Modern Art–Film Stills Archive, New York)*

that any ordinary sexual love would seem incestuous. Consider, as a useful scheme for understanding the metaphorical structures—or species and familial divisions—of the tale, the two interrelated "laws" that Westermarck proposes in his *History of Human Marriage*. First, Westermarck's species "law of similarity" has it that all animals, including humans, tend to mate with those like themselves; they shy away from intercourse with those outside their species—or from bestiality. Second, Westermarck's family "law of dissimilarity" has it that we humans tend to mate with those unlike ourselves; we shy away from intercourse inside the family, or from incest.[55] In the fairy tale, Beauty shies away from intercourse/marriage with the animal that she loves because that animal is so much unlike her as to be outside her species, and she shies away from exclusive love with her father because he is so much like her as to be inside her immediate family. Throughout the middle section of the fairy tale, she vacillates between living with her loving father and living with her beastly lover: She wavers between incest and bestiality.

Yet Beast, who is the extraspecies animal in Beauty's life, is, to all intents, the same as her father, who is the intrafamilial—and, in some versions, essentially the only available—man. This identity between the two male beings—the one too unlike her, the other too like her—is hinted at in the economic bargain informing the tale: the deal whereby Beauty's life is traded to Beast in exchange for, or in

Fig. 12. Father going to the door of Beast's castle, where Beast seems to appear as Father's shadow. Still photograph from Jean Cocteau's Belle et Bête. *(Courtesy of Museum of Modern Art–Film Stills Archive, New York)*

behalf of, the life of her father. And the identity between Beast and Father is also hinted at in the mortal sicknesses of the two rival "suitors," which compel Beauty to choose between attending to the needs of one or to those of the other, hence between killing one or the other (see fig. 12).[56]

In "Beauty and the Beast" father and beast—or human exogamy and bestial endogamy—are, for Beauty, one and the same. This conflation of incest and bestiality is the key to understanding the structural significance and widespread popularity of the tale.[57] It is a conflation that informs even its ancient analogue, Apuleius' "Tale of Psyche and Cupid." In this Roman story, Psyche = Beauty does not know whether the invisible being with whom she makes love night after night is an animal, a human kinsperson, or a god. It is as though Beauty might as well be sleeping with anyone or anything—with her father or with a great viper (which is what Beauty's sisters tell her that she is doing)—as in a grand masquerade where all beings can or do pass for one another.[58] In "Cupid and Psyche" it turns out, however, that Psyche was unwittingly not the lover of beast or father but of a god, Cupid.[59] Psyche = Beauty is a Roman version of the Christian *sponsa dei* or Roman vestal virgin, as if only a god could transcend in one leap both familial and species boundaries. Like Beast, Cupid (and, if you will, also Christ) is, in the instant of his osculatory transformation, a member of two apparently different species. Beast =

Cupid is like the amphibian lover in the "Frog Prince" tales, where the animal groom is both aquatic and land being, as well as both animal and man.[60] Beast's re-creation as man through a kiss from Beauty—a kiss that is at once chaste and unchaste—is like God's re-creation as man through a sexual intercourse with Mary that is similarly ambivalent. (That is, God as lover, like Beast as lover, is both a human being and a husband and He is neither.) The kiss at the center of the animal groom tale would seem to transcend the ordinary questions of "Who is it proper to kiss?" and "In what manner is it proper to kiss him?" as much as does the Christian universalist "kiss of peace" or "holy kiss"[61]—a kiss promiscuously given to all persons alike. It is a kiss which, by virtue of the Christian monachal doctrine, would seem to transcend the distinction between kin and nonkin and hence between chastity and incest.[62]

In more or less psychoanalytic terms, though by no means from a psychoanalytic perspective, I may summarize thus: In "Beauty and the Beast," Beast becomes, in the instant of its osculatory transformation from animal to man, a transitional object between parental and spousal love.[63] Only Beast's miraculous transformation into a human husband allows for the chaste marriage, neither bestial nor incestuous, that gives the fairy tale its Edenic or paradisiacal ending. (In the 1761 ending of the tale, the instant of Beast's transformation into human form in the castle corresponds to the immediate translation of "her father and his whole family" to the castle: husband and father appear together in the castle at the same instant. The last words of this version are "Their happiness . . . was compleat."[64] This family reunion combines or transcends kinship relationships like daughter and wife or father and husband, and thus provides the right ending.) The millennial popularity of "Beauty and the Beast" is attributable to its comedic blending together of human endogamy with bestial exogamy in such a way as to make human exogamy.[65]

However, the success of Beauty's search for a wholly acceptable being with whom to mate—an intraspecies and extrafamilial one—depends on our granting that there is a real difference of kind between human and animal beings. For the difference between incest and bestiality, upon which Beauty's search for the right beast (call him husband)—and hence also for a chaste marriage—depends, vanishes if we hypothesize not only that all human beings are our kin (the traditional view, say, of the United Nations) but also that all animals are our kind (the view of Saint Francis and some pet lovers). The ultimate consequence of this view would be understanding the only alternative to absolute celibacy as bestial incest, or incestuous bestiality.

In the same way, we might understand the notion of our kinship with animals as leading either to vegetarianism or cannibalism;[66] and even the comforting difference between vegetarianism and cannibalism vanishes if we hypothesize a human kinship with vegetables as well as with animals. The consequence of this view would be that the only alternative to starvation is cannibalism.

To choose between celibacy and bestiality/incest and also between starvation and cannibalism is difficult and needful, yet society has imposed its sentence upon us: Eat, drink, and be married. For fully socialized human beings—i.e., for civilized adults—the primordially tragic need to choose between such dire alternatives must appear comedic. That is where the beast story comes to the rescue.

FABLES OF PASIPHAË; OR, VACCINATION AND VACCIMULGENCE

When Iesus Christe shall be come, Princes must be protectours of Christianitie, and Queenes must be nurse mothers.

—*Calvin's Sermons* (1579)

Advt., Wanted, a Child to Wet Nurse, by a Young Woman with a good breast of milk.

—Want ad in the *Morning Chronicle*, April 13, 1784

In the story of "Beauty and the Beast," human endogamy intersects with bestial exogamy so that human exogamy comedically arises from the topos of the animal groom or god groom. In much the same way, the topos of the animal or divine nurse-mother sometimes makes for endogamy. For according to the regulations of many societies, milk kinship results in the same diriment impediments to marriage as blood kinship.[67] And a nurse-mother, or wet nurse, transmits familial kinship—and hence species kind—through her milk just as a consanguineous parent transmits kinship through the blood.

Montaigne, in considering the institution of the nurse-mother in his essay "Of the Affection of Fathers for Their Children" (1578–80), writes that "it is ordinary where I live to see village women, when they cannot feed their children from their breasts, call goats to their aid. . . . These goats are promptly trained to come and suckle these little children. . . ." Animal nurse-mother and human nurse-child become much attached to each other: "If any other than their nurseling is presented to them, [the goats] refuse it, and the child does the same with the goat." Montaigne refers to instances where nurse-mothers even develop, for their non-consanguineous nurse-children, "a bastard affection [*affection bastarde*] engendered by habit more vehement than the natural, and a greater solicitation for the preservation of the borrowed child than for their own." He concludes first, that the "natural affection [of a genetrix for her consanguineous child] to which we give so much authority" has "very weak roots," and second, that "animals alter and corrupt [*abastardissent*] their natural affections as easily as we [humans do]." According to Montaigne, then, both interfamilial and interspecies collactaneous kinship can take precedence over intrafamilial consanguinity. Affective kinship is engendered by custom, not by sexual generation.[68]

Since (as we have seen) consanguineous kinship is in any case always ultimately deniable and since the collactaneous affinity that develops between nurse-mother and nurse-child can change a child's family and even its species, the human or animal nurse-mother sometimes becomes a feared creature. Where the social institution of nurse-mother is widespread, as in some aristocratic and slave-based economies, we find tales and fears of accidental changing by an unwary nurse or purposeful changing by a wily one. How many children born slave-owners have thus became slaves? How many born paupers thus end as princes? Where nurse-mothering is widespread, we also find accounts of foundling and orphan children fostered by kindly wild animals in the forests. Virgil's story of the she-wolf's fostering of Romulus and Remus is a *locus classicus.* In eighteenth-century Germany and France there circulated hundreds of "true accounts" of *enfants sauvages* raised in the "state of nature." In the nineteenth-century United States, the institution of black nurse-mothers (Mark Twain's Roxana) haunted America's white population as a specter undetectably diluting their race (or species).

Collactaneous affinity, like consanguineous, affects not only family kin but also species kind. Having intercourse with a human nurse-mother or her near kin counts as incest. (The story of Ambleth and his foster sister—Shakespeare's Hamlet and Ophelia—is an important example.) Similarly, since "der Mensch ist was er isst [You are what you eat]," as Feuerbach says, the human nursling of a she-animal is part animal, and anyone having intercourse with that nursling commits bestiality.[69] In a prerevolutionary France concerned with an ideal of Nature, the question of the "bastardizing" effect of a child's imbibing human milk from outside the family was debated in much the same terms as the question of the "animalizing" effect of imbibing milk from outside the species. In his *Émile* (1762) Jean-Jacques Rousseau attacked the institution of human nurse-mothers as immoral; in her popular handbook for breast-feeding, *Avis aux mères qui veulent nourrir leurs enfants* (1767), Le Rebours stressed the importance of endogamous breast-feeding and argued that what a child receives from the nurse-mother is generally a dangerously "alien and bastard milk [*lait étranger et bâtard*]."[70] Fear of animalization was similar: Garden writes that "one of the most serious charges leveled at the wet nurses of the Lyons area was that they substituted goat's milk for their own,"[71] and only a few Frenchmen had a more practical view toward animal milk. (Boerhaave's *Traité* of 1759, for example, argues that there was nothing at all wrong with goat's milk.)[72]

Linked with the controversy surrounding the animalizing effects of drinking animal milk—and the vampiric effects of transfusion with animal blood[73]—was the debate in the Romantic era about the animalizing effects of vaccination, or the inoculation of human beings with animal diseases.[74] For example, the arguments informing discussion about the efficacy of drinking cow's milk, or "vaccimulgence" (a neologism of the 1790s whose sound Coleridge seems to have relished)[75] informed the discussion about the efficacy of "vaccination." Dr. Edward Jenner, who had studied the medical implications of the fact that people who worked around cows did not get smallpox, argued in his *Inquiry into the Causes and Effects of the Variolae Vaccinae* (1790) that vaccination with cowpox could prevent smallpox. But many people preferred smallpox to being cowed, or to being "grafted" with a cow, or *vacca.*

Grafting, which in the botanical realm involves an implantation that turns two things into one, has its counterpart in the zoological realm. This counterpart is not quite twinning, as where two friends are as alike as "twinned lambs"—to which actual animals the two kings in *The Winter's Tale,* King Leontes of Sicilia and King Polixenes of Bohemia, are compared. Nor is it quite the relationship that a father and his consanguineous son have when they are "as like as eggs"[76]—to which potential animals Leontes compares himself and his possibly bastard son Mamilius. The real zoological counterpart to botanical grafting occurs where there is identity, or unity, of the two creatures, as when we learn, at the beginning of *The Winter's Tale,* that "Sicilia cannot show himself over-kind to Bohemia. They were train'd together in their childhoods; and there rooted betwixt them then such an affection which cannot choose but branch now."[77] Leontes cannot show himself overkind to Polixenes, because he and Polixenes are as kind as can be. Like the Siamese twins that Montaigne discusses in "Of a Monstrous Child" and Mark Twain depicts in his fabular *Those Extraordinary Twins,* they are grafted together by a process that is the business of the whole *mythos* of *The Winter's Tale* to reveal.

There are two procedures in *The Winter's Tale* whereby two become one. First, there is marriage, as traditionally understood by Pauline theologians. Man and wife become one flesh—as when Leontes' offspring Perdita marries Polixenes' offspring Florizel. Second, there is artful engineering of nature: the gardener's art of grafting plants, say, or the shepherd's art of breeding animals.[78] It is with his shepherdess daughter Perdita—whom he has feared to be a zoological bastard (i.e., an illegitimate or natural child)—that the disguised Leontes discusses botanical bastards (a specific kind of a flowering plant).[79] And he describes the gardener's artful implantation of one *gens* (or species) into another:

> You see, sweet maid, we marry
> A gentler scion to the wildest stock,
> And make conceive a bark of baser kind
> By bud of nobler race. This is an art
> Which does mend Nature—change it rather—but
> The art itself is Nature.
>
> (4.4.92–97)

Implantation or inoculation—the two terms have been near synonyms since Francis Bacon's *New Atlantis*[80]—is the hallmark of civilization when it comes to the culture of plants and animals. In agriculture, *inoculation* once meant "to implant an eye or bud into," "to bud one plant into, unto, on, or upon another," or "to join or unite by insertion (as the scion is inserted into the stock so as to become one with it)." But in eighteenth-century zoological and medical discourse, *inoculation* came to mean "to impregnate (a person or animal) with the virus or germs of a disease . . . for the purpose of inducing a milder form of the disease."[81] The inoculant (or its animal or human source) and the inoculated become as one.

Controversy about medical inoculation involved two questions. First: Should a human being ever be inoculated with a disease from another human being?[82] Where the "inoculant" is a poison, like syphilis, the medical civilization that brings inoculation appeared like "syphilization"—which in the 1840s became a near synonym for "inoculation."[83] Second: Should a human being ever be injected with a disease from an animal (or another kind of animal)? That people could catch animal diseases was well known. (Such human diseases as "chicken pox," "cow pox," and "swine pox" were named after animals.) But could people catch animality itself from animal inoculation?

Animal inoculation thus became the focus of an ideological debate about interspecies miscegenation and the definition of human being. Those who argued for vaccination were called "cow maniacs" and those who argued against it were called "cow phobics."[84] Cow phobics argued that the practice of vaccination would make for the the gradual and imperceptible bestialization of humankind (see fig. 13). For example, Dr. Verdé-Delisle, in his *Physical and Moral Degeneration of the Human Species Caused by Vaccination* (1855), argued that vaccination would mean the end of human beings as we know them.[85] In his *Medical Tracts* (1799) Dr. Moseley asked "Who knows . . . what ideas may rise in the course of time from a brutal

Fig. 13. Colored copper-plate engraving. Cow-Poxed, Ox-faced Boy. *In Rowley,* Cow-Pox Inoculation No Security Against Small-Pox Infection, *1805. (Institute of the History of Medicine, The Johns Hopkins University, Baltimore)*

[animal's] fever having excited its incongruous impressions on the brain?''[86] Several cow phobics insisted that people who were vaccinated soon looked and acted like animals. For example, Dr. William Rowley's engravings (1805) show "how children who had been vaccinated were developing cow's faces,''[87] and so do Dr. R. Squirrel's illustrations (1805).[88] And a child vaccinated in 1801 at the Russian court was afterward called "Vaccinoff,'' as if he were part human and part cow.[89] A few cow phobics, focusing on the fear that vaccination produced mixed species, claimed that it amounted to incest. For example, Dr. Joseph Merry, in his *Conscious View of Circumstances and Proceedings Respecting Vaccine Inoculation* (1806) "stated that cowpox inoculation was comparable to incest, introducing into the human body a disease of bestial origin similar to syphilis.''[90] And Dr. Moseley, referring to the Greek story in which the human Pasiphaë breeds with a bull and gives birth to a Minotaur,[91] asked, "Who knows also, but that the human character may undergo strange mutations from quadrupedian sympathy, and that some modern Pasiphaë may rival the fables of old.''[92]

In a similar vein cow phobics attacked Benjamin Waterhouse, a member of the universalist Society of Friends who had helped to introduce vaccination to the United States; they accused Waterhouse of attempting to homogenize humans and animals in a kind of spiritually incestuous stew.[93] A few cow phobics commented on vaccination as a cannibalistic rite-of-passage that defines communal brotherhood in much the same way as the Eucharist. By the century's end, the vegetarian George Bernard Shaw, in his preface to *Back to Methusaleh: A Metabiological Approach* and his preface to *The Doctor's Dilemma,* attacked inoculation "as a horrible reversion to the most degraded and abominable forms of tribal ritual.''[94]

Where the argument against vaccination of human beings boiled down to a fear of "bovinization,'' it seemed useful to cow maniacs to employ such reassuring analogies to vaccination as milk-drinking or beef-eating. Many cow maniacs tried to point out the relationship between vaccination and imbibing or eating cow. For example, Adams, in his *Answers to All the Objections Hitherto Made Against Cow-pox* (1805), writes of the "nonsense about the danger of inoculating humours from an animal whose milk makes the principal part of our children's food, whose flesh is the source of Old English courage, and whose breath is not only fragrant but salubrious.''[95] A speaker at the Third Festival of the Royal Jennerian Society in 1805 made the same point about the humanness of cow's milk.[96] But whether it was essentially human either to drink cow's milk or to eat cow's flesh was already a hot debate.

At this time in Germany, where most practicing physicians were against vaccination,[97] the novel entitled *Die Kuhpocken* [The Cowpox] focused on a family quarrel in which a cow-maniac father wants to have his child vaccinated and a cow-phobic mother does not. In later years, Thomas De Quincey was struck by how so eminent a thinker as Kant could have seriously questioned Jenner's evidence in favor of vaccination,[98] and have insisted that vaccination was a monstrous "inoculation of bestiality.'' In response to queries from respondents on the subject of vaccination in 1800, Kant went so far as to claim that the practice was morally unjustifiable.[99] And under Kant's influence, his friend and physician Dr. Marcus Herz published his *The Inoculation of Bestiality* (1801)[100] which again defined the question of disease in relation to breeding between species and within families.[101]

KINDNESS AND CHRISTENDOM

Let us backtrack here and reconsider, on the one hand, the relationship of the idea of universal animal kinship to the idea of universal kinship, and, on the other hand, the connection of the idea of universal animal kinship with the purported moral obligation to be kind to kin, or to be our brother's keeper.

Universalism and Particularism

Universalist ideology, we have seen, overtly posits that there is essentially only one human family in the world. For believers in such a religion, "All human beings are my siblings and only animals are others." But the rhetoric of universalism invites us to wonder whether only our kin are our kind even as it recalls the languages of primitive tribes where the word for "human being" and the word for "fellow tribesperson" are one and the same.[102] For the universalist statement turns out easily to mean "any creature who is not my brother is not human" or "only my brothers are human, all others are animals"—a conclusion with catastrophic moral consequences. Particularist ideology, on the other hand, rules out of order this politically dangerous slide by holding that there is more than one brotherhood or tribe of human beings. For particularists there are human brothers, and also others who are not, on account of their otherness, less than human. While the universalist insists that all aliens are animals (he cannot admit the existence of others who are human), the particularist allows "us" to think of and treat some beings from outside our tribe as human rather than as animals.

(Who "we" are matters. The creature that a person from one universalist tribe calls "extrafamilial," hence "nonhuman," may be essentially "familial," hence "human," in the view of someone from another universalist sect; or that creature may be essentially "human," if also "nonfamilial," in the view of someone from a particularist sect. The changes in meaning of such terms pose for the interlocutor—"mon semblable—mon frère"—a sometimes intolerable complication in discussing toleration.)

Greater political consequences follow when universalist theorists or sentimentalists claim to extend the boundaries of kinship to include not only beings considered human but also those considered animal—an extension generally made in order to eliminate the otherwise burdensome problem of distinguishing human beings from animals. The universalist statement "All creatures are our siblings" now turns easily into the view that "Only our brothers are creatures, all others are nonanimal—or both nonhuman and nonanimal." Universalist religion thus not only treats alien humans as nonanimal but also treats alien humans and animals as nonanimal, or as insentient things. It treats both alien humans and animals differently than does particularist religion. As we shall see, the practical moral and political consequences of this difference are far-reaching.

Kindness and Cruelty to Animals

There are many universalist tribes—groups of human beings for whom all extratribal beings are also essentially nonhuman, or for whom the concepts of "human" and "fellow tribesperson" are indicated by two synonymous terms or by one and the same term. Examples abound in the European secularist tradition: the French revolution held out for universal "fraternity," for example, and the English Family of Love claimed that all human beings were their brethren. More significantly for our purpose of understanding long-standing aspects of the Judaeo-Christian tradition, the prime Christian stance is that Christianity itself is a universalist religion and Judaism a particularist religion.[103] In what follows I shall adopt this possibly inaccurate Christian representation of Christianity and of Judaism, as a means to tease out certain practical moral and political implications of universalist and particularist ideologies.[104] I shall discuss the largely secular and nationalist states of Christendom—such states as France and England. Their thinking about and treatment of "others" is connected with the universalist ideology that Christianity tends to embrace, whether in official Pauline or unofficial Napoleonic and later guises.

For both the religious and the secular Christian there are, in regards to animals and kinship, two positions in polar opposition to each other: (1) Animals are akin to humans; that is, they are our brothers, hence are to be thought of and treated as if they were members of our tribe or species. In this view, we are not the "keepers" of animals but their equals. In holding to it, Christianity is what one historian of animal sentimentality calls "the most anthropocentric religion that the world has ever seen."[105] (2) Animals are extraspecies and extratribal beings, and hence, like all essentially nonhuman things, are outside the "covenant."

The position that all living creatures are our brethren is a hypothesis entertained by Christian thinkers such as St. Francis, who preached his doctrine to the birds, and St. Anthony, whose horse, it is said, used to kneel to receive the eucharistic host, and even Renaissance "free thinkers" influenced by Stoic "cosmopolitanism" such as Montaigne.[106] And the universal kinship of all living creatures is a popular notion in secular Christian culture. Thus one widespread old English Christmas carol—a song celebrating the birth of an extraspecies god as an intraspecies human being—includes the following refrain:

> The friendly beasts around him stood:
> "Jesus, our brother, kind and good."[107]

In performing this carol, the members of a chorus of human beings, a species that we define as *speaking animals,* pretend to be members of a choir of domestic animals that speak (goats, chickens, sheep, etc.)—as if the caroling humans were animals or the animals were caroling humans. The notion of the animals' friendship, kindliness, and brotherliness to man is of course pleasant. But it can also be pretty unsettling. For example, universal kinship turns all meat-eating into cannibalism—even into incestuous cannibalism, since the flesh that a meat-eater devours must come from the body of a "brother," that is, from the body of a butchered member of one's essentially human family, or, if you prefer, from a member of one's superspecies family. At the very least, kinship, such as this "Carol of the Beasts" supposes, turns all meat-eating into cannibalism for him—or for Him—whom the animals call their "kind . . . brother."

The significance for Christian thought of the analogy between Jesus and the domestic animal—Jesus is part man and part God and the domestic animal is part man and part animal—is hinted at by the birthplace that Christians generally assign to Jesus: a stable.[108] There being no proper place among human beings for this extraordinary creature to be born—"no crib for a bed"—Mary gave birth to Jesus "away in a manger."[109] Many Christian pet lovers, following or imitating Jesus, wish that they too had been born among animals: "I often think that I should have been happier," writes Barbara Woodhouse, "born in a stable than at St. Columba's College."[110] (In Christendom, popular stories about interspecies sexual generation between human beings and animals or gods often recall Mary's giving birth to the godman Christ. Thus Hogarth's "Medley," which shows a woman giving birth to bunnies in a church, mocks the Enthusiast Methodists' pre-Darwinian belief in interspecies generation.)[111]

Taken together, the two universalist views we have discussed here—that animals are akin to humans and therefore part of the covenant and that animals are not akin to humans and therefore not part of the covenant—can be contrasted with the view of animals informing a "particularist" religion such as Judaism, which allows that there are nonhuman, extraspecies members of the covenant, just as it allows that there are human extratribal members. For Jews, animals and human beings are both within the covenant, albeit they inhabit a different place in it. For Christianity, on the other hand, that only essentially human beings are covered by the covenant and the laws, means that insofar as we are obliged to treat animals kindly, we must also treat them as (if they were) responsible humans. In some realms of Christendom, therefore, animals that have committed a "wrong" are admitted to the ecclesiastical courts dressed in human clothing, and then compelled to sit on a chair or to stand on two legs.[112]

The view of all God's creatures not only having a place in the choir but also inhabiting the same species-place in it is, in a literal sense, totalitarian. And since it denies the various animals a place of their own, it often leads to their relegation, not to human status, but to that of mere "things"—say, only things to eat. (Instead of saying, "All the animals are our kin," we come to say, "Only my kin are animals, the being that I eat is only a thing.") Indeed, the second Christian view of animals has it that, insofar as they are not, like family pets, inside our family—or insofar as they are, unlike family pets, entirely outside our species—they are, like vegetables and stones, outside the covenant of the law. Christianity, especially in its post-medieval context, thus grants to human beings virtually limitless dominion over the world of "things," including animals.[113]

Augustine himself underwrites the Christian withdrawal of protection from all fauna except humankind in his commentary on the "Story of the Legion of Devils Cast Out." In this story, Jesus allows—even induces—thousands of pigs to drown themselves in the sea, as in a holocaust of swine *(marranos)*. "Then went the devils out of the man [who was called Legion] and entered into the swine; and the herd ran violently down a steep place into the lake, and were choked."[114] Augustine recognizes that Jesus was ill-treating a large number of animals and was thus apparently violating certain rules of the Old Testament. So he argues that, according to the New Testament, there is no protection whatever for them: "Christ himself shows that to refrain from the killing of animals . . . is the height of superstition, for judging that there are *no common rights* between us and the beast . . . [God] sent the devil into a herd of swine. . . ."[115] Thus Pope Pius IX was in step with the historical Christian tradition when he refused to sanction the establishment of the Society for the Prevention of Cruelty to Animals (SPCA) in Rome, on the grounds that its ideology implied that humans would have duties toward animals that according to Christianity they do not have.[116] After all, it is not Christianity but Judaism which has laws that specifically include animals in the covenant[117] and that protect nonhuman animals; and it is Christianity, not Judaism, whose doctrines, from Jesus to Saint Augustine to modern popes, deny any obligation toward nonhuman beings.[118]

What are the practical consequences of this difference between Christianity and Judaism regarding the treatment of animals? Jewish doctrine, while disclaiming equal kinship with animals, enjoins humane kindness toward them. There are Old Testament commandments, for example, that one should help even the ass of one's enemy when the ass is under its burden,[119] that one should allow animals to rest on the Sabbath,[120] that one should not muzzle an ox when it treads out the corn,[121] and that one should regard highly the life of one's beast.[122] Christian writers before the modern era usually ignore the biblical passages enjoining specific kindly treatment of animals. Or they interpret them allegorically, to refer not to animals but to humans; for example, Christian clergy said that the "muzzled ox" in the passage from Deuteronomy stands for the "inadequately paid clergy";[123] and Church fathers, citing Paul's letter to the Corinthians, argued that God does not care at all for oxen as oxen.[124]

How some Christian thinkers converted the rules of the Old Testament against being cruel to animals into a license to dominion over them[125] is as interesting as how the Jewish injunction against physical cannibalism became an endorsement of spiritual cannibalism (the Eucharist) or how that against physical incest became an endorsement of spiritual incest (God as the Father of Himself). Many Christians argued that beasts were not fit parties for taking part in a Covenant and therefore could not be part of one.[126] Thomas Aquinas presented another argument: "If any passage in holy scriptures seems to forbid us to be cruel to brute animals," says Thomas, "[it does so] lest, through being cruel to an animal, one becomes cruel to human beings, or because an injury to an animal leads to the temporal hurt of man."[127] Calvin wrote similarly that animals were permitted to rest on the Sabbath only to ensure that their human masters do as well.[128] This Christian view of kindness to animals, a view both Catholic and Protestant, suggests that if we could be sure that cruelty to a species other than our own would do our own species no harm, then cruelty to animals would be permissible. (And what if we were sure that cruelty to another "species," even annihilation of it, would do our "species" some good . . .?)[129]

In the late, Christian secularist eighteenth century, when animal sentimentalists sought scriptural justification for their own feeling that humans should be kind to animals, they could not, generally speaking. find it in the New Testament. The only relevant doctrine they could find in Christendom involved old-fashioned "pagan" customs and political institutions that had long existed in more or less unofficial ideological tension with official Christianity.[130] Only "un-Christian" customs enjoined a polytheistic awe for zoological and botanical creations (the *Tannenbaum* of the German tribes is a good example). Only "un-Christian" political institutions called upon human beings to be kind to creatures outside their own kin or kind group. (Some basically pre-Christian feudal doctrines thus called upon the feudal lord to be kind to his vassals in the same way and in the same words that they called upon the vassal to be kind to his animals.) The doctrines of the Pythagoreans and of the Hindus—that all animals have souls equal with ours—might have been useful to the sentimentalists in their search for authority, but these doctrines seemed too outlandish to be persuasive. So the sentimentalists turned for help to the Jewish view that cruelty to animals is wrong regardless of how we feel about either the human consequences or about the animals' interspecies equality with us.[131]

The sentimentalists cited Hebrew scriptures for their purpose of encouraging kindness to animals. But they mistook the complex and balanced Jewish understanding both of stewardship and of kindness, representing the Jewish position on the relationship between humans and animals—and on the relationship between one group of animals and another—as having, like the sentimentalist position, no reference to real individual and political needs. This misrepresentation of Judaism continues among sentimentalists and ecologists to the present day.[132] Nowadays, indeed, the sentimentalist position has become idealist to the point of wondering whether it might be unkind, not only to eat sentient, nonhuman beings (animals), but also nonsentient beings (vegetables) and even nonliving beings (rocks).[133] It is as though, for thoroughgoing sentimentalists, universal human starvation was the only alternative to being inhumane. *Returning to Eden* is their aptly entitled handbook.[134]

The Pet as Inedible Animal

In Judaism the rules of how to be kind to animals are inextricably connected with the rules of how to be cruel. The God of the Pentateuch tells us not only how to tend the animals but which to butcher and how to butcher them. The *baalei hayyim,* or legislation prohibiting causing animals undue pain, includes explicit consideration of the humaneness of *shehitah* and of the "bird's nest law," which enjoins against slaughtering a mother animal with its young or a young animal before the eyes of its mother.[135] The closeness of the relation in Judaism between being kind to animals and eating them (as in the rule that Jews should not boil a kid in its mother's milk)[136] recalls, once more, the words from *Hamlet,* one is cruel in such a way as to be kind.

In this context, it is important to emphasize that Christianity, unlike many other religions, does not ban the eating of *any* food: It is essentially an omnivorous ideology.[137] Indeed, even cannibalism, which is banned absolutely in most other cultures, is sometimes enjoined in Christianity. In the Eucharist, for example, the celebrant eats or says that he eats—in however extraordinary a sense—a being that is not only partly nonhuman, or divine, but also partly human.

In an essentially Christian culture, the widespread modern institutionalization of sanctioned pethood made a crucial difference to the view of animals and men alike. Christianity had been an omnivorous religion, but in the new order the pet emerged as the one essentially inedible animal. (Indeed, one definition of *pet* that reveals our characteristic "indulgence" toward them is as the animal or nonhuman being that cannot, or should not, be eaten—even in the otherwise virtually omnivorous gourmet centers of Christendom.) In Christianity, the advent of pethood, or of the institution that made human kind and familial kin out of beings that had theretofore been considered extraspecies and extrafamilial, brought with it the feeling, amply illustrated in the literature, that it would be like ordinary cannibalism to eat a pet.[138]

That there should be anything at all it would be wrong to eat was a crucial new position for Christendom. It is an essentially Protestant position; Roman Catholic churchmen frowned upon pet-owning[139]—and also upon the practice of giving "Christian" names, or names "appropriate" to human beings, to animals[140]—because such practices tended to confuse the partly human Christ (and the Eucharist) with the partly human Fido (and pet-eating). For many decades after the introduction of widespread pethood the Romantic and Victorian Christians treated pet-eating and cannibalism as identical; they conflated the new taboo, "Thou shalt not eat a pet"—which the institution of pethood promulgated—, with the old taboo, "Thou shalt not eat a human being"—which traditional Christianity had both promulgated (insofar as the human being on the platter is most anyone) and also broke or transcended (insofar as the human being is Christ). The meat-eater Henry David Thoreau, exhibiting a kind of millennialist spirit, wrote in *Walden* that some day people would scorn our present animal-eating habits much as we scorn the man-eating habits of so-called savages: "Whatever my practice may be, I have no doubt that it is part of the destiny of the human race, in its gradual improvement, to leave off eating animals, as surely as the savage tribes left off eating each other, when they came in contact with the more civilized." (In *The Citizen of the World*, Oliver Goldsmith wrote sarcastically of those who argue for kindness to animals while at the same time eating meat that "they pity and they eat the objects of their compassion.")[141]

The Christian secularist distinction between what animals may and may not be eaten by no means signifies the same as the Jewish laws of Kashrut. In Christian families, the eating of the family lamb is a central event, a sacred and profane Crucifixion, or butchering, and a Eucharist, or act of cannibalism, rolled into one.[142] In Jewish families, what animal may be eaten is indicated not by its human family (is it somehow part of my consanguineous family?) but rather by its animal species (is it a pig?) and its mode of being butchered (is it drained of blood?).

Eating and Intercourse

The anthropologist Edmund Leach, adapting a canonical anthropological thesis that "there is a universal tendency [among human beings] to make ritual and verbal associations between eating and sexual intercourse," argues that "it is a plausible hypothesis that the way in which animals are categorized with regard to edibility will have some correspondence to the way in which human beings are categorized with regard to sexual intercourse."[143] Leach then goes on to find a structural parallel, in the tradition of Lévi-Strauss, between two groups: those beings with whom we are barred from having sexual intercourse, and those that we are barred from eating. Leach makes two or three far-reaching and apparently commonsense assumptions about these two categories of forbidden things. But we shall see, the anthropology of the pet figures centrally in these assumptions, and Leach tends to mistake its complex role in the structuring of taboos on intercourse and eating.

Leach's first assumption is that only humans belong in the group with whom we are barred from having sexual relations. This means that while he considers the taboo on incest he passes over that on bestiality. Yet there are people for whom intercourse with certain kinds of more or less familiar animals, is not absolutely condemned.[144] His second assumption, that only animals belong in the group of beings that can be divided according to which are edible and which are not, means that, while he considers the taboo on eating animals, he does not consider the taboo on cannibalism. Yet there are people for whom eating certain human beings is not only partly allowable but even obligatory.[145]

In any case, the principal examples that Leach uses to support both assumptions contradict an excellent hypothesis of his about pets—namely, that they are not so much animals per se as intermediating beings between animals and humans.[146] Leach contradicts this hypothesis when he says that the quintessential example of a being with whom we may not have intercourse is a member of our own human family—as we have seen, a pet animal might serve just as well. He also contradicts it when he says that the quintessential example of a being we may not eat, is a pet—as we have seen, a human might serve just as well. If, as Leach's original hypothesis suggests, the pet is an "animal-man," it belongs as well or ill in either group; a pet is as much or little a being with whom we cannot have intercourse as it is a being that we cannot eat.

Insofar as we conceive the pet as both human and nonhuman, it stands at an intersection between species (it is my kind and another kind as well); similarly, insofar as we conceive it as being both familial and nonfamilial, it stands at an intersection between families (it is my kin and not my kin). The pet thus stands at the focal *chassé-croisé* of the taboo concerning eating and the taboo concerning incest.

Leach goes some distance toward establishing a structural "homology," or series of correspondences, between eating and sexual intercourse in terms of animals and human beings. What I have been arguing, however, is that neither the phenomenon of pethood nor the problem of the relation between human and animal is easily understood in terms of a structural homology between eating and intercourse—the canonical anthropological viewpoint that Leach represents: Rather, what is here at work is an identity of one with the other.

The family pet stands both at the borderline between family and nonfamily (i.e., at the borderline between those beings with whom it would be incest to have intercourse and those with whom it would not be incest) and at the borderline between animal and nonanimal or between man and nonman (i.e., at the borderline between those beings which may be eaten and those which may not). Pets stand at the *chassé-croisé* between kin and kind.

For a culture where all sex is equally taboo or sacred (say insofar as we are all essentially siblings and hence barred from having intercourse with one another) and where all eating is similarly so (say insofar as we are essentially omnivores), the institution of pethood is, as we shall now see, an especially sensitive barometer of the way human beings grouped as nations are likely to treat one another.

Kindness and Cruelty to Humans

To briefly summarize the Jewish particularist and Christian universalist viewpoints on kin and kind: The Jewish position is that there are human beings both within and without the tribe; both groups are participants in the covenant, however much the ones within the tribe be "chosen," say by divine election. Likewise, the Jewish position is that there are sentient beings both within and without the human species: Both intraspecies and extraspecies beings are protected by the law, however much the former group is "superior" to the latter, say by virtue of being speaking animals.

Christianity, on the other hand, does not allow for either extratribal human beings or for extraspecies sentient beings protected by the rule of the covenant. If one is not essentially akin to a Christian, one is not humankind, and, as an animal, one has no legal right to be treated kindly: One is exploitable along with vegetables and stones.

The disturbing slide from extending kinship to others toward denying kindness to them tends to occur not only when we say we are extending kinship to all extra-species beings. It also tends to happen when we say, somewhat less grandly, that we are extending kinship to all extratribal humans in such a way that they all are brethren in a single group. For the view that all human beings are members of one tribe can turn into a call for turning extratribal human beings into things or for dehumanizing them. Chamfort said that the motto of the French Revolution, "Liberty, Equality, Fraternity," really meant, "Be my brother (*frater*) or I will kill you."[147] And the Elizabethan Family of Love, whose doctrine included the notion that all men are brethren, actually said that "whosoever is not of [our] sect, [we] account him as a beast that hath no soul."[148] Thus the doctrine of universal intra-species human kinship easily justifies treating all beings from outside the Christian tribe (i.e., all beings who feel themselves to be, or who are felt to be, nonconvertible to that tribe) as animals. Was it not, in our century, a millennialist Christian movement, albeit in a radical secular guise, that called beings from outside what its members took to be the one essential tribe "racially inferior"? That movement made the living conditions of those extratribal beings as filthy as it understood those of some animals to be.[149] "Petting" the Jews and Gypsies in slaughterhouses, the Nazis tried to turn them into animals.[150]

By the same token, the doctrine of universal interspecies kinship can justify treating animals as human. Hitler anthropomorphized his pet dog—maybe his pet was the one being he "loved"—just as he tried to dehumanize the Jews and Gypsies.[151]

Such observations do not really help us understand the ideology of a totalitarian and tribal nation torturing and attempting to annihilate nonnationals as both extra-tribal ("We are Aryans, they are Gypsies") and extraspecies ("We are humans, they are animals"); their having such status is not, in any ecological case, a moral or political justification for torturing and annihilating individuals and groups. But such observations do help us understand how it was that Lewis Gomperz, the first secretary for the Society for the Prevention of Cruelty to Animals (SPCA) and a Jew, was concerned not so much with individual family pets—a fanatic obsession in the romantic and nationalist period that was his and is ours still—as with animals in general.[152] Gomperz was interested in animals as part of a tolerant covenant that includes in its purview beings tribal, extratribal, and extraspecies.

Romantic Utopia

Among common minds, aye, among any but very uncommon minds, who enquires whether any one can do that which no one does do—. Add to this all the moral Loveliness of the Disposition of the two affections [sisterly and conjugal], which the better part of our nature feels—tho' only a few speculative men develop that feeling, and make it put forth in its distinct form, in the understanding.—A melancholy Task remains—namely

> *to show, how all this beautiful Fabric begins to moulder, in corrupt or bewildered* (veränderte) *Nature—the streets of Paris and the Tents of the copper Indians, or Otaheitans.—Of this elsewhere, when we must. It is a hateful Task.*
>
> —Coleridge, *Notebooks*, no. 1637

The revolutionary ideology of universal *fraternité*, like the old Christian ideology of universal brotherhood, posits a universal kinship extensible in certain circumstances from one species (that of human beings) to all species. The English intellectuals who visited France during the 1790s, for example, often conceived of animals as potential members of a radical egalitarian community. Samuel Taylor Coleridge thus included animals in the plans that he and Robert Southey formulated for their American pantisocracy. In the human community of pantisocracy we humans would all be brothers and sisters, so that all sexual relations would be incestuous or would transcend mere incest.

And how should we describe sexual relations in pantisocracy if we extend siblinghood to animals as well as to all humans—as bestial, or as transcending mere bestiality?

Coleridge makes the extension of kinship from man to animal not only in the case of his family cat (a familial rather than a species extension) but also in that of less familiar animals. In one version of his poem "To a Young Ass," for example, he writes:

> I hail thee *Brother*—spite of the fool's scorn!
> And fain would take thee with me, in the *Dell*
> Of Peace and mild *Equality* to dwell . . . [153]

The dell of interspecies equality where Coleridge would live together with his animal Brother is the same one he speaks of in his poem "Pantisocracy":

> I seek the cottaged *dell*
> Where Virtue calm with careless step may stray . . .[154]

And one version of "To a Young Ass" makes explicit the link between interspecies equality and the pantisocratic community for which Coleridge yearns:

> I hail thee *Brother*—spite of the fool's scorn!
> And fain would take thee with me, in the *Dell*
> *Where high-soul'd* Pantisocracy shall dwell . . . [155]

For some people, the reason we should be kind to an animal like an ass is that it is protected by the law. For others, the reason might be that a particular ass has been kind to them. (This is the reason God in the Old Testament puts into the mouth of Balaam's ass, whom Balaam would kill with a sword: "Am I not your ass, upon which you have ridden all your life long and to this day? Was I ever accustomed to do so to you?")[156] But for Coleridge, as for Saint Francis, the reason that a man should be kind to an ass is that an ass is kindred with us in a universal interspecies siblinghood.

In the pantisocracy that Coleridge idealizes all intercourse verges on bestiality (or transcends mere bestiality, since animals are his kin, hence his kind). This way of conceiving the matter is, to be sure, outrageous, except perhaps for saints and children. But the matter is restatable in other terms as well: Insofar as the animals are my brothers and sisters in a pantisocracy, all flesh-eating verges on cannibalism or transcends it.

In pantisocracy, and in any other community believing in universal brotherhood, all intercourse verges on the incestuous and the bestial. Can one avoid physical incest and bestiality by means of "spiritualizing" kinship distinctions and species distinctions—that is, by "raising" them high above the physical? To use Coleridge's term, how *high-soul'd* must be one's kindredship with an ass in order to imagine oneself loving it in a wholly chaste, or nonincestuous and nonbestial, way—a way that is not, even unconsciously, physical bestiality or incest? One solution would be an absolute spiritualization of love—of the kind that Coleridge sometimes imagines in terms of "kindred minds."[157] Such a spiritualization of love becomes, in practice, a religious celibacy of the sort promulgated in the Catholic orders. Interspecies equality, for which we say religious celibates such as Saint Francis stood, leads to a spiritual kind of bestiality, and intrafamiliar equality, for which Saint Francis also stood, leads to a spiritual kind of incest.[158]

Celibacy was the solution of Saint Francis to the problem that universal interspecies kinship makes all lovemaking at once bestial and incestuous. But celibacy was anathema to the Protestant Coleridge. And since there is no other way of solving the problem, Coleridge attempts to dissolve it by means of a joke. "To a Young Ass," for example, avoids directly confronting the sexual and nutritional difficulties involved in hypothesizing an interspecies equality by means of its apparently nonsensical ending. In the closing lines of the poem, Coleridge says that, in pantisocracy, the

> . . . Rats shall mess with Terriers hand-in-glove,
> And Mice with Pussy's Whiskers sport in Love.[159]

E. H. Coleridge assures us that the poet "mean[s] . . . [this ending] . . . to have [no] meaning."[160] Yet what kind of messing, or sporting in love, is this? Is it like the love in Shakespeare's *Timon of Athens* between the misanthropic Timon and the vegetable root that Timon must decide either to eat, in order that he might continue to live, or not to eat, in order that he might die?[161] Coleridge bypasses the problem that, in an interspecies pantisocracy where all are kin, it would mean starvation for the terriers not to eat the rats, their natural food, and it would be incest for the pussy to sport in love with the mice. Universal interspecies siblinghood, which makes any ass my brother, turns all flesh-eating into cannibalism just as surely as universal intrafamilial siblinghood, which philanthropically makes any human being my familiar sibling, turns all intercourse into incest. The effect of giving all God's creatures the same "place in the choir"[162] is thus either universal celibacy and starvation or bestiality/incest and cannibalism.

CONCLUSION

In England "by 1700 . . . the symptoms of obsessive [animal] pet-keeping were [already] in evidence."[163] In America in the 1990s the number of pets far exceeds that of children, and the human-animal bond has been presented to us, as we have seen, as an almost universal panacea for psychological and sociological ills.

Pets are especially useful to us here in America, in the age of the small, "nuclear" family, because this age puts unique pressures on the family's kinship structure. In the past, there were family slaves, servants, mistresses, and domestic working animals who provided safety valves for large extended families, which perhaps needed such safety valves less than our smaller families do. The general disappearance of such metakinship institutions has left a lacuna that pets often fill.[164]

Maybe pets provide a better safety valve than metakin of our own kind: One can love a pet more uninhibitedly than one can love a slave, nursemaid, or servant, precisely because in itself the taboo on bestiality (with the pet insofar as it is not a member of the human species) tends to make the taboo on incest (with the pet insofar as it is a member of the family), which we might generally desire, unthinkable. The taboo on bestiality thus makes unnecessary an even more repressive explicit taboo on incest.[165] Fleeing the human for the animal and the sexual for the asexual, one comes upon the family pet with a sigh of relief.

The pet thus represents one solution to the incest taboo. But perhaps it represents more. Animals, after all, are not only sociologically totemic and psychologically transitional objects for human beings; they are also somewhat conscious beings like human beings. They are the same and different. Are animals any less wonderful than extraterrestrial beings like "E.T."? Are the intersubjective barriers to interspecies relationships really greater than the awesome barriers to intraspecies human relationships?

The dominant, anthropocentric ideology of our time—abetted by the Cartesian view that animals are automata—dismisses this question by claiming that animals as such do not really exist. It regards animals as things to be exploited—say as elements of a smoothly running family, old-age home, or farm, or as moral counterparts to human beings (guinea pigs about whom moralists argue whether their lives as "experimental animals" are too painful for human beings to bear).[166] To imagine how animals may be—or may be thought of as—other than mere things would require a leap of the imagination and a feat of historical scholarship beyond the purview of this chapter.[167] Yet it is now clear that we need to reinvestigate several longstanding and influential questions. Did we, for example, look differently upon animals before the advent of an economic system that treats human beings as "living tools"?[168] Did the disappearance of domestic working animals make for a diminution in our emotional life? What happens when people study an ape "growing up human" in a human family or when people join a family of apes?[169] Do closer relations with edible animals and with slaughterhouses encourage a more extended hierarchy of living creatures? What is the cultural significance of the

Enlightenment view that wild animals cannot be owned as property while working animals can be? How, if at all, do we own our pets?[170]

The peculiar institution of pethood generally has the quieting effect of helping to conceal both the sociological urgency of such questions and the articulation of kin with kind that underlies modern nationalist and internationalist ideology. Pethood is in itself a relatively kindly and unthreatening institution. Yet the ideology of pethood comes to the rescue of proselytizing politics by articulating an idealized *chassé-croisé* between kin and kind. In this way pethood helps to conceal even from would-be kindly human beings the brutally inhumane reality of the doctrine of universal (human) brotherhood.

Family pets are generally mythological beings on the line between human and animal kind, or at least thought of as on such a line. Yet sometimes we really cannot tell whether a being is essentially human or animal—when we are children, say, or extraterrestrial explorers. Sometimes we really cannot tell whether a being is our kind or not our kind, our kin or not our kin; we cannot tell what we are and to whom. If there were no beings such as pets, we would breed them, for ourselves, in our imagination.

8

TRIBAL BROTHERHOOD AND UNIVERSAL OTHERHOOD

OR

"In the presence of my enemies"

Thou preparest a table before me
in the presence of my enemies.

—Psalm 23

ENEMIES AND FOES

Thy Friendship oft has made my heart to ake
To be thy Enemy for Friendships sake.

—William Blake, "To H——N"

The doctrine of universal fraternity—"All ye are brethren"—has its difficulties. The doctrine leads either to celibacy or to incest. And it tends toward abstraction—which was the Victorian position against it. James Fitzjames Stephen, in his *Liberty, Equality, Fraternity* (1873–74), thus criticizes the ideology of fraternity in its quasi-secular romantic guise as a hypocritical substitution:

> Love for Humanity, devotion to All or Universum, and the like, are . . . little, if anything, more than a fanatical attachment to some favorite theory about the means by which an indefinite number of unknown persons (whose existence it pleases the theorist's fancy to assume) may be brought into a state which the theorist calls happiness.[1]

Similarly, Aristotle claims that the universal siblinghood set forth in Plato's *Republic* is merely watered-down kinship, saying that "it is better for a boy to be one's own private nephew than a son in the way described."[2]

The doctrine that "all ye are brethren" turns all too easily into the dogma that "only my brothers are human beings, all others are animals." And since the dogma tends toward bestiality and to a cruelty at once inhumane and—one would hope—nonhuman, it is better to be an outsider in a particularist kinship system, where there are human kin and human aliens, than to be an outsider in a universalist kinship system where there are only humankind and animals. William Blake understood this dialectic of brother and other, and fearing its action in the time of the French Revolution, he remarked in his poem "To H——N" that it is sometimes better to be an enemy where some enemies can be human than to be a friend where all enemies are subhuman.

Many Christian thinkers, confronted with such problems and wanting to be neither celibate nor incestuous, asserted that Jesus did not mean what he said. They asserted that the doctrine "All ye are brethren" was meant to apply only to the Perfect (i.e., to Brothers as monks and Sisters as nuns). For the imperfect many, they argued, there could be two kinds of brethren. The first kind we might call "exogamous" and love them in the sexual way; the second kind we might call "endogamous" and shun them sexually.[3]

The fearsome implications of Jesus' other cherished injunction, "Love your enemies," were similarly solved—or so it seemed.[4] Many Christian thinkers, wanting to hate their enemies despite the injunction or considering that politics depended upon our having someone to hate, asserted, once again, that it was meant to apply only to the Perfect. For the imperfect many, they argued, there are two kinds of enemies: "enemies" proper, coming from within our familial group, whom we must love, and "foes" from without our group, whom we may hate. Insofar as we strive to be genuinely political creatures—and not merely supposedly "wishy-washy" or "pluralist" liberals of the sort that detractors of Weimar democracy mocked—perhaps we *must* have foes to hate.[5]

The influential thinker Carl Schmitt, urged to join the National Socialist Party by the philosopher Martin Heidegger, was attracted to the sort of universalist Catholic rhetoric that calls for the inevitability and even political necessity of hatred of foes. He adopted as his own the view that Christ did not mean for us to love our foes.[6] For example, Schmitt argues in his *Concept of the Political* that in Jesus' Sermon on the Mount, which enjoins love, "no mention is made of the political enemy [foe]. Never in the thousand year struggle between Christians and Muslims did it occur to a Christian to surrender rather than defend Europe out of love towards the Saracens or Turks." For foes, or enemies from within one's national kin, Schmitt says, no quarter should be given; we are to show them neither love nor mercy.[7]

Schmitt's suggestion that the Christian tradition is not, in fact, so universalist as to enjoin treating all others as brothers has its basis in much theology and legal practice in Christendom. For example, it was often alleged by Christians (as opposed to Jews) that "no promises made to an enemy of the Christian faith, whether infidel or heretic, need be kept,"[8] and Christendom often adopted specialized Roman rules whereby alien nationals, or others, could be accountable to different laws than citizens, or brothers.[9] (It is as an alien that Shylock is punished in Shakespeare's superficially cosmopolitan Christian Venice.[10]) The stoic or universalist streak in Christian thought ("There is neither Jew nor Greek . . . for you are all one in Christ")[11] counters the legal and political practice of Christendom.

Schmitt's further claim, that it did not "occur to a Christian to surrender rather than defend Europe out of love towards the Saracens," seems calculated to obscure the way in which Christian love leads as easily to murder as to being murdered. For aside from the historical question of whether or not it did occur to Christians to surrender to the "Saracens" (or, in other times and places, to surrender to their foes)—and aside from the theological question of whether Christians should surrender to their foes rather than defend themselves—there is the logical question of whether a Christian's not surrendering to slaughter by others or to slaughtering them means that he will not love them. Unwillingness to surrender to an enemy does not necessarily mean unwillingness to love him, especially according to that Christian doctrine where loving a sacrificial victim (an edible animal or god) is the precondition to killing him, and where being slaughtered is sometimes the sign of divine love. Manuals for Christian slaughterers emphasize over and over again that one should love the animal one butchers. "Hold the lamb gently in your arms as you slaughter it." Not only does the meat taste better that way, as we are told,[12] but slaughtering lovingly also turns butchering into crucifixion. According to the christological interpretation of David's famous psalm, "The Lord is my Shepherd," it is the soon-to-be-butchered speaking animal, the Lamb of God, who, happily surrendering to the slaughterer and apparently looking forward to a dinner "not where he eats, but where he is eaten,"[13] says, "Thou preparest a table before me in the presence of my enemies."[14]

It is in the guise of reintroducing politics to Christian social theory that Schmitt depoliticizes that theory. His easy distinction between enemy and foe, for example, generally boils down to the difference between one tribe and another. This difference, as it is essentially not so much political as biological (or thought of as biological), is fraught with various difficulties surrounding the notion of tribal or racial purity. Among these problems would be, on the one hand, the inevitability of either celibacy or incest where the nation and family are one (or are thought of as one)—a problem, though raised by Plato and recognized by the pro-incest French racialist and nationalist Gobineau, that Schmitt chose to submerge;[15] and, on the other hand, there would be the ever present possibility and fear of secret miscegenation or undetectable internationalization through producing bastards or changelings. Put otherwise, Schmitt's way of distinguishing between endogamous enemy and exogamous friend depends on one's being able to tell, from the start, whether any given conflict is an endogamous civil war (the American Civil War) or an exogamous international war (the War of the States). Yet, as Thoreau hints in *Civil Disobedience,* no wars present themselves that way. All wars are about establishing borders. Real politics takes place not between two camps, but on the line that appears both to separate and to join them. On this line, the being we present to ourselves as both brother and other—the family pet for the nuclear family, say, or the Jew for Christendom—defines the nation as a whole, or defines the species.

Is there a nationalism without racism? The idea that love of nation might not always be linked with love of race or species is comforting to liberal nationalists. Admirable thinkers such as Hannah Arendt and Ernst Cassirer argue in the midst of the mass murders of World War II that there can be a clear distinction between racism and nationalism.[16] And the editors of the *Oxford English Dictionary* maintain the difference between nationalism and racism when they claim that a *nation* is

> an extensive aggregate of persons, so closely associated with each other by common descent, language, or history, as to form a distinct *race* or people, usually organized as a separate *political* state and occupying a definite *territory*. In early examples, the racial idea is usually stronger than the political; in recent times the notion of political unity and independence is more prominent.[17]

Ironically, however, racial theorists argue for the same difference between race and nation as do liberal nationalists. Thus Gobineau inveighed in 1854 against the idea of *patrie* in favor of the idea of race. According to him, patriotic nationalism falsely substitutes, for the ideology of race, the ideology of a nation supposed to transcend any "ethnic admixture, that is, the intermingling of different races." And he complains that, whatever the "original" French tribe or *genus* had once been, it had now "de*gen*erated" into a national hotchpot.[18] Hankering after the "original" consanguineous and genetic affiliation, Gobineau claims to discover racial purity in a few blue-blooded aristocratic classes and regions in France. (Not surprisingly, he also finds such racial purity in his own individual genealogy, which, he says, is "Aryan or Teutonic.")[19]

According to Gobineau, the modern liberal ideology of universal brotherhood or racial equality—"that the most different powers are or can be possessed in the same measure by every fraction of the human race, without exception"[20]—is bent on destroying the "original" division of the world into discrete tribes or unique brotherhoods with vastly differing physical and intellectual qualities. In his book *The Third French Republic,* Gobineau thus calls the French revolutionists' ideal of national *fraternité* an insidious enemy[21] and a "false patriotism," a substitution of mere "politics" for genuine "blood"—which ideal, he claims erroneously, was learned from too great an attention to Jewish (as well as ancient Greek) political theory.[22] And in much the same way he criticizes the liberal universalists for their explicit endorsement of the ideology "All men are brothers."[23]

Although the racialist Gobineau is different from most liberal nationalists in wanting *race* rather than *nation* to be the distinguishing mark of the polis, he is the same insofar as the "original" race that he hankers after is fictive in the same way as any geographically boundaried nation. Just as the racialist links together members of a tribe through a bloodline—or fiction about that bloodline—, so the nationalist links individuals through a polis. Thus there can hardly be a nation—if not also a polis—without an ideology of common political nativity. Plato says that the people of the polis must think of their fellows as sharing a common ground. So the French people are, in "La Marseillaise," children of the same land. (People who are not indigenous—or are merely "naturalized" citizens—do not always share the same rights as "native born" citizens, even in the liberal democracies. Because I was not native born in the United States, for example, I cannot become president of the United States unless the Constitution of the United States becomes more liberal.) By the same token, liberal democrats who claim to be indigenous make much of the common ground. In Europe in the early 1990s, even as the iron curtain seemed to crumble, the *Volk* of a long divided Germany cried out for a unified "*Vaterland!*" just as if they were autochthonous brothers-german reborn.[24]

Enlightenment thinkers, of course, would want to disagree with the view that common nativity is what makes the nation. In *Qu'est-ce que le Tiers État?* (1789), the revolutionary Abbé Sieyès, for example, set down his hope that a nation was "a union of individuals governed by *one* law and represented by the same law-giving assembly."[25] That is also the American dream. The utopian Sieyès was no leader, however. And political history tells that the boundaries of a nation's ground or blood are always already in dispute. "The existence of the nation," writes Renan in *Qu'est-ce qu'une Nation?* (1882), "is a daily plebiscite."[26] The boundaries of a nation are fictive plots bandied about by real-estate agencies of the mind.

WAR OF THE WORLDS

O, wonder!
How many goodly creatures are there here!
How beauteous mankind is! O brave new world,
That has such people in't!

—Shakespeare, *The Tempest*, 5.1.181–83

Before we judge of them too harshly we must remember what ruthless and utter destruction our own species has wrought not only upon animals, such as the vanquished bison and the dodo, but upon its own inferior races. The Tasmanians, in spite of their human likeness, were entirely swept out of existence in a war of extermination waged by European immigrants, in the space of fifty years. Are we such apostles of mercy as to complain if the Martians warred in the same spirit?

—H. G. Wells, *The War of the Worlds*

Can we ever really tell for sure who are our species' insiders and who are on the outs? Do we always know, when we see E.T., what is Extra-Terrestrial and what is merely an Extraordinary Twin? Would interplanetary travellers be better able to distinguish extraterrestrial creatures as essentially human or nonhuman than were the Christian missionaries who thought the extra-European Kodiak islanders to be nonhuman because they supposedly had no incest taboo?

The Christian missionary Brothers were confronted with the notoriously ambiguous quality of kinship terminologies among alien peoples. And, thanks to their universalist doctrine that "all human beings are siblings," the Brothers were quick to project unto others the profane familial incest they themselves feared or the sacred incest they adored. (They feared incest insofar as all sex acts for human beings perfected as Brothers are physically incestuous. They adored incest insofar as the crux of Christianity, Jesus, is his own father and also the father, brother, husband, and child of Mary.) By the same token, the missionaries were prone to exclude from the human species those humanoid creatures whom they suspected of incest. For these outlandish forerunners to secular anthropologists, the only being with whom one might have nonincestuous intercourse would be a monster, as in the folk tale "Beauty and the Beast," or a god, as in the Annunciation, or an extraterrestrial being, as in Roeg's science fiction film *The Man Who Fell to Earth.*

It was as part of the millennial problem of "plurality of worlds" that the old question, "Are the creatures on other planets, if such there be, our brethren?," assumed importance for secular Christian universalists. Leibniz considers human kinship in these terms, and suggests that in the universe "there may be species intermediate between man and beast . . . " and that "in all likelihood there are rational animals somewhere, which surpass us."[27] And Leibniz takes seriously the *Tempest*-like lunar romance of Bishop Francis Godwin: "If someone . . . came from the moon . . . , like Gonzales . . . , we would take him to be a lunarian; and yet we might grant him . . . the title *man.* . . ."[28] Humanoid bipeds from other celestial bodies could well be granted the status of human beings, says Leibniz. And, given the usual conflation of species with family, he says, these bipeds would also have to be "brethren." But could they also be granted the status of being members of our family? "No doubt some [people] would maintain that rational animals from those lands [are not] descended from Adam."[29] If the view that these animals are unrelated consanguineously to Adam is correct, then the lunarians would be of our species but not of our family. The doctrine that all human beings are brothers and sisters would no longer obtain.

The immediate quandary here for Christian universalists pertains to the potential unredeemability or unconvertibility to Christianity of the humanoid lunarians if, as seems likely, they are not descended from Adam. (In America in the 1680s, by analogy, planters in the Virginia colonies argued that black slaves should or could not be Christianized because they were not descended from Adam.)[30] Though such creatures might be of our species, they could hardly be of our family. Christianity, though it usually belittles blood ties, as we have seen, makes them all-important in the case of the link with Adam.

The genealogy of the partly extraterrestrial Jesus—His Adamic consanguinity—has a great importance in the Gospels.[31] Jesus was the "son of man [*adam*]" born pretty much of humankind. And in line with this anthropocentrism, which makes of God an earthling with blood ties to Adam, it is hard to maintain that extraterrestrial creatures could be considered our brethren without making some such potentially blasphemous claim as that God the Father visited the inhabitants of other planets and mated interspecies-wise with one of them (just as with the Virgin Mary) or that there has been interplanetary space travel linking the humanoid lunarians as our consanguineous kin. Thus the quintessentially romantic Chateaubriand, in his *Les Martyrs,* sees Jesus as visiting the inhabitants of other planets.[32] Any further solution to the problem of the "plurality of worlds" so defined would have to internalize the vacillation at the heart of Christian doctrine between the view that kinship is essentially physical and the view that it is essentially spiritual.

When other worlds are presumed uninhabited, the fact of nonhabitation is often explained by the corruption of human life on Earth. Thus Chateaubriand, in his *Génie du Christianisme,* argues that, although God created the other planets as future habitations for "the race of Adam," those worlds have remained only "sparkling solitudes" because man sinned.[33] Other thinkers have argued that extraterrestrial creatures, whether human or not, might be in some theologically technical sense better than us earthlings. In *Night Thoughts,* for example, Edward Young addresses the extraterrestrials thus: "Had your Eden an abstemious Eve? . . . / Or if your mother fell, are you redeemed?"[34] In 1768, the English publisher John Newberry[35] wrote: "I cannot imagine the inhabitants of our earth to be better than those of other planets. On the contrary, I would fain hope that they have not acted so absurdly with respect of [God], as we have done."[36] That the sons and daughters of terrestrials are uncorrupted by us is the eighteenth-century American David Rittenhouse's thought.[37] Dostoevsky's story, "The Dream of a Ridiculous Man," is the "tale of a man who would devote his life to the gospel of universal love after having been conveyed in a dream to a planet peopled by innocent, loving creatures, whom he corrupts."[38]

Kant was an advocate for the view that there is humanoid life on other planets. For him, it is our contemplation of the stars and the creatures on them that links the aesthetic sublime with the moral world. In the *Critique of Practical Reason* Kant thus recalls man's place in the universe.[39] And in *Observations of the Feeling of the Beautiful and the Sublime* Kant considers in a Rousseauist spirit that "it will not be necessary for women to know more of the cosmos than is necessary to make the aspect of the sky touching to them on a fine night, after they have grasped, to a certain extent, that there are more worlds, and on them more creatures of beauty to be found."[40] Kant's *Theory of the Heavens,* the third part of which is "based on analogies of nature, between the inhabitants of the various planets,"[41] considers directly the kinship or likeness among the various planetarians. Kant here suggests that the inhabitants of the solar system are morally or intellectually developed according to the distance of their planets from the sun. Earthlings stand in this scheme somewhere near the middle. The inhabitants of Jove and Saturn are said to be superior to us; and the inhabitants of Mars are supposed to be sinful like us, just as if they too had experienced a fall from grace and a redemption. From Earth, then, we might look out onto the other creatures of the solar system: "From one side we see thinking creatures among whom a man from Greenland or a Hottentot would be a Newton, and on the other side some others would admire him as an ape."[42] Kant compares the differences between inhabitants of the planets (human earthlings and humanoid Saturnarians) with the differences between inhabitants of the Earth (humanoid apes and human earthlings).

We ordinary folk would want to believe that Newton's extraordinary tribe, and Kant's, is our own, but we might also wonder whether we are not, after all, fellows of the Hottentots. In Sayles' cult movie *The Brother from Another Planet* it is suggested that any mirandous "brave new world" we might discover in some distant galaxy would only discover to us again the problems of species and family difference that inform humanoid life on Earth. We are strangers in a strange land, aliens on an alien planet. And like the angels who visited the hospitable Abraham under the burning desert sun, we want only to spend the night in the goodly tents of Jacob.

WAR OF THE SEXES

Surely a man might unmask fraternity without vociferating that he is an egotist and a misanthrope.

—Mr. Harrison, cited in James F. Stephen, *Liberty, Equality, Fraternity*

Monastic brotherhoods and nationalist fraternities sometimes depend for their internal cohesion on thinking of women as extratribal if not also extraspecial. The anthropologist Robert Paul, writing of Tibetan monasteries, quotes the misogynistic French song, "Without these damn women we'd all be brothers / Without these damn women we'd all be happy."[43] A universal siblinghood, however much it may transcend other tribal differences, does not always entail gender neutrality.

Yet Protestant sects and European Catholic orders do sometimes promulgate gender sameness. In America, the incestuous Perfectionists and celibate Shakers, or Shaking Quakers, would rise above ordinary gender distinctions,[44] for example, and many Catholic orders cited Saint Paul in their regulations, saying, along with millennialist and apocryphal works, that "there is neither male nor female, for ye are all one in Christ Jesus."[45] Many people in these traditions emphasize the idea of the holy kiss, which is given without any restriction as to gender.[46] A few stress the idea of the androgynous god, as when they call Jesus a male lover and female parent combined.[47] And some Brothers and Sisters, principally the Franciscan friars, institutionalize in the kinship structures of their societies not only an incestuous conflation of kinship roles—where spouse, child, parent, and sibling become one—but also an androgynous conflation of gender roles—where celibate friars take on, in relation to one another, the alternating roles of mother and son.[48] By the same token, in early stages of their historical development, universalist political movements like the French Revolution, with their ideals of potentially universal siblinghood, often show a similar lofty tendency toward gender homogeneity.

Ideal gender homogeneity, however, is already the seedbed for real gender heterogeneity. Confronted with a world of sexuality and politics, universalist gender-neutral and androgynous sects and movements reestablish strict distinctions. Where the universal siblinghood is defined as essentially a *fraternité,* it becomes misogynous, or fearful of women, who are often regarded as being outside the species insofar as they are outside the family; and where the siblinghood is defined as essentially sororal, it becomes fearful of men, who are similarly regarded as extraspecial.

In 1789, the first year of the French Revolution, though the delegates to the Estates General owed their selection to women's votes, the equal status of women was already under siege. Condorcet saw the potential for a misogyny to arise which, by defining women as essentially nonhuman, would destroy the universal siblinghood hypothesized by revolutionary ideologues. He therefore published his defense on behalf of women's rights. In *On the Admission of Women to the Rights of Citizenship,* Condorcet argued that the women of ancient France had liberties that should be institutionalized for all women. Even as he worked from within the conservative context of the Rousseauist view that women had a reason of their own and that their domesticity should serve the wider polity,[49] Condorcet laid out the logical premise of the universalist position regarding the war between brother and sister Frenchpersons: "Either *no* individual of the human species has any true rights, or *all* have the same. And he or she who votes against the rights of another, of whatever religion, colour, or sex, has thereby injured his own."[50] Condorcet's articulation of this definitive position didn't much help the immediate cause. The French Constitution of 1791, in its *Declaration of the Rights of Man and Citizen,* granted suffrage to some men and denied it to all women,[51] and in 1792 the First Republic granted it to all men, continuing to deny it to women.[52]

Thinkers such as Etta Palm had argued as early as 1790 that "we [women] are your companions, not your slaves."[53] And many women saw themselves betrayed both as human beings and as siblings by the republican Constitution. In 1791, adopting Condorcet's views about species and special rights, Olympe de Gouges published her *Declaration of the Rights of Women.* In 1792, the Prussian Theodor Gottlieb von Hippel wrote in his essay, "On Improving the Status of Women," that "all human beings have the same rights—all the French men and women alike should be free and enjoy citizen's rights."[54] And in the same year Mary Wollstonecraft exclaimed, in her memorable "Vindication of the Rights of Men," that the "nature of reason must be the same in all [people]"[55] and warned the now increasingly fraternal and misogynistic revolutionaries that they must change their constitution, "Else this flaw in your NEW CONSTITUTION will ever shew that man must, in some shape, act like a tyrant."[56]

In England the problem of equality of the sexes was often discussed in terms of whether marriage was a reciprocal and equal relationship like that of brother to sister, which is what the conservative Coleridge eventually came to believe,[57] or a generally univocal and unequal relationship, like that of father to daughter, as argued by the radical William Godwin in his *Political Justice* (1798). Godwin, following out the implications of the French Revolution's near-Kantian redefinition in 1792 of marriage as a civil contract involving "the mutual and exclusive use of each other's genitals,"[58] argued for "the abolition of the present system of marriage" and raised the question "whether the intercourse of the sexes, in a reasonable state of society, would be promiscuous, or whether each man would select for himself a partner to whom he would adhere as long as that adherence shall continue to be the choice of both parties."[59] Godwin's clarion call for an end to marriage—even for a republican promiscuity that recalls Sade's argument for incest in his "Français, encore un effort si vous voulez être republicains"—was heard by thinkers such as Shelley.[60] (Shelley later married the daughter of Godwin and Wollstonecraft.) But, generally speaking, Godwin's work gave rise only to lampooning. (*The Anti-Jacobean Review* [July 1797], for example, mocked "the promiscuous intercourse of the sexes as one of the highest improvements to result from Political Justice.")[61]

One political theorist remarks on the turn against women's rights in France in the 1790s that, "having revolted against the older patriarchy of the father-king . . . fraternal men imposed on women a legally secured definition of politics—this time, the patriarchy of the brothers and honorable husband-fathers."[62] This remark has the ring of truth, but it begs the question as to why men, who called all women sisters (i.e., beings from inside their family and species), came to treat them as animals (i.e., beings from outside their family and species).

Beyond any need or desire to form a gender-neutral siblinghood, which many in the early 1790s certainly had, there is the fear that incest must result from such a siblinghood. As the revolutionary nuns who played an important role in the *doléances* of 1789[63] would have known, universal siblinghood requires either absolute celibacy or incest—either liberty from the flesh, as in their sacred celibacy, or liberty of the flesh, as in the profane incest of the republican Sade and Byron. Small wonder that most mid-nineteenth-century would-be "liberators," faced with these alternatives, preferred the zealous liberty of saints and sinners to the more moderate and inherently misogynistic worship of women "as on a pedestal." (An example is the attitude toward women at work in Auguste Comte's positivism.)[64]

In the same way, we might reconsider the religious bent of the feminist "liberator" Flora Tristan. An early Fourierist in the familist tradition, she defined herself as a "Sister in Humanity" and embraced the views of the Saint-Simonists and Owenites that "sexual equality would be achieved through the emancipation of the working class."[65] Tristan was a precursor of Friedrich Engels, writing in France in the 1840s at the same time that Margaret Fuller was writing in the United States; she demanded of Frenchmen that they "emancipate the last slaves still remaining in French society, proclaim the rights of woman, in the same terms that your fathers proclaimed yours."[66] The religious tendency of her universalist model is all too clear. Tristan, an illegitimate child, compares herself to St. Teresa of Avila, and turning away from the interconnected problems of incest and celibacy to which St. Teresa's *Life* draws our attention, she likens her own role in history to that of the *filius nullius* Christ.[67]

JUDAISM AND CHRISTIANITY

Love thy neighbor. That is, as Jew and Christian assure us, the embodiment of all commandments.

—Franz Rosenzweig, *The Star of Redemption*

Early church thinkers promulgated the philanthropic view that all human beings are essentially members of one brotherhood. But who, or what, are brothers, or human beings, according to a Christian? Many Church Fathers argued that only those who have already become baptized or become active members of religious orders are "brothers." Origen writes: "Learn then what gift you have received from my Father. You have received, by your new birth the spirit of adoption, in a manner to be called sons of God and brothers."[68] Similarly, some authorities claimed that only members of one's own holy order are brothers,[69] and others that only confessors were to be called "brothers."[70] Optatus says that it is impossible for a Christian not to be a brother to other Christians,[71] implying that there are human beings in the world who are, to Christians, other than brothers. The universalist impulse thus led to a remarkable particularism. Paul had written that "there is neither Jew nor Greek . . . for ye are all one in Christ Jesus."[72] But while some Christian thinkers urged their followers to call all human beings "brethren,"[73] others claimed that in the new dispensation the new distinction between Christian and non-Christian was at work.

Jesus' rule, "Call no man your father upon the earth: for one is your Father, which is in heaven,"[74] would seem to collapse all human generations into one generation, almost as in Thomas Paine's radical American *Rights of Man* (1791). Yet the group of human beings that becomes an egalitarian brotherhood merely by assuming a common Father is only replacing an earthly hierarchy, that of the sons against the father, with a heavenly hierarchy, that of the sons against the Father. The sons are unequal to the Father in the heavenly hierarchy just as the sons are unequal to their fathers in the earthly hierarchy. One of the sons in the second hierarchy, whose central ideological figure is the spiritually incestuous Holy Family of Christianity, is Christ Jesus, the theological and structural linchpin of Christianity. Jesus is the ambiguously illegitimate son of no one or of no one human *(filius nullius)* and also the Son who Fathers himself. Christianity's equal liberty (association of the sons, or *liberi*) thus depends upon a liberal Fatherhood that is also a paternal liberty. (In Shakespeare's *Measure for Measure* Duke Vincentio, Vienna's "father friar" and "brother father," wavers between the role of parentarchal duke and that of either fraternal monk or universalist libertine.)[75]

Not surprisingly in view of these difficulties, Christian sects fell off from the egalitarian ideal of "calling no man your father upon the earth." They adapted the pre-Christian mystery cults' practice of calling "father" the director of ritual practices.[76] They called the priest "father."[77] (Eventually some Christian authorities said that bishops were to call one another "brothers" and to call their abbots and priests "sons.")[78] Thus these sects violated both the letter of Jesus' rule, "Call no man your father," and the letter of most Jews' rule, "Call your father your father" (honoring him according to the fourth commandment).

I mean to suggest not that Christian authorities were unduly lax or hypocritical, though perhaps they were. I mean to suggest that there is a historical tendency for any universalist movement, whether religious or secular, to differentiate between tribes and within the tribe. I would characterize this tendency as more inevitable than hypocritical. After all, the full implications of universal and equalitarian siblinghood are utopian and idealist, even unbearable, for most of us. (Exceptions here would be thinkers like the chaste Saint Francis and the incestuous Marquis de Sade.) If the universalist doctrine were realized, Christendom or the national state would become the spiritually or physically libertine society, or antisociety, envisioned by the English Family of Love and by the French libertines—precisely the sort of society that political Christendom is compelled officially to condemn.[79]

The most disturbing aspect of the idea of universal siblinghood is not, as we have seen, the inevitability of a retreat from it, but rather the inevitably inhuman or inhumane practical consequence of making a retreat that is unwitting or unacknowledged.

It is one small step from the most extreme *philanthropeia,* or love of others as brothers, to the most extreme *misanthropeia,* or hatred of them.[80] The doctrine of universal siblinghood according to which all human beings are brothers and all brothers are human beings—a doctrine Dante presents in secular guise in his vision of a United Nations living in perpetual peace[81]—can transform any being whom we would ordinarily call human, but who cannot or will not become a member of our brotherhood, into an animal, or can lead us to treat that being as an animal, just as, in some primitive societies the notions of "human being" and "fellow tribesman" are indicated by the same word and are thought of as being the same.[82] By the same token, an injunction to love as brothers all men—including both exogamous "foes" and endogamous "enemies"—can turn into an injunction to treat all men equally as others.

Moreover, if the human family and the human species are one and the same, interfamilial crossover (marriage) and interspecies crossover (dehumanization and anthropomorphism) are also one and the same. When Paul says, "Love thy neighbor as thyself,"[83] implying that we humans are all brotherly neighbors, he leaves us only those beings outside the neighborhood, or outside the brotherhood, to love in a sexual manner, without incest. But if all human beings are brothers, then we humans can love chastely only animals. Bestiality would be the only way, besides celibacy, to avoid incest. For universalist Christian philanthropists there thus develops an almost overwhelming need to dehumanize certain beings, like Muslims and Jews—a need to treat beings from outside their own neighborhood not only as extratribal but also as extraspecies.

As we have seen, the quintessential alien for Christians is the Jew. And it is against the Jewish particularist doctrine of love that Christendom for millennia has sought to define itself, claiming with Saint Matthew and other anti-Semites that the Jews teach misanthropy and that it is impossible for a Jew to understand universal love.[84] Yet Judaism and Christianity are poles of the same spectrum. Neither the idea of universal brotherhood nor that of total otherhood, which are so important to Christianity, are wholly foreign to Judaism. (Nor are these ideas alien to Islam, as we have seen.)[85] For example, the cancellation of particular kinship bonds that are important in the Christian rite of profession (as described by the Port-Royalists) and in the Catholic doctrine of the "equal forgetfulness of all persons, whether relatives or not" (as described by Saint John of the Cross)[86] plays a role in the Jewish ritual of conversion, which involves a stripping away of kinship bonds followed by a rite of "brothering." For many Jews, the "convert" stands in relation to everyone as both brother and other, or neither. This stance involves quandaries about celibacy and marriage. In some branches of Judaism, for example, "a proselyte is as a new born babe who stands in absolutely no relationship to any preconversion relation. Consequently, a proselyte's brothers and sisters, father, mother, etc. from before his conversion lose their relationship on the proselyte's conversion. Should they too subsequently become converted, they are regarded as strangers to him, and he might marry, e.g., his mother or sister."[87] Christian sects, faced with the same threat to the integrity of the ordinary kinship structure as the influx of either the converted or born again (i.e., the person regenerated *sui generis,* without consanguineous parents, like the foundling or *asufi* discussed by Maimonides), resorted to the injunction to be celibate. Jewish rabbis, on the other hand, developed kinship theories such as the *rikkub* principle and promulgated ways to forbid marriages between the "new born babes" that are converts (or foundlings) without resorting to the Christian alternatives of absolute liberty of and from the flesh.[88]

By the same token, the idea of universal brotherhood is not foreign to Judaism. Postexilic Jewish thinkers said that one day the Temple would be called "a house of prayer for all peoples" and that every individual could become a Jew.[89] (The similar view that "the human world" is one or ought to be one is, as we have seen, the imperialist legacy of Alexander the Great and of Greek Stoic and Jewish Hellenist cosmopolitan thinking.)[90] It has always been a Jewish view that Abraham, by his piety and philanthropy, "made brothers of the whole world."[91] The Old Testament asks, "Have we not all one father? hath not one God created us?"[92] This apparent transcendence of human consanguinity moves the injunction to be one's brother's keeper in the direction of loving neighbors, thence to loving strangers who dwell within the community, thence to loving strangers who dwell in strange lands, and finally to loving everyone.[93] Rabbis Hillel and Meir enjoin Jews to love all mankind, even all creatures.[94] And in 1609 in the cosmopolitan city of Venice, Aaron ibn Hayyim published his powerful defense of the view that the law of "Love thy Neighbor" includes in its purview non-Israelites as well as Israelites.[95]

Across the spectrum of the Judaeo-Christian tradition, then, there is a polar tension, both radical and ineradicable, between the demands of tribal heterogeneity and those of universal siblinghood. Toward universal kinship Judaism—

which began as a practical, particularist tribal religion and has long existed as an other in the Diaspora—is bound to strive; and away from universal siblinghood Christianity—which began in an idealist rejection of particularist kinship—is bound to stray.[96]

WAR OF ALL AGAINST ALL; OR, USURY

Just as the conceptions of human brotherhood and (in a less degree) of human equality appear to have passed beyond the limits of the primitive communities and to have spread themselves in a highly diluted form over the mass of mankind, so, on the other hand, competition in exchange seems to be the universal belligerency of the ancient world which has penetrated into the interior of the ancient groups of blood relatives. It is the regulated private war of ancient society gradually broken up into indistinguishable atoms.

—Sir Henry Sumner Maine, *Village-Communities* . . .

One of the few passages in the Old Testament where the law explicitly allows Jews to treat fellow tribespersons differently than other people is that which forbids a Jew to lend money at interest to a brother and allows him to do so to a stranger: "Thou shalt not lend upon usury *(neshekh)* to thy brother *(ahikha)* . . . Unto a stranger thou mayest lend upon usury."[97] This rule for tribal brotherhoods was taken up by the anti-interest rhetoric of premodern Christendom. One unlikely result, as we shall see, was that the Christian definition of usury came to inform Christendom's definition of nation and to prepare the way for the universal otherhood of modern individualism.

To obey the Mosaic code concerning monetary interest one has to know who is one's brother. For a Jew this means knowing, or at least believing that he knows, who is akin to Jacob, or a fellow "son of Israel" *(b'nai yisrael)*, and who is not. For a Christian, however, all men are brothers, not only the sons of Israel, so there is no need to know any man or woman's genealogy. It makes no difference to a Christian if he is descended from Ishmael rather than from Isaac, say, or from Esau rather than from Jacob. It is enough for a Christian to know that he is a human being, that is, descended from Adam or Eve. Saint Jerome, commenting on Jesus' statement "All ye are brethren" in view of the Deuteronomic rule regarding monetary interest, thus argues that all monetary interest is forbidden.[98] Thomas Aquinas later bolstered this rejection of monetary interest by adapting to Christian purposes the Greek view that all monetary interest, or *tokos*, was perverse. Had not Aristotle argued that monetary interest was unnatural insofar as it made money breed as though it were animal offspring?[99] Although a few Christian jurists inquired into the ambiguity of the term *ahikha* (brother) in the Deuteronomic law—it was said to refer to a child of the same parents, a fellow tribesman, and to a man who participates in the same *logos*[100]—most Christians contended that the prohibition of usury among particular brothers in Deuteronomy had been universalized.

Not all Christians agreed, however, that the Christian condition of universal fraternity made all men the sort of fraternal kin that would outlaw usury. In the medieval period, for example, the Patarenes and Cathars understood Jesus to be rejecting the distinction between kin and nonkin and so they regarded all men equally as both brothers and others. For the Jews there are distinct brothers and others, and for Saints Jerome and Thomas Aquinas there are only brothers, but for the Patarenes all human beings are simultaneously both brother and other or neither brother nor other. Accordingly, the Patarenes insisted that money lending at interest might be allowed for everyone. (By the same token they defended celibacy for everyone, arguing that if all men are kin, celibacy is the only way to avoid incest. The libertine Brethren of the Free Spirit similarly defended incest, arguing that if all men are nonkin, there is no incest.)

The Patarenes were persecuted by the official Church in precapitalist times. But their views on brotherhood became acceptable among Protestants such as John Calvin. Calvin argued that all men are brothers in Christendom insofar as they are equally others.[101] In much the same way Saint John of the Cross, who founded the discalced Carmelites with the *converso* Saint Teresa of Avila, argued for universalizing everyone as a stranger: "You should have an equal love for or an equal forgetfulness of all persons, whether relatives or not, and withdraw your heart from relatives as much as from others. . . . Regard all as strangers."[102] ("But the fury of tyrants does not stop short of their own relations," writes Seneca to his young tutee Nero. "They treat friends and strangers alike.")[103] Maine, in his Victorian study of village-communities, argues in the same vein that in the global village everyone is equally estranged and familial, as if every human being were an indistinguishable atom. It is as if human beings, who had once been bound together in many tribal brotherhoods, now became atoms bound together in a single universal otherhood.[104]

In eighteenth-century France, the old rhetoric of usury that opposes the ideal of universal or national fraternity to interest-based capitalism was especially influential. There the traditional Christian double charge against Jews—that Jews regard Christians as nonbrothers and therefore lend money to them at usurious rates. This was the gist of Foissac's prize-winning anti-Semitic essay *Le cri du citoyen contre les juifs* (1786). Isaiah Berr Bing responded to the charges in a letter of 1787. First, Bing disproved Foissac's thesis that Jews hate non-Jews and treat them as aliens; he showed that, according to Jewish law, all those men who observe the Noachite laws partake of salvation and that Jewish principles "oblige us to love you as brothers, and to perform for you all the acts of humanity that our position permits."[105] Second, Bing pointed out that Deuteronomy allows not for "usury" but for "interest."[106] Finally, Bing linked the problem of fraternity with that of interest, claiming that the intention of the Mosaic code regarding usury was "to draw closer the bonds of fraternity [between the Children of Israel], to give them a lesson of reciprocal benevolence."[107] This lesson of fraternal benevolence, he said, rightfully applies also to nations other than the Jews. Referring specifically to the debate about whether Jews might be allowed to become citizens of France—as if the Deuteronomic term for "brother" were now to define "fellow national"—Bing said that there might exist "two nations . . . who live under the same climate [and] between whom there is a sort of political equality."[108]

Some Frenchmen praised Bing's letter. (Among them was the revolutionary leader Mirabeau, author of *Sur Moses Mendelssohn* [1787], writing in his *Monarchie prussienne.*) And some political thinkers in the nineteenth century also asserted the idea of equal fraternal or national coexistence between two nations. (That was the ideal rallying cry half a century later in bilingual Québec in the New World, where, as we have seen, Lord Durham "found two nations warring in the bosom of a single state.")[109] But most revolutionaries opposed the idea of binationalism or multinationalism. They wanted not two coexistent nations (*convivencia*) but one brotherhood of free and equal brethren: *liberté, égalité, fraternité.* The French revolutionary leader Abbé Grégoire thus argued in his *Essay on the Physical, Moral, and Political Regeneration of the Jews* (1789) that Jews should become equal brothers. And Clermont-Tonnerre argued in his *Opinion relativement aux persécutions qui menacent les Juifs d'Alsace* (1789) that the Jews of France "must constitute neither a state, nor a political corps, nor an order; they must individually become citizens; if they do not want this, they must inform us and we shall be compelled to expel them. The existence of a nation within a nation is unacceptable to our country."[110]

Encouraged by this sort of possibly well-intentioned universalism, Napoleon in 1806 encouraged a convention of Jewish notables (Sanhedrin) held in Paris to become French brethren or citizens. All that they needed to do, Napoleon said, was to repudiate the teachings of Deuteronomy regarding usury. Insisting that Jews and Frenchmen were essentially or potentially brothers who share the same father, Napoleon, apparently referring to Mendelssohn's *Jerusalem,* argued that "the [Jewish] Sanhedrin, after recognizing, as the [National] Assembly has done, that Frenchman and Jews are brothers . . . must prohibit usury in dealing with Frenchmen or with inhabitants of any countries where the Jews are allowed to enjoy civil rights. . . . I am anxious to do all I can to prevent the rights restored to the Jewish people proving illusory—in a word . . . I want them to find in France a new *Jerusalem.*"[111] Among the twelve questions that Napoleon's Commissioners posed were three that connected usury with the idea of fraternal nationhood: "In the eyes of Jews, are Frenchmen considered as brethren or as strangers?" "Does the law forbid the Jews from taking usury from their brethren?" and "Does the law forbid or does it allow usury toward strangers?"[112]

The convention of Jewish notables eventually asserted that "all Frenchmen [are] our brothers"[113] and endorsed the notion that from that time forward the technical term "Jews within France" would become "Frenchmen of the Mosaic persuasion."[114] And it ruled out usury for French citizens, promulgating the view that the term *ahikha* (brother) in Deuteronomy applies to all French citizens regardless of religion.[115]

The convention's rulings about usury and about brotherhood as French nationhood (even united nationhood) found popular expression in Jewish education. Samuel Cahen's *Précis élémentaire d'instruction religieuse et morale* (1820), for example, taught that the term *ahikha* in the Deuteronomic commandment concerning usury means "all human beings who recognize God" and stressed that the obligations of "fraternal charity" apply to Jews and non-Jews alike.[116] Elie Halévy's catechism of the same year taught similarly that "a Jew of today cannot, without . . . transgressing the law of God . . . engage in this illicit commerce towards individuals whose religious opinions, it is true, differ from his, but who are no less strict observers of these great principles, fundamental principles of all . . . civilized peoples."[117] Sometimes this universalist impulse seemed to transcend altogether the differences between Jew and non-Jew. For example, Simon Bloch, editor of *La Régénération,* wrote in 1836 that "we are born into the religion which our parents profess, but we must elevate ourselves to piety *(piété)* by our own efforts. . . . We receive, in the bosom of our mother, our first religious impressions, but we draw piety from a celestial source. Religion is the temple; piety is the goddess."[118] Bloch recognized that consanguineous matrilineality is a key factor in defining membership in the siblinghood of Israel *(b'nai yisrael),* but he transposed it, through the standard French nationalist rhetorical figure of collactaneous matrilineality (as in Jacques-Louis David's revolutionary "Festival of the Unity and Indivisibility of the Republic"), into a potentially universal spiritual affinity that seems to include the controversial hypothesis of diverse gods and goddesses.

Many European Jews believed that they might participate fully in the national or political life of their adoptive countries just so long as they endorsed the doctrine of essential universalism. For example, the Rousseauist historian Isaac D'Israeli, a Jew of Iberian ancestry, came to the conclusion that a particularist Judaism "cuts off the Jews from the great family of mankind."[119] And praising the new spirit of univeralism, D'Israeli arranged to have his twelve-year-old son Benjamin baptized in 1817 into the Anglican Church. (Benjamin Disraeli became Prime Minister half a century later. One might compare Karl Marx's father in Rhenish Prussia arranging to have his six-year-old son Karl baptized as a Protestant in 1824.)

However, Moses Mendelssohn had not been mistaken to warn Jews and Christians alike of intolerance's mask of universalism.[120] And just as *fraternité* was being extended to the newly liberated Jews of France, the Imperial Edict of 1807 "imposed discriminatory regulations against the economic and political opportunities of the Jews" as aliens.[121] Chamfort, pondering the French republican promise of "liberty, equality, fraternity" and the French revolutionary slogan of "fraternity, or death!," commented thus: "'Fraternity or Death?' Yes: be my brother or I will kill you!"[122] "If I had a brother," quipped the Austrian Metternich, "I would call him cousin."[123] Laments the American Emerson: "In France 'fraternity' [and] 'equality' . . . are names for assassination."[124] If a person cannot or will not become a brother then he is not human and may as well be treated as such. The promise of universal brotherhood—even a united nations—turns all too easily into the individual and fatal fraternity of Cain and Abel.[125]

CONCLUSION: UNCOMMONLY COMMON KIN

Who are you?

—Lewis Carroll, *Alice's Adventures in Wonderland*

"Our true nationality is mankind," wrote H. G. Wells in his popular *Outline of History* (1920). This apparently genial sentiment, here cited from *Bartlett's Familiar Quotations,* seems kindly enough.[1] But the universalist sentiment militates inexorably against nonnationals, that is, against humanoid creatures, both terrestrial and extraterrestrial, who are deemed nonhuman or "racially inferior"—as Wells himself had suggested in *The War of the Worlds* (1898).[2]

Many apparently alternate kinship structures informing Western nationalism aim to become a "universal siblinghood"—an outlandish term suggesting an association of men and women that recognizes only one tribe of human beings, with no essential intergenerational, intragenerational, or gender differences. As we have seen, however, steps taken in the direction of universalizing homogeneity ("all men are my brothers") are almost always matched by steps in the direction of heterogeneity ("others are animals"). The present book has examined the attempt to attain universalist kinship structure—or, as some sociologists have called it, "nonkinship"[3]—within a world of nations, and has pointed up some dangers of this impossible essay.

In the 1990s, half a century after two world wars, the veneration accorded to universalism is as disturbing in its political implications as the particularist nationalism that universalism pretends to eschew but actually merely defines by polar opposition. After all, the ideal of universal brotherhood—"All human beings are brothers; none are others"—offers no specifically human mediation between species and family, that is, recognizes no being that is both "other" and "human" besides such partly domesticated borderline creatures, defining kin and kind, as pets, animal nurse mothers, and godmen. And so it makes virtually inevitable the slide away from the purportedly humane ideal.

The direction of the slide is toward a particularist actuality that excludes all nonfamily members from the human species: "Only my brothers are human; all others are animals." This direction, which varies somewhat according to local tribal characteristics and involves diverse traditions about kindness to nonhumankind, is due mainly to distinct combinations of universalism's inability to give up in good faith its ideal conflation of species and family (i.e., its doctrinal reluctance openly to surrender the field to tolerant particularism) together both with politically inevitable apprehensions about universalism's actual end—"incest" or "celibacy" broadly understood—and with consequent internal needs to have or create outsiders and foes. (This perspective on universalism's slide to a brutal particularism complements and politically qualifies several "commonsense" propositions adduced elsewhere to explain why social homogeneity, if it ever exists, always caves in to heterogeneity—for example, that making excessive demands often results in getting less than making moderate demands, or that there is an upper limit to the size of any discrete human group. Thus Aristotle teaches that the universal siblinghood—hence incest—hypothesized in Plato's *Republic* is merely diluted kinship, and sociobiologists contend that human tribes, like dog packs and baboon troops, can get only so big before they divide or break up.)

The dehumanizing universalism that congratulates itself—sometimes sincerely, sometimes not—on its own humane intentions even as it conceals a parochial and imperialist particularism comes to despise and persecute the more realizably tolerant possibilities of openly particularist brotherhoods. Exemplary among these brotherhoods would be the politically disempowered bipartite particularism of Judaism—which distinguishes in good faith between brother Jews and those others who are both nonbrother and human—and the empowered tripartite particularism of Islamic Spain, which distinguishes in good faith both between brother Muslims and others and also between brother non-Muslims who are peoples of the Book and others who are pagan. (These examples of potentially tolerant particularism are Semitic mainly because Judaeo-Christianity's distinctive deprecation of political attachment defines itself against Judaism and Islam; but all particularist ideologies, unlike univeralist ideology, recognize explicitly the existence of human others, and some, like Judaism and Islam, have distinctive rules for tolerating others both human and animal.)

The universalist view that all human beings are equal siblings can take an apparently different ideological turn: not the dehumanizing version that nonsiblings are nonhuman but the twist—at once communalist and individualist—that because all human beings are siblings, every sibling is also a nonsibling. That twist allows various adamist, antinomian, and nationalist sects—including the Brethren of the Free Spirit and certain French revolutionaries and American communalists—to endorse what Shakespeare's quintessential sister/Sister Isabella calls "a kind of incest."[4]

This chaste incest is to the Christian tradition what the noble lie of political autochthony in Plato's *Republic* was to the Greek. It is the affinity reflected in Marguerite of Navarre's and Elizabeth Tudor's "distant-close"[5]—an extraordinary collapse of kin with nonkin whose tragic appearance in ordinary society, expressed in *Hamlet,* occurs only on the margins of kinship, a no-man's-land where sisters become wives and Sisters, or brothers become husbands and Brothers.

The same twist, individualist as well as communalist, allows protocapitalist Calvinist and other individualist sects to combine the particularist rule that a brother must not charge monetary interest with the universalist doctrine that all human beings are siblings, concluding that because every sibling is also a nonsibling, one can charge interest to any human creature. Taken together, the communalist and individualist tendencies inform a decisively modern and latently misanthropic drift from tribal brotherhood to universal otherhood, figured both in Melville's *Pierre* and in Catholic and revolutionary France, whence came to the American Pierre, as his ambiguously illegitimate sibling and lover, the orphan Isabel, "consecrated to God."

It would be intolerable to lay down at the doorstep of the universal (catholic) church or its nationalist secularized counterparts all the ills of the world. Nor would it be fruitful to blame the universalist doctrine, "All ye are brethren"—an inextricable component of Christianity's distinctive beliefs about the coming of the incarnate godman Christ and His spiritually incestuous Family—for the Catholic Inquisition's brutalizing emphasis on blood purity or for the millennialist Nazi dehumanization of others in German Christendom. After all, to some extent blood racism in fifteenth-century Spain was un-Christian: Pope Eugenius IV, for example, was at first distressed about the doctrine of blood purity.[6] And in twentieth-century German Christendom, Nazism presented itself expressly as anti-Christian: the Nazi swastika was non-Christian, the Nordic legends and Aryan mythologies of Nazi rituals effaced the Christian gospels, Fascist laws banned Christmas from the official books, and Nazi ideologists claimed that they sought their Supreme Being only in the *Volk.* ("We do not want any other god, only Germany," said Hitler.)

Yet the soil was decidedly Christian where "the mystical brotherhood" of the German SS flourished, and the Spanish ideology was Christian that encouraged the Catholic Inquisition. By virtue of its sometimes sincere belief and always bad faith, the genius of universalist imperial Christendom often dehumanized "others" through an ingratiating anthropomorphization of extraterrestrial gods and aestheticization of terrestrial creatures. The godman "Christ" and secular Christendom's perennially popular beautiful "Beast" matched the so-called primitives' collapse of the distinction between fellow tribesperson and human being—a necessary distinction for both coexistence and toleration.

Universalism's ideal cancellation of kinship by nonkinship was a necessary, although by no means sufficient, condition for the German extermination of Jews and Gypsies. By the same token, the disjuncture between the creed of universal love and Christendom's widespread practice of hatred and cruelty—to which Spinoza, among others, draws our attention in the case of the Iberian auto-da-fé and the expulsion of Muslims and Jews—comes not so much from hypocrisy as from various combinations of universalism's inevitable and essential failure to acknowledge its never-ending slip to particularism together with ancient racialist obsessions—here Gothic—with who's who in Germania.

The skeptical philosopher's conviction that he does not know who's who—so that all sexual intercourse is possibly incestuous—here complements the perfected Catholic belief that everyone is essentially the same—so that all sexual intercourse is teleologically incestuous. (The parallel psychoanalytic persuasion is that almost all sexual intercourse is oneirologically incestuous.) Peoples' awareness of and concern with the indeterminability of kinship—say, the possible presence everywhere of illegitimate children, foundlings, and changelings—depends on varying economic, religious, and other demographic factors. For example, medieval Europe had myriad oblates and foundlings, while eighteenth-century France had bastards and orphans. But whatever the local level of conscious concern with the indeterminability of kinship, the virtual denial everywhere of literal blood kinship helps to explain the perennial popularity of that psychologically cathartic and sociologically reassuring fictional literature, from Sophocles' *Oedipus the King* to contemporary American television soap operas, where apparent nonkin are revealed to be kin (or apparent kin to be nonkin), as if sure revelation were possible anywhere *but* in fiction.

Through study of the intellectual place of kinship words or names, and hence the poetics or metaphorics of the classification of kin and kind, literary study questions the problematic distinction between literal and fictive kinship—between biology and sociology, say, or nature and culture. We have seen how the conjoined twins in Twain's *Those Extraordinary Twins* might present the only case where literal (consanguineous) kinship is undeniably known; the ambiguously affined siblings in Melville's *Pierre* likewise present the otherwise universal case where literal kinship is undeniably unknown. Genetic and forensic science are ready, now as ever, to take on Tiresias' apparently artless question, "Do you know who your parents are?"—ready, that is, to remove it from the province of skeptical doubt. But the problem of doubt or skepticism about kinship relations—and, for that matter, about their nationalist and racialist counterparts—cannot be tested or proved. Whether there is an unequivocal blood or DNA test to detect lineage or parentage and whether true parentage is other than consanguineous remain unsettled questions even in the 1990s. Like the child in the story of Solomon's judgment, we cannot truly know who our blood parents are—we could all be changelings, switched in the cradle—and there is no unshakeable answer to the question of our true parents. "The family is no more than a lexical area," or, stated differently, incest is a "nominative" crime or "surprise of vocabulary" that "consists in transgressing the semantic rule, in creating homonymy" (Barthes)[7]—in much the same way that the nation, with all its legitimate territorial claim, is the "accumulation of human beings who think they are one people" (Wells).[8]

The extraordinary status of universal kinship or nonkinship as both literal and figural, and also as neither literal nor figural, informs many ordinary nationalist ideologies. In a historical continuum paradigmatically including ancient Rome, medieval Spain, Renaissance England, seventeenth-century France and Holland, Enlightenment Germany, the French Revolution, nineteenth-century America, and twentieth-century Québec, we have seen how the univeralist doctrine interacts with institutional issues involving affinities such as adoption, adrogation, authochthony, baptism, blood brotherhood, brotherhood-in-arms, carnal contagion, catenary lineage, class grouping, collactation, consanguinity, conversion, Eucharist bonding, feudal lordship, foster relationships, fratriarchy, friendship, gossipred, language, marriage, nobility, nomination, and oblation. All these mark societal articulations of universal with particular.

Elizabeth Tudor's "Glass of the Sinful Soul" suggests one hitch between the profane family life of the princess—charged by her royal father with being the illegitimate offspring of an incestuous union—and the subsequent reformation in Elizabethan England of the medieval Christian notion of universal siblinghood, where all men and women are equally kin or nonkin as children of adoption to Christ. This was the modern reformation into the idea of nation not as parentarchal institution but as liberal estate, an idea that still influences the politics of tolerant liberalism, with all its contradictions. Thanks to the orphaned Elizabeth's transmutation of fear of the desire for physical incest into a desire for spiritual incest (and also, of course, to the political exigencies of the time), there developed an ideology of national siblinghood where the "ruler" was simultaneously parent and child, sibling and spouse. Britain's exemplary transformation was part of a movement in early modern Europe toward becoming a polis bereft of ordinary kin, an ideological evolution that reversed the ancient Roman imperial family's and vestals' momentous conversion into the Catholic religious orders. One of these orders adopted the orphan Racine and prompted his *Britannicus*.

In the late eighteenth century many European states, having taken over from the church its role of administering orphanages and foundling homes, began to internalize specific aspects of ostensibly secularized medieval oblation into new rites of political "naturalization" and doctrines of nationhood. Thus Rousseau, founding father of the French Revolution, by way of explaining why he abandoned his consanguineous children, disingenuously confessed that, like a latter-day Hagar, he put them up for state adoption in such a way that he would never recognize them as grown ups. (Had not Plato said that in the ideal polis a person would greet everyone equally as a sibling?) In the 1790s France raised the children whose parents it had executed during the Terror, adoptively sublating them as "children of the nation." These children were quintessentially national *alumni*, like the passengers on nineteenth-century American orphan trains. Their new home helped to define a regime where everyone—liberal, equal, and fraternal—might be reborn from the common ground, as the national anthem stated. They were born, like Adam, from clay. Like the celebrated dragon's teeth that Cadmus sowed in the earth, they sprang up, already naturalized, as humanoid Earthmen and Earthwomen uncommonly kin to none and all.

Yet the ancient promise of universal brotherhood, stranger to no creed, now turns, as always, into the particularist fraternity of Cain and Abel. Human beings, suspended between universal otherhood and tribal brotherhood, and still vertiginous from the last world war, want a general peace. However, we are unlikely soon to change our political being or patterns of loving—as if we were chrysalids about to become butterflies in some Wonderland. Particularist ideologies, which openly recognize that some beings outside of their familial or national neighborhood are human, would seem wonderfully essential to the missing peace—if, as we have surmised, they might also provide a temporary basis for treating human others reliably with Noachic common decency.

Toleration, with all the indeterminacy that it entails, is neither genial nor effortless. If ever we come to live a civilized existence, without the periodic catastrophes of war and genocide, it will be, still discontentedly, thanks to the endless adversities—pains without salve—that our own putting up with others imparts to us as a kind of political gadfly. This toleration goads us, at the very least, to avow openly the rightful coexistence of human terrestrials and national territories that we do not much like or like too much (because they are too little or too much like us). And it drives us to admit in good faith as ours the particularist likes and dislikes that constitute political and sexual being. Toleration likewise requires the never-ending vacillation between acknowledging that our kin are ours even as we know that it is deniable that they are, and denying that our kin are ours even as we recognize that it is probable that they are.

NOTES

Chapter 1

1. See Pitt-Rivers, "Pseudo-kinship."

2. Montaigne discusses societies supposed not to have ordinary incest taboos (*Essais,* in *Oeuvres,* ed. Thibaudet and Rat, bk. 1, chap. 23, p. 113). Charron agrees the incest taboo is a "mere custom" that "mastereth our souls, our beliefs, our judgements, with a most unjust and tyrannical authority" (*Of Wisdome,* p. 310). Cf. Sánchez, *Quod Nihil Scitur* (1581), ed. Comparot, p. 78.

3. Bachofen, in his *Mutterrecht,* made the absolute certainty of the mother-child link the basis of a social anthropology that survives to this day. On the "Fiktions-bedeutung der Paternität," see Bachofen, *Mutterrecht,* in *Werke* 2:57 p. 57. Bachofen also writes that "der Vater ist stets eine juristische Fiktion, die Mutter dagegen eine physische Tatsache: *(mater) semper certa est, etiamsi vulgo conceperit, patero vero is tantum* [*sic*], *quem nuptiae demonstrant* [citing Julius Paulus, *Dig.* 2, 4, 5]" and that "die Mutter is nun stets sicher . . . " (*Mutterrecht,* in *Werke,* 2:102, 118 pp. 102, 118). Engels, *Origin of the Family* (1884), intro. Barrett, p. 71, writes that "in all forms of group family, it is uncertain who is the father of a child; but it is certain who its mother is" and attributes this view to Bachofen. This view of female descent Engels emphasizes also in the preface to the fourth edition (1891). I would take issue here with with the traditional view, summarized in the gist of Alan Grossman's scholium on the terms *mother* and *father,* namely, that "you know who is the mother. You do not know who is the father" (Grossman, "Primer," §42, pp. 119–21).

4. The debate involving politics (Orestes) and family (Clytemnestra) in Aeschylus' *Oresteia* is ended by the argument of the virgin goddess Athena (born from the head of Zeus), who proclaims that since "no mother gave me birth, only the male of the human species is the begetter" (*Eumenides,* 736). In an Orphic hymn (31.10) Athena is called both male and female, and in a Homeric hymn (9.3) she is called a virgin divinity (W. Smith, *Dictionary* [1967] s.v. "Athena"). Athena's argument was backed by Aristotle's view that in human reproduction "the male provides the form and the principle of the movement; the female provides the body, in other words the material" (*Generation,* 1.2 729a). See also Plato's similar view of the male Eros of begetting (*Timaeus,* 73, 90–91) and related evidence collected by Onians (*Origin of European Thought,* pp. 108–9). In Anaxagoras the mother is merely the "breeding ground" and the father is "the seed"; in Diogenes of Apollonia the father, not the mother, provides the offspring (Freeman, *Pre-Socratic Philosophers,* pp. 272, 282). Compare Jesus' claim that only (the male) God is His only parent. In Malinowski's view "the most important moral and legal rule concerning the physiological state of kinship is that no child should be brought into this world without a man—and one man at that—assuming the role

of sociological father, that is, guardian and protector, the male link between the child and the rest of the community. I think that this generalization amounts to a universal sociological law" ("Parenthood," pp. 137–38). Vico states that "with the first human, which is to say chaste and religious, couplings, [humanoid creatures] gave a beginning to matrimony. Thereby they became certain fathers of certain children by certain women" (*New Science*, 1098).

5. On "nominal" incest generally see Crawley, *Mystic Rose*. For a discussion of the purchase of family names in relation to incest, see B. Thomas, "Writer's Procreative Urge in *Pierre*." For incest and kinship by adoption, see chapter 6. On kinship "in law," see chapter 5. One might also include in the list of kinds of pseudokinship or kinship by extension that can make for a diriment impediment to marriage the relationship between trading partners, between feudal lords and their servants, between brothers-in-arms, and so on.

6. Shakespeare, *Hamlet*, 1.2.107–8. For the view that in *Hamlet* there is no one either kin or nonkin but thinking makes him so (cf. *Hamlet*, 2.2.256), see chapter 5.

7. See Matt. 23:8. Christianity's universalist doctrine both incorporates and transcends ordinary kinship by means of an extraordinary unifamilial kinship. Cf. Gal. 3:26–28: "For in Christ Jesus you are all sons [or children] of God. . . . There is neither Jew nor Greek . . . for ye are all one in Christ Jesus." Church Fathers who urge Christians to obey the implications of Matt. 23:8 in such a way as to call all men their "brothers" and to call no one by the name "father" include Ignatius of Antioch ("To the Ephesians," 10.3); Clement of Alexandria (*Stromata*, 7.14.5); Justinian (*Dialogue with Tryphon*, 96); and Tertullian (*Apologeticus*, 39.8–9).

8. Heraclitus, frag. 93.

9. *Notes and Queries* 5:493, series 7 (1888).

10. For the voice, see Makkot 23b; for Judah bar Ila'i, see *Jewish Ency.* 11:439. In a strikingly christological misprision of the Judgment of Solomon, René Girard even fails to notice the tale's literal significance: "The woman who cries, 'Give her the living child and by no means slay it,'" says Girard, "is presented to us as the true mother in the biological sense, which resolves the matter within a family context" (*Things Hidden*, p. 242). Girard says that he has read "the commentaries and that Solomon's judgment is "a possible solution to the dilemma," but, as we have seen, it is not a solution and the commentators say as much. It is worth adding that not only Solomon but also neither of the women could really have known for sure which of them was related consanguineously to the living child.

11. For Hegelian *Aufhebung* (sublation), see esp. chapter 6, the section entitled "Roman Sublation and Christian Oblation," in this volume.

12. One sect practiced gazing at navels as a means of inducing hypnotic reverie. Vaughn (*Hours with the Mystics* 1,272) writes that "they call these devotees Navel-contemplators." For the Holy Umbilical Cord, see Collin de Plancy, *Dictionnaire* 2,45.

13. Browne, *Pseudodoxia Epidemica* 5.5.240.

14. Bryant, *Analysis of Ancient Mythology* 1,245.

15. For the background to this relationship, see chapter 4.

16. *Dormition*, or *koimēmsis*, refers to the specific "dying" of the Virgin Mary: she passes from being the mature Mother of God (the infant Son) to being the infant daughter of God (the mature Father), as discussed in the body of church literature called the *Transitus Mariae* (see Lampe, ed., *Patristic Greek Lexicon*). Cf. Warner, *Alone of All Her Sex*, pp. 88–89. The dormition of the Virgin Mary, though generally considered a doctrine of the Eastern Church, is depicted prominently in such public buildings of Western Christendom as La Martorana in Palermo, Sicily, where Jesus gives new life to Mary's infant soul, which resembles a baby in his arms. It is also depicted in such manuscripts as the (probably) British "Winchester

Psalter" of the twelfth century (British Museum Cotton Manuscripts, Nero CIV, folio 29; see Saxl and Wittkower, *British Art and the Mediterranean,* 24.6,7).

Julia Kristeva, citing Dante's well-known line "Vergine Madre, figlia del tuo Figlio," says that "not only is Mary her son's *mother* and his *daughter* [as in dormition], she is also his *wife.* Thus she passes through all three women's stages in the most restricted of all possible kinship systems" ("Stabat Mater," p. 105). But, as we shall see, Mary is also the *sister* of God the Brother, and it is this Sisterhood which makes equal siblings of all mankind and reaches even beyond the restriction that Kristeva notes.

17. *New York Herald,* August 31, 1868; emphasis mine. For this and other references, see Gillman, *Dark Twins,* esp. pp. 53–73.

18. Wallace, *The Two.*

19. On the freak show, see Clemens, *Pudd'nhead,* pp. 45–46, 232, 295.

20. According to Schweitzer (*Herakles,* p. 19) Kteatos and Eurytos, the twin sons of Molione, were originally Siamese twins. For the view that they were originally regular twins, see Farnell, *Greek Hero-Cults;* Farnell argues that the Siamese-twin version of the legend is Hesiodic and hence comparatively late. In Homer, the twins are normal and mortal (*Iliad* 11.750), though sons of Poseidon; they marry and have sons (*Iliad* 2.621). See Smyth, *Greek Melic Poets,* p. 278.

21. The medical anatomist Serres, in his *Theory of Organic Formation and Deformation, Applied to . . . Duplicate Monsters* (1833), which was influenced by Montaigne's "Of a Monstrous Child" (Montaigne, *Essais,* bk. 2, chap. 30; written during the French civil war), says of a set of Siamese twins with four legs and a single head that "there is a perfect unity produced by two distinct individualities. There are sense organs and cerebral hemispheres for a single individual, adapted to the service of two, since it is evident that there are two *me's* in this single head" (quoted in S.J. Gould, "Living with Connections," p. 75; see also Gillman, *Dark Twins,* p. 61).

22. Mankowicz and Haggar, *Pottery and Porcelain,* p. 202. The Chalkhurst twins died in 1734.

23. For example, see "Tocci Twins," p. 374.

24. For the term "conglomerate," see Clemens, *Pudd'nhead,* p. 213.

25. Clemens, *Those Extraordinary Twins,* p. 123.

26. The "Duplicates" in *The Mysterious Stranger* say, "Although we had been born together, at the same moment and of the same womb, there was no spiritual kinship between us" (*Mysterious Stranger,* p. 334). Twain writes of the identical twins in *Pudd'nhead Wilson* that "one was a little fairer than the other, but otherwise they were exact duplicates" (Clemens, *Pudd'nhead,* p. 43).

27. It is worth noting that Twain performed the part of one of a pair of Siamese twins at a private dinner party at his home in 1906 (see Gillman, *Dark Twins,* p. 181) and that he died in 1910 murmuring inchoately about the "duplicates" Dr. Jekyll and Mr. Hyde (see Fiedler, *Freaks,* p. 270). Twain discussed the female Siamese twins Millie and Christine—a "wonderful two-headed girl"—in his "People and Things" column in the *Buffalo Express,* Sept. 2, 1869; for other references to Siamese twins in Twain's journalistic writings, see note 127.

28. Twain writes that he "took those [Siamese] twins [of *Those Extraordinary Twins*] apart and made two separate men of them [the identical twins or changelings of *Pudd'nhead Wilson*]." In *Pudd'nhead Wilson,* moreover, the identical twins become Siamese (*Pudd'nhead,* pp. 212, 295, 245). For the boundaries of nation and race, see Clemens, *Those Extraordinary Twins,* pp. 216–17.

29. Anthropological studies at once phonetic and sociological might focus on the sig-

nificance of the term "Siamese twins"—or the lack of one—in languages throughout the world, but few do. An exception is Naden, "Siamese Twins in Mampruli Phonology"; on the Gur language of the Mampruli in the northern part of Ghana and its more general significance, see also Rattray, *Ashanti Hinterland.*

30. Dreiser, "Mark the Double Twain."

31. Plato, *Republic,* 524; *Theaetetus,* 185; *Greater Hippias Major,* 300. In the ideal realm, the music of the spheres rules supreme. There "the two [identical twins] become one" conglomeration and knock . . . out a classic four-handed piece on the piano in great style" (Clemens, *Pudd'nhead,* p. 49). Twain's "combination consisting of two heads and four arms joined to a single body and a single pair of legs" becomes a single harmony (*Pudd'nhead,* p. 208).

32. For the biological aspects of Siamese twins, whose connections with each other can vary widely, see Newman et al., *Twins,* p. xiii, and David J. Smith, *Psychological Profiles of Conjoined Twins.* For Pauline marriage, see 1 Eph. 5:31–32.

33. The term *cuckold* probably stems from the cuckoo bird's habit of laying eggs in the nests of other birds: the black-billed cuckoo occasionally lays a few eggs in the nests of as many as six other species of birds, and the bronzed cowbird lays its eggs in the nests of as many as fifty-two other species; moreover, the brown-headed cowbird, or "black vagabond," which has no nest of its own, lays all its eggs in the nests of other species. (See Terres, ed., *Audubon Society Encyclopedia,* pp. x, 939, 940; Friedmann, *Cowbirds;* and Payne, "Clutch Size.") It is worth considering that not raising one's own children—or raising others' children—is the "natural" state of affairs for some animals—though perhaps not for human beings. Consider the ornithological term *stepmother,* which refers to the bird that hatches another bird's egg. (For these references I am thankful to Professor Sarah Lawall.)

34. There were strict Roman laws that aimed at avoiding substitution and changelings. The existence of such laws suggests how easy it can be for babies to pass for one another in cases of institutionalized fraud and private deception (see Boswell, *Kindness,* pp. 110, 114).

35. On the prevalence in the eighteenth and nineteenth centuries of the institutions of the nursemaid and foundlings, see Laslett et al., *Comparative Bastardy.*

36. Boyeson, in "World of Arts and Letters" (1894): 379.

37. For "a single drop," see Frederickson, *White Supremacy,* p. 130.

38. Nazi ideology claimed that if an Aryan woman had even one sexual encounter with a Jew, none of her children would ever be Aryan. Julius Streicher, for example, stated that "one single cohabitation of a Jew with an Aryan woman is sufficient to poison her blood forever. Together with the *alien albumen* ["the sperm of a man of alien race . . . which is partially or completely absorbed by the female and thus enters her bloodstream"] she has absorbed the alien soul. Never again will she be able to bear purely Aryan children. . . . [T]hey will all be bastards" (quoted in Reynolds et al., *Minister of Death,* p. 150).

39. Clemens, *Pudd'nhead,* p. 12; emphasis mine.

40. "Children are everywhere thought to be of the same substance as their parents," says one anthropologist, "because they are produced by them: 'like breeds like' in every system of thought" (Pitt-Rivers, "Kith and the Kin," p. 92). Cf. Langland, *Piers Plowman,* Text B, p. ii, l. 28; cf. also Aristotle, *Politics,* 1.2 (1255b): "From good parents comes a good son." On legal fictions, see Fuller, *Legal Fictions.*

41. Williamson, *New People,* p. 63. According to Abraham Lincoln, there were 405,751 "mulattos" in 1850, nearly all from black slaves and white masters (*Speeches and Writings,* p. 400). Slavery should be outlawed, he said, because miscegenation was bad.

42. Williamson, *New People,* p. 47, citing the French Count de Volney.

43. See note 3.

44. Clemens, *Pudd'nhead,* p. 12.

45. Clemens, *Pudd'nhead,* p. 12.

46. Clemens, *Pudd'nhead,* p. 13.

47. There are those who would say, though, that Jocasta did recognize Oedipus early on; see Vellacott, "Guilt of Oedipus."

48. In other places—imperial Rome, medieval Christendom, and eighteenth century France, for example—the number of foundlings raised as foster children was sometimes more than one quarter of the child population. See chapter 5.

49. Clemens, *Pudd'nhead,* p. 21.

50. See Gillman, *Dark Twins,* p. 74.

51. In *Puddn'head Wilson,* alcohol and tea join blood and milk as consubstantial liquids that confer familial identity: Rowena loves the teetotaler Angelo, but she will not marry him because his identical twin brother Luigi drinks to excess and makes Angelo drunk (Clemens, *Pudd'nhead,* p. 292).

52. Abraham Lincoln said that "there is a natural disgust in the minds of nearly all white people, to the idea of indiscriminate amalgamation of the white and black races" (in a speech on the Dred Scott decision at Springfield, Illinois, on June 26, 1857; see Lincoln, *Speeches and Writings,* p. 397). One folk etymology of Lincoln's key term *amalgamation* links it with *ama* ("together") and *gamos* ("marriage").

53. On lactation and kinship: Crawley, *Mystic Rose* 2,230; Koran, ch. 4 ("Women"), p. 75; and chapters 5 and 7 below. For the racialist argument that blacks are non-human, see section 3. For collactation by goats and other animals: Montaigne, *Essais,* in *Oeuvres,* ed. Thibaudet and Rat, bk. 2, chap. 8, pp. 379–80.

54. Clemens, *Pudd'nhead,* pp. 266–67.

55. Hanging Luigi means ending the history of "those extraordinary twins" (*Pudd'nhead,* p. 294). Says Luigi: "If I had let the man kill [Angelo], wouldn't he have killed me too. I save my own life, you see" (*Pudd'nhead,* pp. 91–92).

56. Joyce, *Ulysses,* p. 404.

57. The following story is told in the midrashic Tosafot, or gloss, to the Talmudic tractate Menahot (37a): "At King Solomon's court one day an evil spirit Asmodeus presented the great judge with a difficult case. A two-headed man had married a woman who had borne him seven sons. Six of them resembled the mother; but *one* resembled the father in having two heads. After the father's death, the two-headed son claimed two shares of the inheritance. He argued that he was two men. His brothers, however, contended that the two-headed son was entitled to one share only. Solomon, in his wisdom, ruled that the son with two heads was only one man. He rendered judgment in favor of the six brothers." (See Jellinek, *Beit ha-Midrash,* pp. 151–52, and *Jewish Ency.* 11:439, s.v. "Solomon.") There is also the Twain-like case of the slave who claimed he was his master's son (Jellinek, *Beit ha-Midrash,* pp. 145–46).

58. Clemens, *Pudd'nhead,* p. 6.

59. Galton admitted that fingerprints do not reveal racial grouping (Galton, *Finger Prints,* pp. 1–2, 26).

60. Clemens, *Pudd'nhead,* p. 193; *Extraordinary Twins,* pp. 150–51.

61. Cf. Gillman, *Dark Twins,* p. 6.

62. Hooper, in a recent *Wall Street Journal* article, "Identification by DNA Said to Be Better" (p. 1), reports that "the British scientist [Alec Jeffreys] credited with devising 'DNA fingerprinting,' a way to identify individuals from their genetic materials, has come up with an alternative method that he says should cause less controversy in courtrooms [than the earlier method devised in 1985]."

63. T.V. Smith and E.C. Lindeman, *Democratic Way,* p. 19.

64. Paine, *Writings* 2, 304–5.

65. The *Oxford Latin Dictionary,* s.v. "Liberi," defines the word as "sons and daughters, children in connection w[ith] their parents."

66. For "noble lie": Plato, *Republic* 414c-415d. For autochthony and the Theban tale: Plato, *Laws,* 663e; Odysseus' stories among the Phaeacians. Thebes is the birthplace of Sophocles' Oedipus. Plato's *Menexenus* proposes to unify the people of the state by turning autochthony into an ideology of real estate: "Their ancestors were not strangers, nor are these their descendants sojourners only, whose fathers have come from another country but they are the children of the soil, dwelling and living in their own land. And the country which brought them up is not like other countries, a stepmother to her children, her their own true mother" (Plato, *Menexenus,* 237b in *Collected Dialogues,* ed. Hamilton and Cairns; trans. Jowett). See also, "Are we not told that men of that former age were earthborn and not born of human parents?" (Plato, *Statesman,* 269b in *Collected Dialogues,* ed. Hamilton and Cairns; trans. Skemp; cf. 271a–272b). In the Bible, of course, *Adam* means something like "of the earth."

67. "With everyone he happens to meet, he will hold that he's meeting a brother, or a sister, or a father, or a mother, or a son, or a daughter, or their descendants or ancestors" (*Republic* 463). Cornford, in a note to his edition of the *Republic* (pp. 161–63), argues that "Plato did not regard the . . . connections of brothers and sisters as incestuous." Cf. *Republic* 414d. The Platonic argument that incest is politically necessary to a liberal republic runs counter to other arguments; since Gibbon's *Decline and Fall* one common view has been that incest was actually the cause of republican *decline* in Rome (Fowler, esp. chap. 3, "Incest Regulations").

68. The Marquis de Sade, in the wake of the republican French Revolution, argues like Plato that for a people to become a genuine republic where all people think of and treat other people as equal citizens they must refuse to recognize any difference between consanguineous and nonconsanguineous kin, thus practicing either incest or celibacy. See Sade, "Francais, encore un effort si vous voulez être républicains," in *Philosophie,* esp. pp. 221–22.

69. For the history and theology of the legal definition of "spiritual incest" as a Brother or Sister having sexual intercourse with anyone at all, see such documents as Council of Rome (A.D. 402), can. 1,2, in Hefele, *Conciliengeschichte* 2,87; Pope Gelasius I (A.D. 494), letter to the bishops of Lucania, c. 20, in Gratian, *Decretum,* causa 27, q. 1, c. 14; Council of Macon (A.D. 585) can. 12, in Hefele, *Conciliengeschichte* 3,37; Gratian, Decretum, causa 30, q. 1, c. 5.10; and Oesterlé, "Inceste," in *Dictionnaire de droit canonique,* ed. Naz, 5:1298–1314; and [Joseph]-Eugène Mangenot, "Inceste," in *Dictionnaire de théologie catholique,* ed. Vacant et al., 7:1539–56.

70. Among English writers in Christendom it is the practice to capitalize kinship terms when they refer to affiliations that directly involve the Holy Family (Father and Son), the associations of friars or nuns (Brotherhood or Sisterhood), and the radical siblinghood of humankind that this family and these associations imply. Thus "Child" or "Son" is capitalized when referring to Christ as the offspring of God the Father but not when referring to him as the child of Joseph. Likewise, "Sister" is capitalized when referring, for example, to Shakespeare's Ophelia as a would-be nun—her new family would be modeled upon God's relationship to Mary as her Father, Brother, Son, and Spouse—but not when referring to Ophelia as the consanguineous sister of Laertes. Where we take for granted the usual Christian distinction between holy and profane or between spiritual and literal, the traditional practice of capitalizing essentially Christian kinship terms is helpful in clarifying whether a term is being used one way or the other. But difficulties arise in writing about apparently non-Christian or purportedly secularist universalist siblinghoods. These siblinghoods sometimes ignore the usual Christian distinction, or they reflect that distinction so transparently

as to become the brunt of Christian critiques, as did the fraternalist ideology of the French Revolution.

71. On the incestuous relationship between Jesus and Mary: Heuscher, *Psychiatric Study,* p. 207. On the Athanasian doctrine that Father and Son are not literally the same, so that the Son is the Father of Himself: the Council of Nicea in A.D. 325. For Tamar: Matt. 1:3 and Gen. 38:26–30. Rahab, a harlot, and Bathsheba, an adulteress, are included among the human ancestors of Jesus, whose mother was a kind of harlot, since His conception was extramarital. On the genealogy of Jesus: Santiago, *Children of Oedipus,* p. 50, and Layard, "Incest Taboo," pp. 301–2. For the magi: Rankin, "Catullus," p. 119, and Moulton, *Early Zoroastrians,* esp. pp. 204, 249–50. For the Persians: Antisthenes of Athens, *Fragmenta,* cited in Rankin, "Catullus," p. 120, and Sidler, *Inzesttabu.* See Nietzsche's remark that "an ancient belief, especially strong in Persia, holds that a wise magus must be incestuously begotten" (*Die Geburt der Tragödie,* sec. 9, in *Werke* 1:56–57; trans. Golffing, pp. 60–61). For further discussion of fourfold kinship within the Holy Family, see chapter 4.

72. The father in Stendhal's *Cenci,* Francesco Cenci, "taught [his daughter Beatrice] a frightful heresy, which I scarcely dare repeat . . . that when a father has carnal knowledge of his own daughter, the children born of the union are of necessity saints, and that all the greatest saints whom the Church venerates were born in this manner" (Stendhal, *Cenci,* pp. 181–2). For Saint Albanus, see Rank, *Inzest-Motiv.* For Saint Julian, see Bart and Cook, *Flaubert's Saint Julian,* and Berg et al., *Saint Oedipus.* For the story of Gregory, see Hartmann von Aue, *Gregorius.* On the motif in general, see Harney, *Brother and Sister Saints.*

73. See chapters 2 and 4.

74. The alliance between Clément Marot and Anne d'Alençon, parodied by Rabelais in his chapter entitled "The Island of Ennasin" in *Pantagruel* (bk. 4, chap. 9), is a good example of kinship by *alliance.*

75. In Goethe's *Elective Affinities,* Ottilie expresses "a strong belief in the existence of some law of male and female friendship and kinship higher than our actual marriage would in every case now imply." Goethe's Lotte says that "it seems to me that these things are related to each other not in the blood, so to speak, so much as in the spirit" (cited in Dixon, *Spiritual Wives,* pp. 361, 365). "Blood Relations of Choice" or "Chosen Kin" is how Brodsky translates *"Wahlverwandschaften"* (*Imposition of Form,* pp. 88–89). However, since we do not know who *are* our blood kin, electing who we *call* blood kin is the usual situation for human beings.

76. For the phrase "cult of fraternity," see Sandell, "'A very poetic circumstance,'" ch. 2. See also Durbach, "Geschwister-Komplex," pp. 61–63; Rank, *Inzest-Motiv;* Praz, *Romantic Agony,* pp. 111–12; and Schelly, "A Like Unlike." For Byron, see his "Manfred" (1817), "The Bride of Abydos" (1813), and "Cain" (1821), esp. 1.1.187–89, 1.1.380; and Van Der Beets, "Note." Engels reports Marx's view from Marx's letter to him about Richard Wagner; Engels himself writes that "not only were brother and sister originally man and wife, sexual intercourse between parents and children is still permitted among many peoples today" (Engels, *Origin of the Family,* p. 100; cf. p. 102). For Vico, before the establishment of specifically human "institutions," men lived in what might be called a "nefarious promiscuity of things and women" in which "sons often lay with mothers and fathers with daughters" (*New Science,* paras. 16, 17).

77. For Paul, see 1 Cor. 6:12. For Noyes, see his letter of January 15, 1837 to David Harrison (quoted in Dixon, *Spiritual Wives* 2, 55–56).

78. *Bible Communism,* p. 27. Noyes continues: "The sons and daughters of God, must have even a stronger sense of blood-relationship than ordinary brothers and sisters. They live as children with their Father forever, and the paramount affection of the household is . . . *brotherly* love. . . . A brother may love ten sisters, or a sister ten brothers, according to

the customs of the world. The exclusiveness of marriage does not enter the family circle. But heaven is a family circle; and . . . brotherly love . . . takes the place of supremacy which the matrimonial affection occupies in this world" (*Bible Communism;* quoted in Dalke, "Incest in Nineteenth Century American Fiction," p. 88).

79. Ellis, *Free Love,* pp. 187–88. Noyes counters the usual objection to "amative intercourse between near relations"—that "'breeding *in and in*' deteriorates offspring"—with his own genetic theories. Cf. Noyes' monograph on eugenics, "An Essay on Scientific Propagation." For other references to Noyes' writings, see Kern, *Ordered Love;* on the doctrine of free love in nineteenth century America, see Spurlock, *Free Love.*

80. Morison, *Oxford History of the American People,* pp. 524–25.

81. See Emerson's discussion of those Americans who "adopt the word *l'humanité* from [Pierre] Le Roux [follower of the Christian socialist Saint-Simon] and go for *'the race'* [of humankind]." Emerson, who has in mind people such as Orestes Brownson, Christopher A. Greene, Elizabeth Palmer Peabody, and George Bancroft, writes that "the world is waking up to the idea of Union and already we have Communities, Phalanxes, and Aesthetic Families, & Pestalozzian institutions" (Emerson, *Journals,* 8:251; cf. Bercovitch, "Emerson," pp. 3–6). Here Emerson is referring to the schools for waifs, strays, and orphans pioneered by the Swiss Rousseauist Johann Heinrich Pestalozzi. For the term *familism* see also *Tait's Magazine* 15 (1848): 705, where the author remarks that there was a strong "propensity to group" that "embraces love, friendship, ambition, and a fourth passion, called familism." On the American feminist Fourierist Margaret Fuller, who in 1844 published her *Women in the Nineteenth Century* in the tradition of Mary Wollstonecraft, see chapter 8.

82. Fiedler, introduction to *Quaker City,* p. xxx; cf. p. vii. Fiedler writes that what "sets Lippard's book apart from those on which he modelled it is a peculiar emphasis, somehow characteristic of America, on the sanctity in a world otherwise profane of the brother-sister relationship." The booksellers sold 60,000 copies in 1844.

83. Lippard, preface to *Quaker City,* p. 2. On the Brotherhood of Union, see Fiedler, introduction to *Quaker City,* p. viii.

84. See Easton, *Hegel's First American Followers;* Webber, *Escape to Utopia.*

85. The Quakers were directly associated with the doctrine of Familism; see, e.g., Hallywell's *Familism as it is revised by the Quakers* (1673). On the theme of brother and sister in *The Philadelphia Story* see Cavell's remarkable *Pursuits of Happiness.*

86. For the romantic topos, see B. Thomas, "Writers' Procreative Urge." Damon, "Pierre the Ambiguous," puts *Pierre* squarely in the group of literary works about incest; Mogan, "*Pierre* and *Manfred,*" p. 231, discusses the incestuous crime of Byron's Magian hero in relation to *Pierre.* The topos was already an important theme in American letters, including W. H. Brown's *The Power of Sympathy,* and such works as Hawthorne's *Marble Faun* (1860), where Miriam and her mysterious follower are likely Astarte and Manfred revisited, and his "Alice Doane's Appeal," which tells the story of Leonard Doane's love for his sister. Consider also Anna Lewis's poem "Child of the Sea" (1848), in which it is revealed that the affair is not after all incestuous. See also Dryden, "Entangled Text," and F. G. See, "Kinship of Metaphor."

At first the ordinary romantic topos was not quite recognized by the critics. Damon, "*Pierre,*" p. 110, writes that "the nineteenth century literally did not know enough even to guess what the book was about." Some early readers could think of *Pierre* only as a hoax (note in Melville, *Pierre,* ed. Hayford et al., p. 393).

87. The conjunction of the Catholic *sponsa* with incestuous urges is not uncommon in American literature. Irwin (*Doubling,* pp. 129–30) focuses on Christian mother-son incest in the writings of William Faulkner: "The fecundation of Mary by God is a supplanting of

Joseph . . . and since Jesus, the son, is himself that God, then he is, in a sense, the son who has impregnated his mother, and Jesus' birth, as befits the birth of a god, is incestuous."

88. Melville, *Pierre,* ed. Murray, foreword Thompson, p. 37. Subsequent notes refer to this edition unless otherwise noted.

89. Melville, *Pierre,* p. 58.

90. In Disraeli's *Alroy* (1833), as Murray points out, the hero does not know any love, "save that pure affection which doth subsist between me and this girl, an orphan and my sister" (introd. to *Pierre,* p. lxvi).

91. See Brodsky, *Imposition of Form,* p. 244.

92. Melville, *Pierre,* p. 391. The portrait of Beatrice Cenci also appears in Hawthorne's *Marble Faun.*

93. Melville, *Pierre,* p. 109; emphasis mine. Shakespeare, *Hamlet,* ed. Jenkins, 1.5.9.

94. Melville, *Pierre,* pp. 232 and 141.

95. Melville, *Pierre,* p. 116. Melville's Billy Budd is "a foundling" who, when asked "Who was your father?," has to answer, "God knows, Sir" (Melville, *Billy Budd,* p. 298). Budd's "entire family was practically invested in himself" (*Billy Budd,* p. 297).

96. Melville, *Pierre,* p. 310.

97. For a nineteenth-century American view of bundling, see Stiles, *Bundling.* On Dutch aspects of American bundling, see Irving, *History of New York, by Knickerbocker,* p. 211. On bundling among the Indians, see Stiles, *Bundling,* pp. 50–63. On bundling in the Catholic religious orders, see Bayle, *Dictionnaire,* ed. Desoer, 6:508–10, art. "Fontevrault"; and see chapter 4, section on "Social Anthropology of Universalist Orders." *Webster's [Dictionary]* is quoted by Stiles, *Bundling,* p. 13. Stiles, *Bundling,* p. 15, quotes Caesar from the American edition of Logan's *Scottish Gael,* p. 472. The attack of current British sleeping practices is quoted from *Reviewers Reviewed,* pp. 34–35.

98. Isabel, in describing her relationship with Pierre, says: "I am called woman, and thou, man, Pierre; but there is neither man nor woman about it. Why should I not speak out to thee? There is no sex in our immaculateness" (Melville, *Pierre,* p. 178).

99. Melville, *Pierre,* p. 171.

100. This community would be a political utopia that perfects the communities of Mettingen in Brown's *Wieland* (which has only four people, a pair of sibling-pairs) and at Saddlemeadows. Cf. Wilson, "Incest," p. 7.

101. Melville, *Pierre,* p. 304.

102. Melville, *Pierre,* p. 305.

103. Melville, *Pierre,* p. 57.

104. Melville, *Pierre,* p. 318.

105. Gesenius, *Lexicon,* p. 45.

106. Melville, *Pierre,* p. 310.

107. Melville, *Pierre,* p. 342. On the abiogenetic hypothesis that living things sometimes arise from such "lifeless" matter as the earth—without either human or divine parents—see Nigrelli, ed., "Modern Ideas on Spontaneous Generation"; Thomas Huxley, in his *Address to the British Association for the Advancement of Science* (1870), wrote that "to save circumlocution, I shall call . . . the doctrine that living matter may be produced by not living matter—the hypothesis of *Abiogenesis*" (Huxley, "Spontaneous Generation," p. 356).

108. Nationalists in France, whence came Isabel, refer to "enfants de la patrie" [children of the fatherland] who die and are reborn or regenerated from the earth. In "La Marseillaise" fighters for liberty are "produced from the earth": *la terre en produit de nouveau.*

109. Lucy becomes convinced that she has been called "to do a wonderful office" toward Pierre (Melville, *Pierre,* p. 350), and, thinking him to be the complete incarnation of

all her family—her "brother and mother . . . and all the universe to me" (Melville, *Pierre,* p. 351), Lucy "nun-like" (Melville, *Pierre,* p. 350) comes to the Church of the Apostles to serve Pierre as a kind of nun just as Isabel serves him as a kind of Sister.

110. In his *Dictionary,* Bayle reports that Mohammed, though he forbade incest for his followers, allowed it to himself by a special privilege.

111. Isabel says to her brother, Pierre, the Brother of all men, "Were all men like to thee, then were there no men at all,—mankind extinct in seraphim!" (Melville, *Pierre,* p. 186).

112. The universalist viewpoint I am pursuing militates against assuming the importance of the subject's place in a consanguineous family, an importance that the narrative structure of biography often assumes (see chapter 6). However, much could be said about Melville's relationship to his mother, sister, grandfather, and so on. Among the facts: Melville "purposed to write his spiritual biography in the form of a novel." "Melville used the name of his mother [Mary], not that of his wife, on the birth certificate of his son." Melville's "vigorous and pious" sister Augusta, presumably a model for Isabel in *Pierre,* copied his manuscripts. Melville longed for the prenuptial state after his marriage. Isabel is also modeled on Melville's cousin. Pierre's grandfather is like Melville's (Melville, *Pierre,* pp. xxiv, xxxvii, xxxviii, 430; ed. Hayford et al., pp. 339, 397, 399; Mumford, *Melville*).

113. One hero of American-German socialists of mid-century America, the cosmopolitan Stoic Epictetus, says that the common origin of all human beings rules out essential class differences: "Slave, do you not want to help your *adelphos,* who has Zeus for father, who is born of the same germs as you and is of the same heavenly descent?" (Epictetus, *Discourses,* cited in Zeller, *Philosophie der Griechen,* pp. 299–303). Epictetus agrees with the gist of Christian monachism. (For sociological treatments of interclass equality within the orders, see Campbell-Jones, *In Habit,* esp. pp. 196–99, and Séguy, "Sociologie," p. 347.) Paul too held that "there is neither bond nor free, for ye are all one in Christ" (Gal. 3:28).

114. "The earliest American fictionalists," speculates Dalke, "unconsciously used the incest theme to express their deepest anxieties about class upheavals" (*"Had I known,"* p. 88). Cf. Wagenknecht's remark that, "judged by their fiction, the Founding Fathers might appear primarily devoted to incest" (*American Novel,* p. 2). For a Marxist condemnation of incest in American literature as "a neurotic practice of the decadent upper class," see Zelnick, "Incest Theme."

115. In Sophocles' drama, which is partly about class upheaval during the period of egalitarian democratization under the Greek tyrants, Oedipus ascribes Jocasta's outburst at learning he is not the biological son of Polybus and Merope (*Oedipus,* 1077–79, trans. Grene) to her fear that he may have been born from the lower classes. It is thus in an egalitarian and democratic spirit that Oedipus speaks the words I quote.

116. Pierre marries outside his aristocratic class, earning his mother's ire, and inside his family, committing incest. ("But you, Pierre, are going to be married before long, I trust, not to a Capulet, but one of our own Montagues; and so Romeo's evil fortune will hardly be yours. You will be happy" [Melville, *Pierre,* p. 39].)

117. For this term, sometimes translated as *scission* or *diremption* in American descriptions of Hegelian *Aufhebung,* see the nineteenth-century American translation of Hegel's *Logic* by Lt. Governor Henry C. Brokmeyer (Missouri), a frequent contributor to *The Journal of Speculative Philosophy* and writer about the Civil War.

118. I here use the term *brotherhood* and the term *fraternal kinship* immediately below as near synonyms for *siblinghood.*

119. *Documentary Source Book of American History,* doc. 50; emphasis mine.

120. *Documentary Source Book of American History,* doc. 47.

121. *Documentary Source Book of American History,* doc. 43; emphasis mine.

122. Quoted in Decker, *The Declaration of Independence,* p. 149; emphasis mine.

123. Durham, *Report* 2:16. In "O Canada," the anthem of my birthplace, Canada is called the "home and native land" of all citizens, English- and French-speaking, whether they are indigenous or foreign-born. Whether such genuine "natives" as the Mohawk Indians of Québec should call Canada their homeland remains ambiguous. See chapter 3.

124. Decker, *Declaration of Independence,* p. 160.

125. Elizabeth, *Letters,* ed. Harrison, p. 47.

126. Gentili, *De jure belli libri tres* (1598), bk. 3, chap. 18, trans. Rolfe, pp. 387–88; cited in Nelson, *Usury,* p. 156.

127. Mark Twain, in his "People and Things" column in the *Buffalo Express* (Sept. 2, 1869) and in an 1869 article in *Packard's Monthly* (August 1869) entitled "Personal Habits of the Siamese Twins," takes up the question of Siamese twinning and civil war: "During the war they . . . both fought gallantly—Eng on the Union side and Chang on the Confederate." Siamese twins come to stand for the interdependencies of masters and slaves, say, or the Northern pole and the Southern. One pundit, considering the case of the Siamese twins Chang and Eng, wonders "what General [William Tecumseh] Sherman would do if one [of the twins] were disloyal and had to be sent South, while the other remained loyal." (*Alta* [1864]: 1). Another pundit notes that "whether the Chang part was for the North, or the Eng part for the South, or *vice versa,* is not yet made public" (*Downieville Mountain Messenger* [April 28, 1866]; see Gillman, *Dark Twins,* pp. 57–58). Montaigne, in his "Of a Monstrous Child," speaks of the one-headed multilimbed Siamese twin ["ce double corps et ces membres divers, se rapportans à une seule teste"] in terms of the French civil war (Montaigne, *Essais,* in *Oeuvres,* ed. Thibaudet and Rat, bk. 2, chap. 30, p. 691).

128. Delany, *Condition,* pp. 12–20, 209; cited in Sollors, *Beyond Ethnicity,* p. 48.

129. Clermont-Tonnerre, *Opinion relativement aux persécutions qui menacent les juifs d'Alsace.* Cardoso, *Excelencias,* translated in Yerushalmi, *From Spanish Court,* pp. 469–470. See chapter 2.

130. See, for example, Calhoun's speech on the Reception of Abolitionist Petitions (delivered in the U.S. Senate, February 6, 1837); in Levy, ed., *Political Thought,* pp. 307–10.

131. Fitzhugh's *Cannibals All!* is cited in Morison, *Oxford History of the American People,* p. 512. Cartwright, *Prognathous Species,* follows Nott, *Types of Mankind,* in arguing that natural history and the bible "prove . . . the existence of three distinct species of the genus man," including the "white" and the "black" (cited in *Ency. Phil.* 7:60).

132. Cited in Sollors, *Beyond Ethnicity,* p. 38. In *Mardi* (2, 224), the inscription is to be found over the arch of the "tutelary deity of Vivenza."

133. Lincoln, *Speeches and Writings 1832–58,* pp. 396, 402.

134. Sewell, *Selling of Joseph* (1700), argues "that all Men, as they are the Sons of Adam, are Coheirs; and have equal Rights unto Liberty" (in Ruchames, ed., *Racial Thought,* 47). Martin Luther King, in his speech delivered before 200,000 people at the March on Washington (August 28, 1963), accented the traditional church focus on "all of God's children" and argued that racism ends in the familial unity of humankind: "I have dreamt that one day . . . the sons of former slaves and the sons of former slave owners will be able to sit down together at the table of brotherhood. . . . I have a dream that one day . . . little black boys and black girls will be able to join hands with little white boys and white girls and walk together as sisters and brothers" (quoted by Oates, *Let the Trumpet Sound,* p. 261).

135. Acts 17:26; cf. John 3:16. The passage from Acts was often quoted by black American preachers. It can also be found in numerous other texts: Frederick Douglass' "The Meaning of July Fourth for the Negro" delivered in Corinthian Hall, Rochester, on July 5, 1852 (Foner, *Douglass,* pp. 199–200); Delany's *Condition, Elevation, Emigration, and Destiny of the Coloured People of the United States;* and William Wells Brown's *Clotel; or The President's Daughter,* a novel of 1853 about the children that Jefferson fathered with the slave woman

Sally Hemings. Not surprisingly, the text sometimes became the occasion for discussing the evolutionary connection between men and animals, as in the anonymous 1833 essay entitled "Are the Human Race All of One Blood?" (pp. 361–62). W. E. B. DuBois uses Acts prominently at the beginning of *Darkwater* (1920) when he writes "I believe in God, who made of one blood all nations that on earth do dwell." See Timothy Smith, "Slavery and Theology," and, for the references to Delany and to Brown, Sollors, *Beyond Ethnicity*, pp. 58–65.

136. Young, "The Mother of Us All," p. 408. Cf. Hubbell, "Smith-Pocahontas Story," and Sollors, *Beyond Ethnicity*, esp. p. 264, n.7.

137. Sollors, *Beyond Ethnicity*, pp. 85 ff., discusses the analogy between Americanization and Christian rebirth.

138. See chapter 8.

139. Hawthorne, *Passages* 2, 24.

140. Lincoln said that "we have besides these men—descended by blood from our ancestors—among us half our people who are not descendants at all of these men, they are men who have come from Europe—German, Irish, French, and Scandinavian—men that have come from Europe themselves, or whose ancestors have come hither and settled here, finding themselves our equals in all things. If they look back through this history to trace their connection with those days by blood, they find they have none, they cannot carry themselves back into that glorious epoch and make themselves feel that they are part of us, but when they look throughout that old Declaration of Independence they find that those old men say that 'We hold these truths to be self-evident, that all men are created equal,' and then they feel that that moral sentiment taught in that day evidences their relation to those men, that it is the father of all moral principle in them, and that they have a right to claim it as though they were blood of the blood, and flesh of the flesh of the men who wrote that Declaration, (loud and long continued applause) and so they are. That is the *electric cord* in that Declaration that links the hearts of patriotic and liberty-loving men together, that will link those patriotic hearts as long as the love of freedom exists in the minds of men throughout the world" (speech at Chicago, Illinois, delivered on July 10, 1858, in Lincoln, *Speeches and Writings, 1832–58*, p. 456; emphasis mine).

141. Concerning sexual history: The universalist doctrine that all human beings are siblings, insofar as it requires that people who want to avoid incest must either become celibate or commit bestiality, would encourage some white Americans to think of blacks as somehow animals. On celibacy, see chapter 3; on bestiality, see chapter 7.

142. See Fiedler, *American Novel*.

143. This motto was approved for the seal of the British and Foreign Anti-Slavery Society on October 16, 1787.

144. Montgomery, "The Rainbow," in *Works* 2, 361; capitalization and italicization are from this edition.

145. For "man and brotherism," see *Pall Mall Gazette*, March 27, 1865, p. 3.

146. Quoted by Williamson, *New People*, p. 66.

147. W. D. Jordan, *White Over Black*, p. 178.

148. See Dominguez, *White By Definition*.

149. In the West Indies, for example, there were mulattos (half whites), sambos (one quarter whites), quadroons (three quarters whites), and mestizos (seven eighths whites). Analogously, in bilingual and biracial Canada, there were not only English Protestants and French Catholics but also *métis*. Such nonuniversalist, intermediating terms and the gradations of political rights that they seem to require are disconcerting to those who would insist, in unmediated and idealist fashion, on the equal creation of all men. But these terms can have certain practical benefits. As they force recognition of the reality of racial miscegenation

(which the United States did not), they allow for the conceptual distinction of race from class and hence make less credible the racialist hypotheses that confuse the legal boundaries of class and race with those of species. It would be worth comparing the situations of English and French America with that of Spanish and Portuguese America. In this vein, Sequera, in his 1988 consideration of the apparent union of the "ethnic" groups of Venezuela—the European Spanish, Indian-American, and African "elements"—into the modern "nation of Venezuela," writes that this union of internal ethnic groups, which many people call "mestization," should be understood in relation to the United States. Sequera argues that "Nuestra 'identidad' no es, en este momento, el simple resultado de la confluencia de tres etnias—nunca lo fue, en términos reales—como se nos enseñó y aún se enseña en los colegios, sino algo más complejo, puesto que hemos sido víctimas voluntarias de un mestizaje ideológico por medio del cual casi hemos erradicado del mapa a Centroamérica, en un desesperado intento por compartir fronteras con Estados Unidos" (Figueroa, Prologue by Sequera, *Folklore Venezolano,* p. 13).

150. Quoted in Takaki, *Iron Cages,* pp. 46, 49–50; cf. Gillman, *Dark Twins,* pp. 82–84.

151. That is, whereas racialists presumably want to keep the blood pure, nationalist liberals want to be chaste (literally, nonincestuous). Yet, the liberal maxim "All men are brothers" requires a lifting of the incest taboo in much the same way as the racialist rule "Marry only your brother." For Gobineau, see chap. 8, nn. 15–22, in this volume.

152. Few abolitionists were willing to follow Noyes' ideas to the point where they accepted equality within the family (hence incest) or even miscegenation within wedlock, however.

153. Quoted from a letter of 1837, in Parrington, *Romantic Revolution,* p. 336. On Perfectionism from this perspective, see also Garrison, *Life.*

154. "Out of mannes nacion Fro kynde thei be so miswent, / That to the likenesse of Serpent Thei were bore" (Gower, *Confessio Amantis,* 1.55).

155. Schiller, "An die Freude," *Sämtliche Werke,* 1:133–36. Beethoven's music for Schiller's ode was first conceived in 1812, when Percy Bysshe Shelley was also making "the earth one brotherhood" ("Prometheus Unbound," 2.2.95).

156. See Douglas, *Implicit Meanings,* p. 289. Douglas considers how in some tribes "the contrast between man and not-man provides an analogy for the contrast between society and the outsider" (*Implicit Meanings,* p. 289; cf. Needham, *Primordial Characters,* p. 5).

157. On this role of Lady Liberty, see Paulson, *Representations,* p. 16. Sollors, *Beyond Ethnicity,* p. 84, argues that the American Statue of Liberty recalls the alma mater tradition.

158. "Unions occur unseen (*aphonós*) and in the dark between whatever man happens by with whatever women happens by" (Strabo, *Geography,* 11.5.1). See also Tyrrel, *Amazons.*

159. Mme. de Tencin denied throughout her life that d'Alembert was her son. See chapter 6.

160. For the view that it was not only a simple moral failing that led Rousseau to abandon his five illegitimate children but also a political program, see chapter 6.

Chapter 2

1. The *Koran* grounds this series of divisions and is consistent with the well-known Pact of Umar I, which established special regulations for Christians and Jews living in Muslim lands. "There is to be no compulsion in religion. Rectitude has been clearly distinguished from error. So whoever disbelieves in idols and believes in Allah has taken hold of the firmest handle. It cannot be split. Allah is All-hearing and All-knowing" (Sura 2:256, trans. in Stillman, *Jews of Arab Lands,* pp. 149, 157–58). Cf. Sura 109:6: "To you your religion, to me my religion." On Sabianism as a fourth religion of the Book, see chap. 8, n. 85.

2. Cf. Bernal, *Black Athena*, p. 241: "If Europeans were treating Blacks as badly as they did throughout the 19th century, Blacks had to be turned into animals or, at best, subhumans; the noble Caucasian was incapable of treating other full humans in such ways"—a statement with whose irony I would agree if, for "noble Caucasian," Martin were to have substituted "man of Christendom who held that all humans are brothers and that one must love one's brother."

3. Kahana, *R. Avrhaham b. Ezra*.

4. Stillman, *Sources*, p. 76.

5. E.g., Paret, "Sure 2,256," pp. 299–300.

6. In the eleventh century, e.g., the Christian monarch Alfonso VI (sometimes called "El Bravo") offered limited protection to Muslim subjects in León and Castile (*Ency. Brit.* [11th ed.] 1:734); Alfonso designated as his heir a son by Zaida, daughter of the Muslim king of Seville. The tolerance toward Muslims and Jews of the "Muslim-Christian" king Frederick II of Sicily is another example.

7. Boswell, *Royal Treasure*, p. 327; cf. Lewis, *Jews of Islam*, chap. 1, and Burns, "Christian-Islamic Confrontation in the West."

8. On the fate of the *çala* in Christian states, see Roca Traver, "Un siglo de vida mudéjar en la Valencia medieval," p. 127; and Boswell, *Royal Treasure*, pp. 262–63. On the classification of Muslims as animals, see Tilander, *Los Fueros de Aragón*.

9. For the Iberian origin of the term *caste*, see Gilman, *Spain of Fernando de Rojas*, p. 113. The origin of race, from the Portuguese *raça*, is obscure, but many scholars have connected it with various cognates of "generation."

10. For the Iberian term *nação*, see note 55.

11. On these statutes *(Sentencia-Estatuo)* see Sicroff, *Controverses des statuts;* Yerushalmi, "Spinoza's Words"; and Netanyahu, "Américo Castro."

12. Yovel, *Spinoza*, p. 17. In fact, the pope had complained as early as 1437 that certain conversos in Aragon were being excluded from public office (Révah, "La controverse sur les statuts de pureté de sang," p. 265).

13. In Gratian's *Decretum* and Thomas Aquinas' *Summa Theologica* there is no invidious discrimination between Old Christians and New Christians, since "baptism into the faith" was regarded as a "regeneration of man and rebirth" (Cohen, *Martyr*, p. 290n; see Aquinas' *Summa Theologica*, pt. 3, quest. 65, pp. 2375 ff., esp A 2 [p. 2376]). Kamen, *Inquisition*, p. 122, writes, "In theory canon law limited the extent to which the sins of fathers could be visited on their sons and grandsons." Compare the specifically Jewish position that Marranos, or compelled Christians, were still Jews. For example, the Spanish Jew Isaac Abravanel, who served Ferdinand and Isabella, argued in his *Ma'ayene ha-yeshu'ah* that religious conversion cannot bring about the ethnic assimilation of the Jews: "The Ingathering shall be for the Children of Israel who are called Jacob, and also for the Marranos who are of their seed" (Commentary on Isaiah 43:7; cited in Netanyahu, *Don Isaac Abravanel*, pp. 203ff).

14. Dahn, *Lex Visigothorum*.

15. Some writers have said that the "Visigoths" were more tolerant toward "others" than were the Catholics (Salvian, *De Gubernatione Dei*). Visigoth law codes contained discriminatory provisions against Jewish converts (Baron, *Social and Religious History* 3, 33–46), but in comparison with Catholic codes, such provisions appear moderate. Such practices as the Catholics' expelling Jews—and selling into slavery anyone found practicing a Jewish ceremony (Grayzel, *History of the Jews*, pp. 302–303)—generally ceased with the Muslim conquest and began again only with the Catholic reconquest.

16. His history of the Gothic or Teutonic "brotherhood," the *Historia Gothorum, Wandalorum, Sueborum*, is included in the nineteenth-century German national series *Monumenta Germaniae historica (Scriptores, Auctores antiquissimi, Chronica Minora* II).

17. Sicroff, *Les controverses des statuts,* pp. 36–41.

18. Sicroff, *Les controverses des statuts,* pp. 36–41; see Oropesa, *Luz para conocimiento de los Gentiles* (1465) and Alonso de Cartagena, *Defensorium Unitatis Christianae* (1449–50).

19. *Germa* means "brother" in some Iberian dialects. *Germania* was the name for the agrarian revolt in Valencia in 1520 that included forced conversions to the Christian "brotherhood" of Muslims living under Spanish Christian domination—*Mudéjares,* so called from the Arabic word for "allowed to remain" (Cagigas, *Los Mudéjares* 1, 58–64). Cf. Ferdinand's and Isabella's *hermandad,* which was a system of "brotherhood" practiced by Castilian towns as a type of police force.

20. See Peñalosa's fixation on imputing *limpieza* and *hidalguía* to the "true" Spanish person: "Among the Spaniards is found the most ancient nobility of any nation, retaining always the blood of their first progenitor, Tubal [who, according to Peñalosa, came to Spain in the year 2,163 B.C.]" (Peñalosa y Mondragón, *Libro de las cinco excelencias del Español,* cited in Yerushalmi, *From Spanish Court,* p. 386).

21. See Yerushalmi, *Assimilation and Racial Anti-semitism.*

22. In Fascist Italy, "racial laws" followed fast on the doctrine of the "Aryan Italian." The *Manifesto degli scienziati,* for example, proclaimed on July 14, 1938 "that the Italian population was Aryan in origin, that a pure Italian race existed to which Jews did not belong, and that this race had to be defended from possible contamination." In Italian East Africa, laws prohibited conjugal relationships between Italian citizens and subjects, in order "to prevent the growth of interracial marriage and to furnish the Italians with . . . awareness of their racial dignity and superiority." On August 20, 1938, the Rome journal *La Difesa della Razza* [*Defense of the Race*] called for the sword to protect Italian "Aryans" from Jewish and African contamination. See Toscano, "Jews in Italy," pp. 39, 40.

23. Rafael de Tramontana y Gayango—Marquis of Guadacorte and President of the Fundación Gayangos—is quoted in Abercrombie, "When the Moors Ruled Spain," p. 92. Juan Antonio Llorente—General Secretary of the Inquisition from 1789–1801—pointed out in his *Memoria Histórica* that "you will find hardly a book printed in Spain from the time of Charles V to our own day in which the Inquisition is not cited with praise" (cited in Kamen, *Inquisition,* pp. 44, 133). The English term *blue blood* is a translation of the Spanish *sangre azul;* aristocratic Christian families of Castile, claiming to be of a light complexion that made their veins appear relatively blue, apparently used the term to characterize themselves as having never been "contaminated" by Jewish or Muslim blood (OED, "Blood," 1:8); cf. the royal Egyptian Queen Cleopatra's proud reference to her "bluest veins" in Shakespeare's *Antony and Cleopatra,* 2.5.29.

24. The relevant statute was instituted by the archbishop Juan Martínez Silíceo in Toledo in 1547, ratified by Pope Paul V in 1555, and upheld by Philip II in 1556 (Yerushalmi, *From Spanish Court,* p. 15).

25. Kamen, *Inquisition,* p. 121.

26. On the *sambenitos,* see Kamen, *Inquisition,* p. 122.

27. Baroja, *Los Judios en la España moderna y contemporanea* 2, 304. The *Tractatus bipartititus de puritate et nobilitate probanda* was published in the seventeenth century.

28. "Pouco sangue Judeo he bastante e destruyr o mundo!" (Costa Mattos, *Breve discurso contra a heretica perifidia do Iudaismo* [1623], trans. Yerushalmi, *From Spanish Court,* p. 416).

29. Cervantes, *Don Quixote,* bk. 2, chap. 63, p. 1515.

30. Cervantes, *Don Quixote,* bk. 1, Prologue.

31. Cervantes, *Don Quixote,* bk. 2, chap. 6; trans. Cohen, p. 506. On Cervantes' *converso* ancestry and his mockery of lineage in general, see Madariaga, "Cervantes y su tiempo," and Castro, *Cervantes y los casticismos.*

32. Yovel, *Spinoza,* p. 112. As Gilman points out, there are no direct references to *con-*

versos in the *Celestina* (Gilman, *Spain,* p. 366). However, Calisto describes Melibea as possessing "limpieze de sangre e fechos" (Rójas, *Tragicomedia de Calisto y Melibea* [*The Celestina,* act 12]). In the "Preface," Rójas speaks of his "fellowmen" or *socios* (cited by Yovel, *Spinoza,* p. 90).

33. Is it *essentially* blood kinship with Adam and Eve or spiritual kinship in God that makes us universal kin? Disagreement about this matter informs the difference between King Alfred's and Queen Elizabeth's translations of a well-known passage from Boethius's *Consolation of Philosophy: "Unus enim rerum pater est."* King Alfred the Great, ruler of the West Saxon tribes in the ninth century, translates the line in such a way as to stress blood, or tribal, kinship. Boethius's "one father of all things" becomes for Alfred "the father and mother of the race"; divine kinship becomes human ancestry. Five years after the defeat of the Spanish Aramada in 1588, Queen Elizabeth offered a more accurate interpretation of Boethius's words: "All humain kind on erthe / From like beginninge comes: / One father is of all, / One only al doth gide [guide]" (Boethius, *Consolation,* 3.6; in Elizabeth, *Poems,* ed. Bradner, p. 32. For the historic proverb and King Alfred, see Friedman, "'When Adam Delved . . .,'" pp. 220–21).

34. Yovel, *Spinoza,* p. 17. See Cohen, *Martyr,* on Spanish laws forbidding Jews, Muslims, and *conversos* to emigrate to the New World and on Spanish intolerance towards the Indians of the Americas. When Pernambuco (in Brazil) was conquered by the Dutch, Jewish New Christians, like other people in the New World freed from Spanish domination, established relatively free communities (Wiznitzer, *Jews of Colonial Brazil*). Greene writes about the Iberian epic *Os Lusíadas*'s "imperialism and nationalism" (*Descent,* p. 220). *The Lusiads,* composed by the Portuguese poet Luiz de Camoëns in the years following 1556 and published in 1572, concerns the discovery by Camoëns' kinsman Vasco da Gama of the sea route to India; it includes nationalist views of such major events in Portuguese history as the massacre in 1510 of every Muslim in Goa (*Ency. Brit.* [11th ed.] 12, 160). Cf. Greene's discussion of *La Christiada* (1611) by Diego de Hojeda, the Spanish-born Dominican monk of Peru (*Descent,* pp. 231ff).

35. Marquillos de Mazarambros, who instigated the first discriminatory statute, calls the *conversos* "children of incredulity and infidelity" (see Gilman, *Spain,* p. 191n, citing Benito Ruano, "El memorial contra los conversos").

36. *"El humor o error nacional"* (Laínez's Letter to Araoz [1560], cited in Kamen, *Inquisition,* p. 126).

37. *"El humor español"* (cited in Rey, "San Ignacio de Loyola"). The harsh judgment of Spain which this view of history would entail might be tempered with the observation that, during this period of the emergence of modern nationalism, few other states were called upon to live peaceably with strangers in their midst. There were no Muslims in England, for example, and the Jews had been expelled from that country centuries earlier.

38. See Calderón, in Pérez de Ayala, *Política,* p. 186.

39. See Cossio, *Los Toros* 4, 765; Ortiz Cañavate, "El toreo en España"; Conrad, "BullFight," p. 114; and Marvin, *Bullfight,* p. 139.

40. Campos de España, "España y los toros," cited in Marvin, *Bullfight,* p. vii.

41. See Dryden's "scaly Nations of the Sea profound" (in his translation of Virgil, *Georgics,* 3:806).

42. The Old Irish *braido/brado* ("wild," "savage") is apparently linked etymologically with the Spanish and Portuguese *bravo.* Cf. Provençal *braidin* ("fiery," "spirited" [horse]) and Latin *rabidus* ("mad," "fierce"). See Storm, "Mélanges étymologiques," pp. 170–71.

43. Hemingway, *Dangerous Summer,* p. 26.

44. Marvin, *Bullfight,* p. 96.

45. The verb *lidiar* means something like "to out*mano*euvre [sic] the bull," but it also

retains the sense of its Latin etymon *litigare* ("to dispute, to sue at law"), as at an auto-da-fé. According to Bollain, *lidiar* also concerns "the most efficient preparation for death of the bull" (Bollain, *El Toreo,* p. 16). In this context it is worth noting that bullfighting in Spain has a role much like that of rodeo bronco-busting in the United States, fox-hunting in England, and cockfighting in Bali (see Lawrence, *Rodeo;* Bouissac, *Circus and Culture;* Howe, "Fox-hunting"; and Geertz, "Deep Play").

46. Pitt-Rivers, "El sacrificio del toro."

47. See Gilpérez Garcia and Fraile Sanz, M. *Reglamentación Taurina Vigente,* article 131, cited in Marvin, *Bullfight,* p. 140.

48. Marvin, *Bullfight,* pp. 34–35, 61, 96, 75, 203–4.

49. In *Teshuba* 3.5 Maimonides, who left Spain as a young man, says that the pious gentiles have a share in the world to come, and in *Edut* 11.10 he says that, in certain circumstances, pious idolaters have such a share (Maimonides, *Moreh Nebukhim* [*The Guide of the Perplexed*]; cited in Strauss, *Spinoza's Critique,* p. 273n).

50. Lopez's assertion, "being spoken by a Jew, as it was," writes Camden, "was but onely laughed at by the people" (*Historie of the Life and Reigne of that Famous Princesse Elizabeth,* p. 105; cited by Mullaney, "Brothers and Others," p. 72). For an extended treatment of the problems of conversion, race, and religion in *The Merchant of Venice,* see my *Money, Language, and Thought,* chap. 3.

51. On the "Festival of the Christians and Moors," see Abercrombie, "When the Moors Ruled Spain."

52. Maurice Cranston, "Toleration," in *Encyclopedia of Philosophy* 8:142, citing Eliot's "Idea of a Christian Society."

53. Cf. Latin: *tollo, tollere, sustuli, sublatum.* "Sublation," a longstanding term in European philosophy, is the literal translation of Hegel's *Aufhebung;* see my *Money, Language, and Thought,* chapter 4.

54. Quoted from *Encyclopedia of Philosophy,* s.v. "Toleration," 8:143.

55. In Portuguese: *nação.* In Spain, the Marranos became the *nación* and in France the *nation.* See Vincente de Costa Mattos, *Breve discurso contra a heretica perifidia do Iudaismo,* esp. pp. 148ff. Gomes de Solis's work, *Alegación,* calls the new Christians the *hombres de la nación,* and in 1649, King Philip IV called them the *gente da nação;* see Saraiva, *Inquisição et Cristãos-novos* The Latin term *natio* plays a role in the Marrano Baruch Spinoza's *Ethics;* in Proposition 3:46 (in *Improvement of the Understanding*) Spinoza writes: "If someone has been affected with joy or sadness by someone of *a class or nation different from his own,* and his joy or sadness is accompanied by the idea of that person as its cause, under the universal name of the class or nation, he will feel love or hate not only to that person, but everyone of the same class or nation."

56. The Hebrew term *anusim* recalls discussions of coercion and rape in the Talmud. On the ambivalent religious practices of the Marranos, see Roth, "Religion of the Marranos."

57. Yerushalmi, *From Spanish Court,* p. 49. On Cardoso's writing in general, see Yerushalmi, *From Spanish Court.*

58. The *Philosophica libera* was published in Venice in 1673. I leave aside here such works as translations and commentaries by Elijah de Medigo and the neo-Platonic *Dialoghi d'amore* (1535) by Judah León Abravanel ("Leone Ebreo"), son of the Spanish-born Isaac Abravanel.

59. On the Marrano community in Amsterdam, see Silva, "Literatuurlijst."

60. Cardoso, *Excelencias,* p. 389, cited in Yerushalmi, *From Spanish Court,* p. 375. Cardoso refers to such bullfights as that of June 19, 1630, where twenty bulls and three men were killed. And he probably refers also to the national Fiesta Agonal of October 13, 1631. There tigers, bears, bulls, horses, greyhounds—and "other less important animals which

might enhance the laughter and entertainment" of the spectators—were thrown into the ring (Yerushalmi, *From Spanish Court,* p. 96). The bull was triumphant among these creatures, goring most of them to death, after which King Philip IV himself donned his cape and killed the exhausted bull. In his earlier days as a Spanish Catholic, Cardoso had attended this Fiesta. Pellicer de Tovar includes in his *Anfiteatro* a sonnet by Cardoso that depicts King Philip as a "Christian Mars" and the bull as gratefully accepting his wounds at the hand of the monarch (Pellicer de Tovar, *Anfiteatro,* fols. 3–11, 43v, cited in Yerushalmi, *From Spanish Court,* pp. 96–98, 375).

61. Cardoso, *Excelencias,* p. 374, cited in Yerushalmi, *From Spanish Court,* pp. 469–470.

62. Strauss, *Spinoza's Critique,* p. 53.

63. Gilman, *Spain,* p. 193.

64. *Concerning Heretics.* This is an anonymous work attributed to Châteillon, together with excerpts from other works by Châteillon and David Joris; cf. Kamen, *The Rise of Toleration,* p. 75. For Sánchez, see Limbrick, introd. to Sánchez, *Quod Nihil Scitur.*

65. Montaigne's essays show the influence of his mother, the Marrano Antoinette de Louppes (Lopez), and of his father, for whose sake he translated the Spanish schoolman Raymund de Sabunde's *Theologia naturalis* (1569).

Davis, in her excellent study of rites of violence, argues that the French Catholic killers at the Saint Bartholomew Day's massacre in 1572 did not think of the Protestants that they killed as "a foreign race" (as Estèbes had claimed in *Tocsin pour un massacre*) but merely as people who engaged in polluting, divisive, and disorderly actions (Davis, "Rites of Violence," p. 160n). Davis also remarks, however, that the Catholics considered their Protestant victims to be "nonhuman" and that they forgot that their victims were "human beings" (Davis, "Rites of Violence," pp. 175n, 181). It seems to me that such dehumanization often amounts to racism—especially in the context of universalist or catholic ideology. Where "all human beings are siblings," and every creature's status as human being is questionable, the anthropocentric butchering as animals of thousands of creatures called Huguenots—or "kin of Hugo"?—amounts to the same thing as the racist massacre of them as nonsiblings. (On the sixteenth-century interpretation of Huguenot as "kin of Hugo," see Richard, *Untersuchungen,* pp. 46–48.)

66. See esp. *Theologico-Political Treatise,* ch. 20. Strauss, *Spinoza,* p. 16, writes that Spinoza "was the first philosopher who was both a democrat and liberal. He was the philosopher who founded liberal democracy, a specifically modern regime."

67. See Basnage de Beauval's *Tolérance des religions* (1684), a Huguenot defense of religious toleration, and Bayle's *Dictionnaire Historique et Critique,* a discussion of Luke 14:23, where the master tells his servants to force the guests to enter.

68. Spinoza, *Theologico-Political Treatise,* p. 6. When Spinoza's former student Albert Burgh converted to Catholicism and tried to convince Spinoza to forgo philosophizing and likewise convert, Spinoza sent him an uncharacteristically angry letter. Spinoza reminded him how Burgh's own Netherlandish ancestors had been tortured during the period of the Spanish Duke of Alva's "Blood Council" and how the convert Judah the Faithful had recently been burnt alive by the Inquisition (Spinoza, *Correspondence,* pp. 415, 417–18). Judah the Faithful, says Pollock, was Don Lope de Vera y Alcarón de San Clemente, who was burnt to death in Valladolid on July 25, 1644 (see Pollock, *Spinoza,* chap. 2, last note).

69. Matt. 5:43.

70. Spinoza, *Theologico-Political Treatise,* p. 169.

71. See Strauss, *Persecution and the Art of Writing,* pp. 174–75.

72. Spinoza, *Theologico-Political Treatise,* p. 250.

73. Spinoza, *Correspondence,* p. 366. Orobio was a fellow resident in Amsterdam who criticized Juan de Prado. Prado had been "excommunicated" together with Spinoza in 1656. Orobio's *Certamen Philosophicum Propugnatae Veritatis Divinae ac naturalis,* published in

1684 with Fénelon's *Traité de l'existence de dieu*, was a Cartesian response to Spinoza. Orobio's *Prevenciones divinas contra la vana idolatria de las gentes*, published in French under the title *Israel vengé* (1770), was used as ammunition by French atheists against Christianity. On Orobio and the Marranos generally, see Kaplan, *From Christianity to Judaism*.

74. Cf. Strauss, *Persecution*, p. 168.

75. "Economy," says Cardinal John Henry Newman in *Arians of the Fourth Century* (p. 65), means "setting [the truth] out to advantage," as when "representing religion, for the purpose of conciliating the heathen, in the form most attractive to their prejudices," and the *disciplina arcani* is a "withholding [of] the truth" in the form of allegory, by which the same text may express the same truth at different levels to different people. Economy is necessary to "lead children forward by degrees" and may employ similes and metaphors. Newman maintains, for example, that "the information given to a blind man, that scarlet was like the sound of a trumpet, is an instance of an unexceptionable economy, since it was as true as it could be under the circumstances of the case, conveying a substantially correct impression as far as it went" (pp. 72–73).

76. Spinoza, *Theologico-Political Treatise*, p. 53. Cf. Strauss, *Persecution*, p. 190.

77. Kamen, *Rise of Toleration*, p. 223.

78. Spinoza, *Theologico-Political Treatise*, p. 264.

79. Spinoza, *Theologico-Political Treatise*, p. 6.

80. Had Manasseh not been absent from Amsterdam on his mission to England, says Israel Abrahams, Spinoza would probably not have been "excommunicated" by the Jewish community in 1656 ("Menasseh ben Israel," *Ency. Brit.* [11th ed.] 18, 112). On the English translation of Spinoza's *Treatise*, see Elwes' introduction to Spinoza, *Theologico-Political Treatise*, p. xxxiii. On Marranos in seventeenth century England generally, see L. Wolf, *Crypto-Jews*.

81. Rosenzweig, *Star of Redemption*, p. 216.

82. Lessing, the friend of Moses Mendelssohn and himself the German translator of Manasseh ben Israel's work, idealized the Pact of Umar in his Enlightenment drama. In the middle scene of the middle act of this play, Nathan, the Jewish descendant of Solomon the Wise, is asked by Saladin, the Muslim ruler, which of the three "religions of the Book" is genuine. The three religions are here squared off against one another in a way recalling the twelfth century Spanish Jew Yehuda Halevi's *Kuzari*, a work written in Arabic and indebted to the fifth century Islamic philosopher Abu Hamid Muhammad al-Ghazali's *Courteous Refutation*. For the influence of al-Ghazali on Halevi, see David Kaufman, *Geschichte der Attributenlehre*, esp. pp. 119–40. See also *Lessing und die Toleranz*.

83. Strauss, *Persecution*, p. 192.

84. See Laslett's catalog of Locke's library in his edition of Locke's *Two Treatises*.

85. See Locke's "commonplace books," which contain essays on the "Roman Commonwealth" (concerning religious liberty and the relations of religion to the state) as well as one entitled "Essay concerning Toleration" (1666).

86. For Orobio's disputation with Limborch, see Schoeps, *Israel und Christenheit*, esp. pp. 97–113. Philip van Limborch's *Historia Inquisitionis* (1692), esp. vol. 2, fols. 158, 322–23, is indebted to Orobio. Since reviews in the *Bibliothèque universelle et historique* are generally unsigned, attributions are uncertain (Colie, "Locke in the Republic of Letters").

87. Using the pseudonym "Philanthropus," Locke published *A Second Letter Concerning Toleration* (1690) and *A Third Letter for Toleration* (1692). Work on *A Fourth Letter for Toleration* was interrupted by his death in 1704.

88. Locke, *First Treatise*, para. 144–46; cf. *Second Treatise*, para. 110.

89. Américo Castro attributes Inquisitorial fanaticism to the *converso* heritage of some of the early inquisitors (Castro, *Structure of Spanish History*, pp. 521–44).

90. Guyon, *Réflexions sur la tolérance*, is a good example of the arguments that blame

anti-Semitism on the Jews: "Alors que les autres dieux méditerranéens se manifestaient les uns aux autres une déférence de bon goût, le Iaveh israélite apporta brusquement l'intolérance dans ses relations avec ses confrères. Cette invention s'est, par un bien mauvais tour du destin, retournée contre la race élue; son esprit d'intolérance, passé par la suite dans le christianisme, s'est exercé contre les Juifs dispersés et leur a valu d'éternelles persécutions dont ils peuvent, avec quelque mélancolie, trouver le prototype dans leurs propres antécédents" (p. 87). Not surprisingly in this context, Guyon argued in 1933 for the practice of universal incest, basing his views on the universalism of mankind; see the doctrines of the London-based pro-incest Guyon Society as formulated in Guyon's *Sex Life and Sex Ethics,* and my *End of Kinship,* p. 246.

91. See Netanyahu, "Américo Castro."

92. The history of defining nationhood and class in terms of the biblical "book of generations" includes Noah's cursing Ham's son Canaan as a slave ("Cursed be Canaan; a slave of slaves shall he be to his brothers" [Gen 5:1, 10:25]). On the one hand, Christian apologists for serfdom and slavery argued that their "white" serfs and slaves were descendants of the children of Ham (Friedman, "'When Adam Delved . . .'," p. 228, cites Hugo von Trimberg's *Renner* [c1300] and Heinrich Wittenweiler's *Ring* [c1400]). On the other hand, people in Europe and the United States argued that all whites should be free. Canaan, they argued was "black" (Ham presumably copulated with a raven on the ark—as suggested in the *Sachsenspiegel* [c1200]) and God gave Africa to Ham's descendants (which is the gist of *Cursor Mundi* [c 1300] and the "T-map" in Isidore of Seville's *Etymologiae* [1472], bk. 14, chap. 1). They rarely mention that Canaan was not African—other sons and grandsons of Ham, like Egypt and Sheba, were (Gen. 10:6–7)—and they conclude that only blacks should be slaves.

93. For example: In the ancient Hebrew Commonwealth members of all nations (Ammonites and Moabites excepted) had access to the rights of citizenship (Deut. 23:3).

94. Universalist Christians have sometimes criticized Jews in the *diaspora* for "exclusiveness." William Smith, for example, writes that "the liberal spirit of the Mosaic regulations respecting strangers presents a strong contrast to the rigid exclusiveness of the Jews at the commencement of the Christian era. The growth of this spirit dates to the time of the Babylonian captivity" (Smith, *Bible Dictionary,* p. 664). Smith neglects to mention that the liberal spirit of the Mosaic regulations respecting strangers—which he contrasts with Jewish exclusiveness—presents a strong contrast to the intolerance of Christians. Although for Christians exclusiveness and intolerance are theoretically always the same ("Be my brother—i.e., be included in my brotherhood—or I will kill you"), for Jews they are distinct ("If you are not my brother I will still keep faith with you"). Particularist tolerance may not be possible without a polity (that is the gist of Spinoza's remarks on the subject); universalist tolerance may not be possible at all.

95. Penn writes in his "Journal" that "the utmost they [the Hebrew Commonwealth] required from strangers . . . was an acknowledgement to the Noachical precepts" (*Select Works of William Penn* 1, ii; and *Constitution and Select Laws,* p. xvi; cf. Penn, *Considerations Moving to Toleration*). Among the original regulations associated with the Noachic covenant with all men: "Whoever sheds the blood of a man, by man shall his blood be shed; for God made man in his image" (Gen. 9:6). For Locke, see his *Letter Concerning Toleration,* ed. Romanell, p. 43.

96. On *hidalguía,* a sort of nobility which is sometimes distinguished from blood purity, see Gilman, *Spain,* pp. 146–47.

97. Other texts generally cited included the Golden Rule (Matt. 7:12), the parables of the tares (Matt. 13:24–30; Mark 4:26–29), Jesus' discussion of the bruised seed (Matt. 12:20), and Rabbi Gamaliel's exhortation to the Jews (Acts 5:38–39). Another important text was Paul's statement that God "made of one blood all nations of men for to dwell on the face of the earth" (Acts 17:26, cf. John 3:16).

98. Cited in *Ency. Brit.* (11th ed.) 11, 4.

99. Cited in Kamen, *Rise of Toleration*, pp. 65, 78.

100. Luis de Granada's *Libro de la Oración* (1554) is cited in Kamen, *Rise of Toleration.*

101. See *Racovian Cathechisme;* and concerning Socinianism generally, see Williams, ed., *Polish Brethren.*

102. *Homines sunt Protestantes: humani ab illis nihil alienum;* in Conring's introduction to the 1649 edition of the writings of George Witzel and George Cassander, cited in Kamen, *Rise of Toleration*, p. 124.

103. Williams, Letter to the Governor of Massachusetts (1651), cited in Kamen, *Rise of Toleration*, p. 189. Cf. Williams, *Bloody Tenent Yet More Bloody.*

104. The Corinthian sect's acts of incest were not "deed[s] done secretly out of weakness but . . . ideological act[s] done openly with the approval of at least an influential sector of the community" (Adela Yarbro Collins, "Function of 'Excommunication' in Paul," p. 253; see Cor. 5:1).

105. "I will love you if you will be my kin" also became the French Revolution's nationalist promise to the Jews.

106. Mendelssohn, *Jerusalem*, pp. 106–8.

107. The "barbarism of universalism" has not disappeared. Consider, for example, the popular contemporary Russian nationalist ideologue Igor Shafarevich. In his "Russophobia," published by the Union of the Russian People, he writes that "one of the most wonderful phenomena and enigmas of our Earth [is] belonging to one's [own particular] people." And he insists that everyone fully human should convert to this nation. Confusing thus the polis with the family (nation), Shafarevich writes of Jews in the Soviet Union that they must "make the choice between the status of aliens without any political rights and citizenship based on the love of the Fatherland." If a person does not share the same Ur-father—or have the same autochthony from an indigenous Fatherland—as do "genuine" Russians, then he is an "alien"—without political rights. According to the usual universalist principle of Christendom, every alien is nonhuman ("nothing human is alien to me"). So aliens in Shafarevich's universalist Russia would be worse off than strangers in a particularist Commonwealth. Not surprisingly, the nationalist Shafarevich makes the old accusation (like Schopenhauer's) that Jews regard non-Jewish human beings as animals. "Well known are the pronouncements from Talmud," he writes, "which from many points of view explain that a person of another religion cannot be considered human . . . [that non-Jews] are animals with human faces, etc., etc." (quoted by Liah Greenfield, "Closing of the Russian Mind," pp. 33–34 passim). As I show in chapter 7, there are universalist Christians who say that those who are not essentially (or potentially) Christians are nonhuman. And there are particularists who, as Locke reminds us in his *First Treatise of Government*, not only guarantee specific political rights to human strangers but also reserve special protections for animals. (See, for example, Locke's discussion in the *First Treatise* ed. Laslett [37–39] of how, though man was intended to be a shepherd, he was not permitted to take "a Kid or Lamb out of the Flock to satisfie his hunger.") Cf. Shafarevich's *Socialist Phenomenon*, with its foreword by the nationalist exile Aleksandr I. Solzhenitsyn.

108. Locke, *Toleration*, p. 57; emphasis mine. Compare Oliver Cromwell's statement that "I had rather that Mahommedanism were permitted amongst us than that one of God's children should be persecuted" (*Ency. Brit.* [11th ed.] 7, 493). Still, Cromwell was not tolerant towards Roman Catholics and Anglicans.

109. Locke, *Toleration*, p. 43.

110. Locke, *Toleration*, p. 43. Cf. John Milton's forceful advocacy of the separation of church and state in his *Treatise of Civil Power.*

111. See p. 9 of that essay.

112. Locke, *Toleration*, p. 52.

113. Locke, *Toleration*, p. 12.

114. See, for instance, Marcuse, "Repressive Tolerance."

115. Locke, *Toleration*, p. 39.

116. Christian sects, both from primitive times and of the seventeenth century, were frequently accused of "lustfully pollut[ing] themselves in promiscuous uncleanliness," as Locke remarks. Locke says the accusation about the primitive Christians was false (*Toleration*, p. 39), though modern historians might disagree, claiming, for example, that the Corinthian sect's incestuous deeds were "ideological act[s] done openly with the approval of at least an influential sector of the community" (Collins, "Excommunication," p. 253). There were similar incestuous heretics in seventeenth century England. Abiezer Coppe, a member of the Ranter sect, argued that through the intermediation of Jesus sin was "made to disappear"; promoting "sexual license," he praised the Pauline state beyond "good" and "evil," or chastity and incest (Carey, Foreword, in Nigel Smith, ed., *Collection of Ranter Writings*, p. 7, and Cohn, "The Cult of the Free Spirit," p. 68).

117. Locke, *Toleration*, p. 13.

118. In 1689, William and Mary were crowned as joint sovereigns in England; one of the new sovereigns' first bills was the "Act of Toleration," which granted freedom of worship, on certain conditions, to Dissenting Protestants. See James Tyrrell's letter to Locke of May 6, 1687, in which Tyrrell says that "your Discourse about Liberty of Conscience would not do amiss now to dispose people's minds to pass it when the Parliament sits" (cited in Laslett, Introd., p. 67, from Locke, *Two Treatises*).

119. It is important to ask ourselves whether the cruelty of the relatively tolerant American colonists toward the black slaves within their own borders—and of the relatively tolerant Dutch toward the indigenous peoples of Africa and the Malay Archipelago—was based on greed, as Locke might have liked to believe, or on a racialism that follows from a potentially intolerant universalist creed conflating family and species ("only my brothers—my gener*ation*, or *race*—are men worthy of humane treatment"), or on both. In this context, we might consider specifically Christian arguments for apartheid ("apartness") and Muslim arguments for slavery in Africa (cf. Bernard Lewis, *Race and Slavery in the Middle East*).

Chapter 3

1. Darwin, *Descent of Man*, pt. 1, chap. 2, p. 59.

2. See Wright, "Quest," p. 64; and Ross, "Hard Words"; cf. Gen. 4:20: "The man called his wife's name Eve, because she was mother of all living things."

3. Acts 17:26; cf. John 3:16.

4. See Shevoroshkin, "Mother Tongue."

5. For the early history of the idea of "Adamic" language, see D. Katz, *Philo-Semitism*, esp. chap. 2.

6. Cited in Wright, "Quest," p. 48. On the racialist quality of German romantic linguistics in the eighteenth and nineteenth centuries, see Bernal, *Black Athena*, pp. 224–72.

7. The Québec government's Parent Act of 1964 introduced a few supposedly nonconfessional schools—there had been none earlier—and various governments since then have tried to secularize the entire school system.

8. In the early 1960s Montréal's publicly supported primary schools were run either by the Protestant or the Catholic school board. Jews counted as Protestants and paid taxes to the Protestant school board. In previous decades some members of the Jewish community, including the Bundists, had favored the assimilationist tendencies of this arrangement.

9. On the issue of the routing of Montréal's "ethnic minorities" into French language schools in the wake of Bill 101 (1977), see *Vivre la diversité en français*. For an analysis of

"immigrant anglicization" and the Commission des écoles catholiques de Montréal in the 1980s (as seen from the viewpoint of the Italian community), see Taddeo and Taras, *Le débat linguistique*. On Montréal's efforts concerning the trilingual education of "Néo-Canadians," see Behiels, "The Commission des écoles catholiques de Montréal."

10. See Vallières, *Nègres blanc d'Amérique;* Mezei, "Speaking White"; and Lalonde, "Speak White."

11. For the term, see Etiemble's study of the Anglo-Saxon influences on the French language of France, *Parlez-vous franglais?*

12. The language of public signs was already a specific legal issue in Québec in 1910 when the Lavergne law required the use of French in limited commercial circumstances. Articles condemning the CNR are collected in Bouthillier and Meynaud, eds., *Le Choc des langues*. Rumilly, *Histoire de Montréal*, 5:177–78, discusses the Ligue's petition. Marc Levine's "Language Policy" is a good source for these and other references.

13. Laporte, "Queen Elizabeth?," p. 74; my translation.

14. Léger, writing in *Le Devoir* (1959), cited in Levine, "Language Policy," p. 4. Cf. Léger, *Francophonie*.

15. Eventually the Consumer Protection Act (1971) and the Companies Act (1973) were passed.

16. Levine, "Language Policy," pp. 6–7.

17. Bouthillier and Meynaud, *Le Choc des langues*, p. 32.

18. This phrase is Camille Laurin's. For an overview of Bill 101, see Plourde, *Politique linguistique du Québec*. Bill 101's provisions 58 and 69 concern the language of signs.

19. Levine, "Language Policy," p. 2.

20. Gouvernement du Québec 35; Laurin et al., *Politique québécoise de la langue française*, Gouvernement du Québec 21.

21. Levine, "Language Policy," p. 2.

22. For the view that Bill 101 meant a French *reconquête* to follow the English conquest of the eighteenth century, see Miron, "Chus Tanné," pp. 178–79.

23. For one use of this racialist term, see Lochner, "La minorisation des anglophones au Québec," p. 522.

24. For this term, see "Les autochtones et nous: vivre ensemble."

25. Colonel John Winslow, at that time commander of the English colonial militia, wrote in his *Journal* that "we are now hatching the noble and great project of banishing the French Neutrals from this Province. . . . If we accomplish this expulsion, it will have been one of the greatest deeds the *English in America* have ever achieved. . . ." (Cited in Clarke, *Expulsion of the Acadians*, p. 29.) For an account of the expulsion of the Acadians, see Casgrain's *Un Pèlerinage au pays d'Évangeline* and Parkman's *Montcalm and Wolfe*.

26. See Clarke, *Expulsion*, p. 31.

27. Clarke, *Expulsion*, p. 3.

28. For *cajun*, see Barbaud, "Parlerons-nous Cajun?" According to Miron, "bilingualism in institutions is merely the antechamber of assimilation" (Miron, "Chus Tanné," p. 194; my translation).

29. "Je suis un chanteur à deux pattes / qui jappe ses belles chansons / pour une race en voie d'extinction" (Jacques Michel, "If I was a cat"; quoted by Roy, "Ce dur Désir de chanter," pp. 30–31).

30. "Le jour s'en vient où tous les Westmount du Québec disparaîtront de la carte" (*Manifeste du Front de libération du Québec*, p. 360). Westmount was a wealthy, predominantly English section of Montréal. For an overview of the October Crisis of 1970, when the federal government of Canada used force to respond to the kidnappings of James Richard Cross and Pierre Laporte, see the various essays collected in *Focus on Québec: 1970 and Its Aftermath* (*Québec Studies* 11 [1990–91]).

31. Gendron ("Evolution de la conscience linguistique," p. 437) traces the focus on "culpabilité linguistique" at least as far back as 1841.

32. Barbaud, complaining about the "état de diglossie" in Québec ("Parlerons-nous Cajun?," p. 67), gives myriad examples of joual informed by English syntax.

33. Godin, "Le joual politique," p. 57; cited in Gauvin, "From Octave Crémazie to Victor-Lévy Beaulieu," pp. 38–39.

34. "Quand je lisais: *Glissant si humide,* je croyais que c'était du français, je comprenais parce qu'en même temps je lisais *Slippery when wet,* alors que c'est de l'anglais en français, c'est l'altérité. Pendant dix ans j'ai emprunté des centaines de fois sans tiquer au sujet de la signalisation: *Automobiles avec monnaie exacte seulement/Automobiles with exact change only—Partez au vert/Go on green,* etc., et je constate que des milliers d'usagers en font autant, jusqu'au jour où j'ai ressenti un étrange malaise, presque schizophrénique. Je ne savais plus dans ce bilinguisme instantané, colonial, reconnaître mes signes, reconnaître que ce n'était plus du français. Cette coupure, ce fait de devenir étranger à sa propre langue, sans s'en apercevoir, c'est une forme d'aliénation (linguistique). . . ." (Miron, "Décoloniser la langue," p. 12).

35. Godbout, *Réformiste,* p. 195; cited in Gauvin, "From Octave Crémazie to Victor-lévy Beaulieu," p. 43.

36. Cholette, *Commission de surveillance,* p. 18.

37. Cited in Plourde, *Politique linguistique du Québec,* p. 61.

38. Richler, "Oh! Canada!," p. 44.

39. Le Cours, "Bourassa."

40. See esp. Proulx, "Bilinguisme," on Bill 178.

41. Lochner, "La Minorisation des anglophones au Québec," p. 528. Lochner points out that between 1976 and 1983 the English population lost 2.5 per cent of its membership every year. He argues that the anglophone population, which numbered around 13 per cent of the population in 1976, will continue to diminish rapidly to stabilize itself in the year 2003 at between 6 and 7 percent of the population.

42. See Miron's complaint that Meech Lake lacks such provisions (Miron, "Chus Tanné," p. 193).

43. The banner's overtones are both Anglo-American and ecclesiastical, hence unsuitable for the quintessentially nationalist Saint Jean-Baptiste Day Parade, where it was first unfurled. See Roy, "Ce dur Désir de chanter," p. 31. Cf. the Christian will to build the church on the "rock" that is the apostle Peter (Matt. 14:17–19).

44. Miron, "Chus Tanné," p. 185, complains about "cette langue franco-bilingue" in the Métro. Such signs as "Bouton de cas d'urgence" are especially irksome to him.

45. See Beauchemin, "Le Français au Québec," esp. p. 147.

46. Woehrling, "Réglementation linguistique de l'affichage et la liberté d'expression," makes useful comparisons to other countries, including Switzerland (on which see Marti-Rolli, *La Liberté de la langue*).

47. Cited in Levine, "Language Policy," p. 14.

48. Québec is often tolerant of its minorities in the generous yet "haughty" way that Kant describes in his criticism of Muslim *convivencia* ("What Is Enlightenment?," p. 9). Thus the province decides on a year-by-year basis whether it will grant to any group other than English-speaking Protestants the "privilege"—as opposed to the "right"—to receive funds to support non-Protestant or non-Catholic schools.

49. On the "laïcisation de la société" in Québec, see Maheu's *Les Québécois,* pp. 175–83.

50. Coleman, "Class Basis of Language Policy."

51. In the 1940s, as most of us learned only in the 1980s, Paul de Man wrote collabo-

rationist academic journalism in a Belgium verging on Nazi totalitarianism. Coming to the United States in the 1950s, he preferred not to wake the furies. He never acknowledged the collaborationist writing of his past. He lied outright even about his wartime whereabouts. (De Man invited me to a German restaurant in New Haven in the spring term of 1973. During dinner I looked up from my roast lamb and asked him, "Where were you during the war?" De Man answered, "In Zurich, mostly." He answered in such a way that I could not believe him.) It was partly to explain how this academic journalist "outfoxed the fox" by a clever "distancing of all the facts" that led me to publish "The Lie of the Fox" in 1974. For my interpretation in 1974 of de Man's view of Rousseau's dictum, "Commençons donc par écarter tous les faits," see my "Lie of the Fox," p. 123, n. 34.

52. Shek, "Diglossia and Ideology."

53. Mezei, "Speaking White."

54. Monette, "Mon Francais mais Montréal," p. 115. Monette, author of *Traduit du jour le jour,* points out that "la plupart des écrivains majeurs sont des transfuges."

55. As of 1988 there were several anthologies of French Québec poetry in English translation but none of English poetry in French. However, there is a growing interest among French critics in Irving Layton, Bellow, Kerouac, and other writers somehow from Québec (Morisset, "La face cachée de la culture Québécoise").

56. See the bilingual journals *Vice Versa, Montréal Now,* and *Ellipse,* among others (Morisset, "La face cachée de la culture Québécoise, p. 537).

57. On code switching in Québécois literature, see Hodgson and Sarkonak, "Deux hors-la-loi québécois." On code switching among immigrants in the United States, as "the alternate use of two languages—including everything from the introduction of single unassimilated words up to complete sentences or more in the context of another language"—see Haugen, "Bilingualism," esp. p. 21.

58. Poulin, *Volkswagen Blues,* p. 110. The first French sentence may be translated: "He himself translated the phrase in hesitant French." The second may be translated: "When you're looking for your brother, you're looking for everybody!" Saul Bellow, the character in Poulin's novel, is modeled on Saul Bellow, the eminent Montréal-born novelist now a member of the Committee on Social Thought at the University of Chicago.

59. "An epigraph [epigram] is a poem in which . . . our attention and curiosity are aroused with reference to one particular object, and more or less held in suspense in order to be gratified at a single stroke" (Hudson, *Epigram,* pp. 8–10, citing Lessing).

60. "The true inscription is not to be thought of apart from that whereon it stands or might stand" says Lessing (Hudson, *Epigram,* pp. 8–10).

61. I photographed all signs discussed in this article in Montréal in 1973.

62. In this and the following examples, the translative medium is underlined.

63. Prévost, *Dictionnaire.*

64. See Van Rooten, *Mots d'heures: gousses, rames.*

65. See Aristotle, *Nicomachean Ethics* 5.5.14.

66. On the ventriloquistic language of commodities (*Warensprache*), see Marx, *Das Kapital,* in Marx and Engels, *Werke,* 23:66–67, 97 (trans. Moore and Aveling, *Capital,* 1:52, 83); and Marx, "Auszüge aus James Mills Büch *Elémens d'économie politique,*" in Marx and Engels, *Werke,* 3(1):545–40.

67. Turgot, in his "Valeurs et Monnaies," discusses the *langage de commerce.* Marx, in an ironic discussion of the relative form of value and the fetishization of commodities, hypothesizes that commodities speak a *Warensprache* when they come into contact with each other, and that they sometimes speak through the mouths of economists (*Capital* 1, 52, 83). Neither Turgot nor Marx, of course, intends to signify the words of advertisements, but these words are part of the alienated language of material values which men have borrowed, so to speak,

from the formal relations between commodities. For a discussion of the *Warensprache* contemporary to the period when these signs were photographed, see Faye, *Colloque de Cluny*, p. 191.

68. Cf. Ellul, *Propaganda.* Bilingual advertisements may help to teach French to the English and English to the French, but this is hardly their principal social or economic effect.

69. See, for example, the Québec Food Regulations Act of March 15, 1967.

70. See, for example, Sheppard, *Law of Languages in Canada,* who ignores interlinguistic mediation.

Chapter 4

1. Elizabeth I, *Poems,* p. 3. During the period of religious and political upheavals the greatest danger to Elizabeth's life probably occurred in 1554, when Sir Thomas Wyatt headed a rebellion in Kent and Elizabeth was summoned to London and sent to the tower for two months, after which she was sent to live at Woodstock (cf. *Letters,* p. 4).

2. Partridge's review ("Good Queen Bess") of Stanley and Vennema's *Story of Elizabeth I of England* is an example.

3. "But thou, which hast made separation of My bed, and did put thy false lovers in My place and committed fornication with them, yet, for all this, thou mayst come unto Me again, for I will not be angry against thee. Lift up thine eyes, and look up, then shalt thou see in what place thy sin had led thee, and how thou liest down in the earth" (Elizabeth, "Glass," Folio 36v; in Shell, *Elizabeth's Glass*).

4. The poem also appeared in 1538 and 1539; in 1547 and 1548 it appeared as part of Marguerite's *Marguerites.*

5. For the view that Anne Boleyn entered the service of Marguerite of Navarre (then Duchess of Alençon), see Ames', introd. to Elizabeth, *Mirror,* p. 31. The two queens knew each other as early as Queen Claude's coronation in 1516; both attended a banquet in France in 1518 and the Field of Cloth of Gold in 1520 (Ives, *Anne Boleyn,* pp. 38–42).

6. Anne Boleyn and Marguerite of Navarre had a well documented correspondence in 1534–35. In October of 1535, moreover, the English were anxious to interest the French envoys in the young princess Elizabeth (Ives, *Anne Boleyn,* pp. 41, 341).

7. Ames (introd. to Elizabeth, *Mirror,* p. 31) writes that "we may conclude that the copy . . . had belonged to her mother, who may have obtained it from her former friend and mistress [Marguerite of Navarre]."

8. Salminen (*Miroir,* p. 253) says Elizabeth used the edition of December 1533 printed by Antoine Augereau in Paris; but Prescott, p. 66, "Pearl," says Elizabeth used the edition of 1539.

9. For the letter, see *Letters,* pp. 5–7. On Elizabeth's needlework, see Neale, *Elizabeth,* p. 12.

10. Elizabeth asks her stepmother to "rub out, polish, and mend (or else cause to mend) the words (or rather the order of my writing) the which I know in many places to be rude, and nothing done as it should be" (Elizabeth, letter to Catherine Parr, December 31, 1544, Folio 4r, in Shell, *Elizabeth's Glass*).

11. On Bale, the ardent reformer and nationalist scholar and playwright, see my "Bale and British Nationalism," *in Elizabeth's Glass.*

12. This portrait is ascribed by some to H. Holbein. Yet it was probably designed in 1547 by an unknown artist (Ames, introd. to Elizabeth, *Mirror,* p. 7—Holbein died in 1543 of the plague.) John N. King says that Elizabeth's kneeling before Christ with the Bible in hand suggests Protestant learning (*Tudor Royal Iconography,* pp. 209–10) and he draws our attention to Bale's *Illustrium majoris Britanniae scriptorum,* which contains a comparable woodcut

showing Edward VI as a studious king standing at a lectern (*English Reformation Literature,* p. 6); we will see that the kneeling woman's relationship to Christ is considerably more complex than that.

13. For the publication history, see my *Elizabeth's Glass.* Elizabeth also gave members of her family other holograph translations as gifts—an English translation of John Calvin's *Institution Chrétienne* to Catherine Parr (1545); Latin, French, and Italian translations from Catherine Parr's *Prayers, or Meditations* to Henry VIII (1545); and a Latin translation of Bernardino's Ochino's *De Christo Sermo* to her brother, the ten-year-old (King) Edward (1547).

14. I offer a detailed discussion of these matters in *Elizabeth's Glass.*

15. On dormition, see chapter 1, note 16.

16. Elizabeth herself suggests that "the part which I have wrought in it" was "as well spiritual as manual"; letter to Catherine Parr, December 31, 1544, Folio 3v.

17. Jenkins, in his edition of Shakespeare's *Hamlet* (3.4.6), notes that, in the Folio, Hamlet, about to enter his mother's chambers, calls out thus for her: "Mother, mother, mother." For the reference to Melville, see his *Pierre,* foreword Thompson, p. 178.

18. Deut. 25:5–6.

19. Blackmore ("Hamlet's Right to the Crown") argues that "the diriment impediment to marriage with a deceased brother's wife was part of English church doctrine since earliest times and was retained by the English secular authorities until the nineteenth century." As Jones (*Hamlet and Oedipus,* p. 68) points out, "Had the relationship [between Claudius and Gertrude] not counted as incestuous, then Queen Elizabeth would have no right to the throne; she would have been a bastard, Catherine of Aragon being alive at her birth"; on *Hamlet* and the relationship between Henry VIII and Catherine of Aragon, see also Rosenblatt, "Aspects of Incest Problems."

20. In *Henry VIII,* Shakespeare goes to extraordinary lengths to allay anxiety about Elizabeth's possibly illegitimate birth. "Ann Bullen," for example, is noticeably absent from the christening scene. Yet, towards the end of the play, the porter suggests that Ann Bullen, who had been Henry's mistress (just as her sister Mary had been), was a "fornatrix," so that the Princess Elizabeth may be a bastard. When the porter in Shakespeare's play cries out, "what a cry of fornication is at door!" (5.3.34–35) he refers in part to the crowd of common people; yet the smallest "fry" in the play is Elizabeth herself who, from the Catholic viewpoint, is born of a "fornatrix."

21. Much relevant material is included in Edward Fox's *Collectanea statis copiosa,* a basis for *A Glasse of Truth* (1532). It included such treatises as Cranmer's *Determinations . . . that it is unlawful for a man to marry his brother's wife* (1531); see Ives, *Boleyn,* pp. 165, 167.

22. For Luther's marriage to the Cistercian Sister Catherine von Bora, see More, *Tindale,* pp. 48–49. For Rome's condemnation of Bale's marriage to Dorothy, see Pits, *Relationum Historicum,* pp. 53–59; cf. Harris, *Bale,* pp. 22–23. For Elizabeth's work with Ochino, see Craster, "Unknown Translation."

23. The notion that bastards, especially those born from bigamous or incestuous relations, cannot inherit the throne, gained some support from the fact that the children of Edward IV, as the offspring of a bigamous union, were unable to inherit. The issue was hotly contested in the 1540s. Upon the execution of Anne Boleyn, for example, "parliament was required to establish the succession on the new basis of Henry's new queen Jane Seymour . . . it also empowered the king to leave the crown by will if he had no legitimate issue, but the illegitimate son, the Duke of Richmond, in whose favor this provision is said to have been conceived, died shortly afterwards" (see "English History" in *Ency. Brit.* (11th ed.) 9, 522d, 531d).

24. Two other counts were the rumor that the king was impotent (Paul Friedmann, *Anne Boleyn* 2, 280, n. 1; Dewhurst, "Alleged Miscarriages of . . . Anne Boleyn") and the view that

the marriage was declared invalid *ab initio* on the ground of Anne's precontract with Lord Percy.

25. See Friedmann, *Boleyn* 2, 287, 351.

26. On the indictments of Anne for sibling incest and for an account of the trial, see Friedmann, *Boleyn* 2, 262–63, 278–81. The charge of adultery and incest with her brother, Lord Rochford, was made on May 2, 1536; Lord Rochford's wife was a principal witness for the prosecution.

27. See Friedmann, *Boleyn* 2, 262–63, 278–81.

28. The idea of Elizabeth as divine was fostered from the beginning, even as Anne Boleyn was associated with Saint Anne. There was a merging of Mary's son (Jesus Christ) and Anne's (expected or "annunciated") son in literary propaganda of the period (Ives, *Boleyn,* p. 284).

29. In 1536 an Act of Parliament ordered every man who had married his mistress' sister to separate from his wife and forbade all marriages with mistresses' sisters in the future. For Cranmer's views on the matter, see Lingard, *History of England* 5, 74; 540–42, and Friedmann, *Boleyn* 1, 43; 2, 323–27; 351–55. On the view that the "marriage was declared invalid ab initio . . . on the ground of the affinity established between Henry and Anne by Henry's previous relations with Mary [Boleyn]," see "Elizabeth Queen of England" in *Ency. Brit.* (11th ed.) 9:282.

30. For Henry's views of Mary's illegitimacy, see *Calendars of Letters, Despatches, and State Papers,* 1534–35, p. 57 [*Letters and Papers . . . of Henry VIII* 7, 214], cited in Ives, *Boleyn,* p. 271.

31. Fowler, "Incest Regulations," pp. 12–13. For the Latin term, see Rabanus Maurus, *Concilium moguntium.*

32. *Jacob's Well,* ed. Brandeis, 162/15.

33. Jonas of Orleans, *De institutione laicailli, Patrologiae (Latina)* 106:183–84.

34. In A.D. 868, Church Councils ruled that "we will not define the number of generations within which the faithful may be joined. No Christian may accept a wife . . . if any blood relationship is recorded, known, or held in memory" (Worms, can. 32, in Mansi, ed., 15, 875). Pope Julius I specified the seventh remove as the limit of diriment impediment to marriage (*Decret.,* no. 5, *Patrologiae* [*Latina*] 8:969). But there were enormous practical problems of record keeping, and even where the numerical degree was both agreed upon and ascertainable, there were controversies about the correct method of counting. Thus Stephan of Tournai notes that "the counting begins with the brothers according to some and with the sons of the brothers according to others" (*Summa,* 255). Cf. C. E. Smith, *Papal Enforcement,* chap. 2.

35. Eph. 5:31–32.

36. 1 Cor. 6:16, Cf. Gen. 2:24.

37. On the kinship relationship engendered by sexual intercourse, see Gratian, *Decretum,* 35, q. 5, 10; Bandinelli, *Summa,* p. 203; Stephan of Tournai, *Summa,* 250; Balbus, *Summa,* 168; and Feije, *De impedimentis,* chap. 14.

38. Bale, "Conclusion," in Elizabeth, *Godly Medytacyon,* Folio 40r; included in my *Elizabeth's Glass.*

39. For an interpretation of Sophocles' *Antigone* in this context, see Benardete, "Reading," 5. 2, p. 11, and Hegel, *Phänomenologie, Werke* 6, chap. 6.

40. There would have been many bibliothecal sources, including a French metrical account of how Anne was brought to trial and executed, the *Histoire de Anne Boleyne Jadis Royne d'Angeterre* of 1545 and perhaps early versions of the picaresque mid-century Spanish *Crónica del Rey Enrico,* in which one "Marguerita"—maybe a figure of Marguerite of Navarre—is the bawd who procures Smeton for Queen Anne. (The *Histoire* is discussed in Ascoli, *Grande-Bretagne;* M.A.S. Hume's edition of *Crónica del Rey Enrico,* pp. 55–59 and 68–76, refers to Marguerita; cf. Ives, *Boleyn,* pp. 69–70, 375.)

41. The terms "aunt-mother" and "uncle-father" are used in *Hamlet* (ed. Jenkins, 2.2.371–72). "You are welcome," says Hamlet to the twinlike Rosencrantz and Guildenstern, "but my uncle-father and aunt-mother are deceived." Insofar as Hamlet's mother, Gertrude, is married to Hamlet's uncle, she is his aunt. Similarly, Elizabeth's mother, Anne Boleyn, was her aunt. Henry had had sexual intercourse with Mary Boleyn, Anne's sister; and since, by the Doctrine of Carnal Contagion, a lover is, to all intents and purposes, a wife, Anne was Elizabeth's aunt.

42. The quotation is from *Ency. Brit.* (11th ed.) 9:282.

43. For a review of the evidence concerning the Elizabeth-Seymour liaison, see William Seymour, *Ordeal by Ambition*, pp. 215–19, 225–26. Thomas Tyrwhitt, who was sent to Hatfield to extract the truth from Elizabeth, complained that Thomas Seymour, Mistress Ashley, Thomas Ashley and Elizabeth told a story so close in so many details that they were "all in a tale" (*Letters*, p. 9; cf. Elizabeth's letter to Edward Seymour Lord Protector, January 28, 1549). Elizabeth's relations with Thomas Seymour was the occasion for sending a syntactically complex letter to him soon after she left the Lord Admiral's household, in July 1548: "You needed not to send an excuse to me for I could not distrust and not fulfilling your promise to proceed for want of good will, but only that opportunity serve not. Wherefore, I shall desire you to think that a greater matter than this could not make me impute any unkindness in you, for I am a friend not won with trifles, nor lost with the light" (*Letters*, p. 8).

44. *Letters*, p. 11.

45. Duff, "Die Beziehung Elizabeth-Essex," p. 469.

46. The centuries-old tradition of ascribing to Elizabeth various feelings about her family circumstances continues unabated. Prescott, for example, would determine "how [Elizabeth] respond[ed] to the executions of her mother, Anne Boleyn, and her stepmother, Catherine Howard" (Prescott, "Pearl," 68). Only seers such as "Fathers Confessor" and psychoanalysts can claim to see into the heart of a person, of course, but where religious or psychological speculation arises from texts such as the "Glass," it would seem to be particularly compelling—at least where the rhetorical quality of such speculation is recognized.

47. There is some evidence that there was a copy of the book in Henry VIII's household (Henry VIII, *Letters and Papers* 14[1], 369). In 1544 Marguerite had not yet a great literary reputation; she was known as an author pretty much only by her *Miroir*, its companion religious poems, and a farce (Prescott, "Pearl," p. 68).

48. Henry VIII opened negotiations to marry Marguerite of Navarre (then Duchess of Alençon) while he was entertaining the idea of divorcing Catherine of Aragon. Margaret's reply to Henry, made through Wolsey, makes clear that she refused to marry Henry because he committed crimes against Catherine of Aragon. See "Margaret, The Pearl of Navarre" (Edinburgh, 1868); Ames, introd. to Elizabeth, *Mirror*, p. 32.

49. Ames "Introduction," p. 33, in his edition of Elizabeth's *Mirror*.

50. Elizabeth, "Glass," Folio 7v, 8r, 23v.

51. Elizabeth, Letter to Catherine Parr, December 31, 1544, Folio 3r. It reads there as "mother, daughter, sister, and wife by the scriptures she proveth herself to be." The passage from the letter is also in *Letters*, p. 6.

52. In the sixteenth century, replacing consanguinity with a sort of amatory *alliance* of friendship had, in carnival style, begun to affect many levels of society. And, in conjunction with an amatory neo-Platonic view of the Christian exhortation to love (1 John 4:11–12), it seemed to encourage spiritual libertinism. Jourda, in his edition of Rabelais, suggests that the social structure of Rabelais' Island of Ennasin (bk. 4, chap. 9) parodies such spiritual *alliances* as that between Marot and Anne d'Alençon; the Rabelaisian "Island of Alliances" is associable with the ideas of the one-time Franciscan Tourangeau de Tours. (See Telle, "Île," p. 166, and Telle, *L'Oeuvre de Marguerite*, pp. 299–312.)

53. Chaucer, *Boece*, Bk. 2, Prosa 3, ll. 30–31, in *Riverside Chaucer*, p. 410.

54. Elizabeth I, *Englishings*, p. 54 (= Boethius, *Consolation* 3.6). The Latin reads: *Omne hominum genus in terris simili surgit ab ortu. Unus enim rerum pater est.* When the great Alfred, King of the West Saxons, translated these famous lines in the ninth century, he changed "the Father of all things" into "the father and mother of the race." Friedman comments: "Seemingly Alfred preferred the concept of a concrete biblical blood cousinry to the abstract mystical brotherhood of men, sons of a spiritual father, which his text, influenced by late Stoicism, intended" ("'When Adam Delved,'" pp. 220–21).

55. D. Cox, "The Lord's Prayer," in Farr, ed., *Selected Poetry* 2, 503.

56. For the proverb and its corollaries in England and elsewhere, see Friedman, "'When Adam Delved.'"

57. The inability of King Edward, Elizabeth's half-brother, to disinherit his two half-sisters Princess Mary and Princess Elizabeth, hence to devise the succession to Jane Grey in 1552, would seem to prove this rule.

58. Bale, "Conclusion," esp. Folio 42v–46r.

59. With this double meaning, compare Latin *sacer.* Cf. Freud, *Totem,* p. 18.

60. *Measure for Measure,* 3.1.138–39. John Bale, in his edition of *A Godly Medytacyon,* translates the name Elizabeth as meaning *Dei mei requies,* or "the rest of my God." Bale, "Epistle Dedicatory," Folio 9r.

61. *Sponsa Christi* is a technical term used as early as Tertullian. See Jerome's theory of the virgin as the bride of Christ and of "spiritual matrimony" (letter 107, in *Epistulae* 55:298; cf. Dumm, *Virginity,* p. 74). For a modern version of the theory, see Pius XII, "Sponsa Christi." For an anthropological view of the institution of women marrying gods in Christianity and in other cultures, see Westermarck, *History of Human Marriage* 1:403–6, and M. E. Harding, *Woman's Mysteries.*

62. Millin, *Antiquités nationales,* 3.28.6, on an inscription at Ecouis; my translation.

63. Millin, *Antiquités nationales* 3.28.6. Luther retells such stories of incest in his *Tischreden.* See Montaiglon's note to the thirtieth tale in his edition of the *Heptaméron,* by Marguerite of Navarre (4, 281–83), and Saintsbury's note in his edition (3, 214–16).

64. Millin, *Antiquités nationales* 3.28.6; my translation.

65. *Pericles, Prince of Tyre,* 1.1.65–72. The vital solution in *Pericles* requires a resurrection—wife and daughter, believed dead, are reborn from their living deaths in a religious institution and a brothel. Dramatically, the solution to the riddle of Antiochus involves assigning to Pericles and Marina the roles of Antiochus and Antiochus' daughter; beyond the resurrection of the two women, the plot enacts a final rebirth, as Pericles calls it, of the father, Pericles, from the daughter, Marina ("Thou that beget'st him that did thee beget"; 5.1.197). In this atonement, Pericles foreshadows kinship relations in all the other romances.

66. Taylor, *Shakespeare's Darker Purpose,* p. 69, compares the riddle in *Pericles* with similar riddles in two of Shakespeare's sources, Gower's *Confessio Amantis,* and Twine's 1594 translation of Apollonius of Tyre, *The Patterne of Painefull Adventures;* however, *Confessio Amantis* and *The Patterne of Painefull Adventures* do not play up the spiritual nunnish-or-monkish quality of simultaneous parenthood, spousehood, and childhood in the same way that *Pericles* does. For Medrano: see Rank, *Inzest-Motiv,* pp. 334–35.

67. Rabelais, *Gargantua* and *Pantagruel,* bk. 4, chap. 9, p. 468.

68. Elizabeth, "Glass," Folio 26v.

69. Elizabeth, "Glass," Folio 43r–v. For the poet-lover in this fourfold kinship role, see also such passages as: "O my father, brother, child, and spouse" (Fol. 21r); "O what a sweet rest it is, of the mother, and the son together" (Fol. 26r); and "Now then that we are brother and sister together, I care but little for all other men" (Fol. 29r).

70. Elizabeth, "Glass," Folio 13r, 19r–v.

71. Elizabeth, letter to Catharine Parr, December 31, 1544, Folio 3r.

72. The sinful soul is a fourfold traitor: the *child* who leaves the home of the father (as in Luke 15); the *parent* who fails to watch out for the child (3 Kings); the *sibling* who betrays the brother (Numbers 12); and the *spouse* who is adulterous.

73. Bale, "Conclusion," Folios 39r, 40r.

74. Salminen dismisses the fourfold kinship in Marguerite's poem as a strange mysticism: "Marguerite se perd dans un mysticisme étrange: elle décrit l'union de l'âme avec le Sauveur à l'aide des métaphores se rapportant aux relations de parenté. Elle se nomme tour à tour la soeur, la mère, l'épouse et la fille de Dieu" (Salminen, *Miroir,* p. 81).

75. Bale, "Conclusion," Folio 39r. A late instance of the Elizabethan fourfold kin topos occurs in Aemilia Lanyer's reference in *Salve Deus Rex Judaeorum* (1611) to a new Cynthia (i.e., Elizabeth Tudor). Of her it is said that Jesus is "Her Sonne, her Husband, Father . . ." (Travitsky, *Paradise of Women,* pp. 99, 101); see also Lewalski, "Of God and Good Women."

76. For example: the *Marguerites, Dialogue, Triomphe de l'Agneau,* the *Coche, Prisons,* and perhaps the *Comedies.* Such works as *The Mirror of the Sinful Soul* and *Discord Being in Man by the Contrariness of the Spirit and the Flesh and Peace [Being in Man] by the Spiritual Life,* both written in 1531, are in some measure commentary on the doctrine of liberty in Saint Paul's *Letter to the Romans.*

77. Marguerite, *Heptameron,* trans. Saintsbury, 3, 192. (All references below are to the Saintsbury translation.)

78. Marguerite, *Heptameron* 3, 200.

79. Some literary historians have sought to use the story that Marguerite of Navarre was raped in the 1520s to elucidate her tales about rape, including those where religious Brothers rape lay women and Sisters (e.g., tale 72), but they generally ignore the definition of spiritual incest as "sexual relations involving any Brother or Sister" and so they tend to overlook the relationship between the physical incest described in the *Heptaméron* and the transformation of physical incest into spiritual incest depicted in *Le Miroir de l'âme pécheresse.* On the presumed rape of Marguerite by Admiral Bonnivet, see Brântome, *Dames galantes,* pp. 422–23; cited by Cholakian, *Rape,* p. 9.

80. Marguerite, *Heptameron* 3, 201.

81. Marguerite, *Heptameron* 3, 195.

82. *Love's Labour's Lost* 1.1.13; cf. David ed., 50–51, 206–7. Shakespeare very probably knew the work of Marguerite; her influence is suggested by his depiction of academic celibacy in *Love's Labour's Lost,* whose Princess is modeled on Marguerite of Valois, her grandniece. See esp. Lefranc, *Découverte* and *Sous le Masque de Shakespeare,* and the endorsement of Lefranc's position in David's introduction, p. 39.

83. Here too there are biographical dimensions to Marguerite's concern with incest, as to Elizabeth's: for example, Marguerite's love for her brother, King Francis of France, is the subject of her greatest poetry; Saintsbury remarks that "it has been asserted that improper relations existed between the brother and the sister," though the historical evidence is not conclusive on the side of either chastity or incest (*Heptameron,* Introduction, p.56).

84. See Ames, "Introduction" to Elizabeth, *Mirror,* p. 42.

85. Lefranc, *Idées religieuses,* p. 15; Ames, p. 42.

86. Calvin, *Contre la secte phantastique et furieuse des Libertins qui se nomment spirituels* (1545), in *Opera omnia* 7, 145–248. For the antinomian beliefs of the Libertines, see Walker, *Calvin,* pp. 293–94. Calvin's attack offended Marguerite, and he wrote an ambiguously apologetic letter to her on April 28, 1545 (*Opera omnia* 12, 65). On Marguerite and the Libertines, see also G. Schneider, *Libertin,* esp. pp. 81–84, and Dagens, "Miroir."

87. OED 2:1053b; cf. Matt. 19:21.

88. Vautrollier, *Commentarie,* p. 85, marg. (ed. 1577). Luther resented the canonical

imposition of celibacy laws, which, given the conflation of intent and act that characterizes his notion of faith, were impossible for almost all human beings—including such monks and priests as himself—to fulfill.

89. OED 7:684c.

90. Freud, *Totem,* p. 7.

91. "Anyone who has transgressed one of these prohibitions," writes Freud, "acquires the characteristic of being prohibited" (*Totem,* p. 22; cf. p. 35). Cf. Freud, "Taboo on Virginity" and "Obsessive Acts and Religious Practices."

92. Cf. Luke 14:26.

93. Marguerite, *Miroir,* 11. 267–68. A translation of Matthew 12:50: "Quicumque enim fecerit voluntatem Patris mei . . . ipse meus frater, et soror et mater est." Elizabeth's translation of this reads: "Those that shall do the will of My father, they are my brethren and my mother" ("Glass," Folio 17r).

94. For a monk or nun it is official Church doctrine and law that concubinage or marriage to anyone is incest. See chap. 1, note 69.

95. Murray, introduction to Melville, *Pierre,* p. 55.

96. Freud, *Civilization,* pp. 48–49.

97. "Il Canto di Frate Sole," in St. Francis, *Scritti,* p. 168; trans. as "The Canticle of Brother Sun," in *St. Francis, Omnibus of Sources,* ed. Habig, 130–31.

98. Bonaventure, *The Soul's Journey into God* (thirteenth century), p. 93. Franciscans generally regarded Bonaventure, the "Seraphic [or Franciscan] Doctor," as the principal theological and philosophical spokesperson for their order.

99. See also Harney, *Brother and Sister Saints,* for a useful discussion of sibling saints mentioned in that work.

100. *Ency. Brit.* (11th ed.) 18:691. See also Athanasius, *Vita Antonii.*

101. *Ency. Brit.* (11th ed.) 18:691.

102. Three days later, Scholastica died; in the course of time, Benedict joined her in a single grave. See St. Gregory the Great, *Vita S. Benedicti.*

103. Hartmann von Aue, *Gregorius.*

104. St. Leander, *Regula,* 5, 331–32, *Patrologiae* (Latina) 72:874–94.

105. St. Leander, *Regula,* quoted from Montalembert, *Monks* 2:188.

106. St. Damasus, *Epigrammata,* trans. Joseph F. M. Marique, cited in Harney, *Brother and Sister Saints,* p. 20.

107. Lioba the Sister, seeking to serve her cousin Saint Boniface, wrote him thus: "God grant, unworthy as I am, that I might have the honor of having you for my brother." She closed the letter with the following suggestive verse: "May the Almighty judge, who made the earth / And glorious in His kingdom reigns, / Preserve your chaste fire warm as at its birth, / Till time for you shall lose its rights and pains" (Harney, *Brother and Sister Saints,* p. 93; see also Willibald of Mainz, *Vita S. Bonifacii*). Saint Boniface asked that at her death Saint Lioba be buried in his grave, but the monks of Fulda did not carry out his request.

108. Cited in Harney, *Brother and Sister Saints,* p. 114.

109. Bernard's *Sermones in Cantica Canticones* demonstrate "la dernière libération de l'âme" [the ultimate liberation of the soul] of which Paul preaches (Viller, *Dictionnaire de spiritualité* 1:1475; cf. Bugge, *Virginitas,* p. 90).

110. Curtius, *European Literature,* p. 122.

111. Victoria Lincoln, *Teresa,* pp. xxv, 10. Cf. Egido, "Historical Setting."

112. Teresa, *Life,* IV, in *Complete Works.*

113. Teresa, *Way of Perfection,* IV, in *Complete Works* 2, 17; cf. VIII, 2, 39.

114. *Complete Works* 2, 415; and (from *Way of Perfection,* XXXVII) 2, 161.

115. John of the Cross, *Precautions* V, in his *Collected Works,* trans. Kavanaugh and Rodriguez, p. 656.

116. Victoria Lincoln, *Teresa*, p. 24; cf. Teresa, *Life*, II, in *Complete Works*, and *Book of her Life*, 2.6, and *Rules for a Brotherhood*. To the list of saintly siblings mentioned in this chapter might be added Heloise and Abelard. They lived together, first in spiritual incest of a physical kind (he a Brother, she a laywoman), then as secret husband and wife. After Abelard was castrated at the command of Heloise's uncle, she became a Sister and they lived as "brother and sister"—to quote the letters between them (Leclerq, *Monks and Love*, esp. p. 119).

117. Knowles and Hadcock, *Medieval Religious Houses*, esp. pp. 104–5, 194–95, 202.

118. The oldest manuscript of Marguerite Porete's powerful *Mirror of Simple Souls* (the Chantilly manuscript; see Dagens, "Miroir," p. 288, and Hauser, *Études sur la réforme française*) was found at Fontevrault. Porete's Antinomian influence on Marguerite of Navarre and the Siblings, or Brethren, of the Free Spirit will be discussed in a later section.

119. Marguerite of Navarre, *Heptameron*, 3rd day, tale 30, 3, 202. Robert d'Arbissel, the founder of the abbey of Fontevrault, was himself accused of sleeping in the same bed with nuns; see Bayle, *Dictionnaire* 6, 508–10.

120. Ency. Brit. (11th ed.) 18:691.

121. On Ochino and his stay in England from 1547 to 1553, see Benrath, *Ochino*, pp. 172–99. On Elizabeth's 1547 translation, see Craster, "Unknown Translation," p. 723. The autograph manuscript is at the Bodleian Library (Swaim, "New Year's Gift," pp. 261–65).

122. For Bale's defense of his marriage, see *Scriptorum illus. Brit.*, p. 702; for Rome's condemnation, see Pits, *Relationum Historicum*, pp. 53–59.

123. Bale, *Image of Both Churches*, p. 537.

124. "Rabelais's [bastard] children were granted the unusual privilege of an official legitimization by the Pope Himself" (Screech, *Rabelaisian Marriage*, pp. 19–20; cf. Lesellier, "Deux Enfants naturels").

125. For Luther's critique of religious celibacy, see his "Exhortation to All Clergy," esp. pp. 40–52, and "Exhortation to the Knights."

126. "Love needs no laws," said Luther, sweeping away "those stupid barriers due to spiritual fatherhood, motherhood, brotherhood, sisterhood, and childhood"—or so Brain, *Friends and Lovers*, p. 94, has it. (Cf. Luther, "Persons . . . Forbidden to Marry," in *Works* 45, 8, and "Estate of Marriage," *Works* 45, 24.) Luther did not sweep away all barriers to marriage, however. He stressed the distinction between figure and letter, or spirit and body, and thus redefined the incest taboo in terms of a literal, or corporeal, principle.

127. *Ancrene Riwle* 106/21; as quoted, from the Corpus Christi College (Cambridge) manuscript, by Kurath and Reidy, eds., *Middle English Dictionary*, s.v. "Incest," from the *Dictionary's* photostatic copy of the manuscript in Ann Arbor, Michigan.

128. Rolle, *Form*, 413.

129. [Lacy], *Ten Commandments*, 4068.

130. 2, 4068–71. See also W.H. Black, ed., *The Life and Martyrdom of Thomas Beket*, line 757.

131. Chaucer, *Parson's Tale*, l. 908, in *Works*, p. 258.

132. On the Benedictine orders and the idealist policies that More promulgates in *Utopia*, see Chambers, *More*, p. 136; More, *Utopia*, pp. 281–82. Although More generally admired the Catholic orders and their doctrines (Hexter, *More's Utopia*, pp. 85–90), he was not a monk or friar but a husband and legitimate father.

133. Ency. Brit. (11th ed.) 5:603.

134. More, *Tindale*, pp. 48–49. Cf. Luther's references to the "incestuous celibacy" of the papists in *Tischreden* 2, 138–391.

135. Matt. 23:8. Christianity proposes a universalist doctrine that both incorporates and transcends ordinary kinship with an extraordinary unifamilial kinship. Cf. Gal. 3:26–28: "For in Christ Jesus you are all sons of God . . . There is neither Jew nor Greek . . . for ye are

all one in Christ Jesus." Church Fathers who urge Christians to obey to follow out the implications of Matt. 23:8 in such a way as to call all men their "brothers" and to call *no one* by the name "father" include: Ignatius of Antioch, "To the Ephesians," 10.3; Clement of Alexandria, *Stromata,* 7.14.5; Justinian, *Dialogue with Tryphon,* 96; and Tertullian, *Apologeticus,* 39.8–9.

136. 1 Cor. 5:1.

137. Adela Collins, "Excommunication," p. 253.

138. 1 Cor. 10:23; and Lev. 18:8. For this position on Corinthians, see Allo, "Saint Paul," p. 121; and Conzelmann, *1 Corinthians,* p. 97.

139. Rabelais' chapter entitled "Pantagruel reaches the Island of Ennasin [cf. "Essene"], and of the strange Relationships there" is bk. 4, chap. 9.

140. For a general account of the Brethren see Cohn, "Cult." In Italy St. Bernardine of Siena was affected by the Brethren. (See Bernardine of Siena, *Opera omnia,* 1:34, 536; 3:109; 4:544; 6:248). In Germany the Brethren influenced the philosophical Meister Eckhardt and the polygamist Anabaptists of Münster. (Telle, "Île," p. 169; and see *Ency. Brit.* (11th ed.) 8:886). The sixteenth-century Belgian David Joris, a prominent member of a local sect influenced by the Free Spirit, had an effect on Bosch, whose painting, "The Garden of Earthly Delights," is said to depict the Free Spirit's "incestuous orgies" (Fraenger, *Millennium,* p. 42).

141. This was the motto of the Libertine Brethren (Telle, *Marguerite,* p. 297).

142. 2 Cor. 3:17.

143. Women were especially prominent as the theorists of the Brethren of the Free Spirit. Thus the poet Bloemardinne (or Hadewijch) of Brussels wrote much of the spirit of liberty and supposedly impious sexual love. The "Homines Intelligentes" of Brussels were a local sect of the Free Spirit; their leaders, Giles Cantor and William Hilderniss, were condemned by Pierre Ailly, bishop of Cambrai, in 1411. See Pomerius, *De origine monasterii Viridivalli,* p. 286, and McDonnell, *Beguines and Beghards,* p. 502.

144. Similarities between the literary work of Marguerite of Navarre and Marguerite Porete include verbal parallels, as in the use of the term "Distant Close" or *loingpres* (see Dagens, "Le 'Miroir des simples ames' et Marguerite de Navarre") and historical connections (see Bédier, "La tradition manuscrite," Eekhoud, *Les Libertins d'Anvers,* and Frederichs, "Luthérien français").

145. The relation between Lutheran doctrine and Spiritual Liberty as they converge in the court of Navarre is the subject of Frederichs, "Luthérien français," and Eekhoud, *Libertins d'Anvers.*

146. See Perrens, *Libertins.*

147. See Doiron, "Middle English Translation."

148. Champneys, *Harvest.*

149. Gardiner (*Great Civil War* 3, 380) writes of the Levelers that "they have given themselves a new name, Viz. Levelers, for they intend to set all things straight and raise a parody and community in the kingdom." Cromwell attacked the Levelers in his speech to parliament in September 1654, quoted by Carlyle in his *Cromwell's Letters and Speeches,* Speech number 2: "Did not that Leveling principle tend to the reducing of all to an inequality." On the Ranters, see below.

150. Parrington, *The Romantic Revolution,* pp. 335–36, suggests that the basis for American Perfectionism lies not so much in French Romanticism or German Idealism as in a medieval utopian past that extends to the religious utopianism of the 1650s and to Roger Williams and beyond, and that Noyes and his Perfectionists were at one with the Diggers and Levelers of Commonwealth times.

151. The sociologist Lewis Yablonsky describes a ritual (now called "the funeral of the

hippie movement'') involving San Francisco hippies. Their ''explicit idea was to bury the hippie image, as they put it, produced by the mass media, and to signal the birth of *The Free Man*'' (Yablonsky, *Hippie Trip,* p. 290). The Love Children wanted to be Brethren of the Free Spirit.

152. In the 1520s Lollards and other groups tracing their origin to Wycliffe came together with Lutherans and thus became a sect to be contended with (*Ency. Brit.* [11th ed.] 8:529).

153. *Ency. Brit.* (11th ed.) 1:905, referring to an event of 12 April 1549.

154. Athanasian Christian Church doctrine claims that the Virgin Mary had no biological part in the making of her baby. In the first half of the ninth century, Theosterikos said the following about the iconoclast emperor Constantine V: ''Taking in his hand a purse full of gold and showing it to all he asked, What is it worth? They replied that it had great value. He then emptied out the gold and asked, What is it worth now? They said, Nothing. So, said he, Mary (for the atheist would not call her Theotokos), while she carried Christ within was to be honoured, but after she was delivered she differed in no way from other women'' (*Vitae Nicetae,* in *Acta Sanct.,* Ap. I, app. 23; cited in Edward Martin, *History,* p. 62).

155. Telle, ''Île,'' p. 169.

156. Regarding the view that Ann Askew may have been one of Catherine Parr's ladies in waiting, see *Writings of Ed[ward] VI,* p. 238.

157. Cited in McDonnell, *Beguines and Beghards,* p. 497.

158. See Hartmann's interrogation at Erfurt by Walter Kerling, recounted in von Dollinger, *Sektengeschichte* 2, 386.

159. Leff, *Heresy* 1, 378–79.

160. Leff, *Heresy* 1, 378–79.

161. ''Amye, amez et faites ce que tu vouldrez'' (Marguerite Porete, *Miroir,* ed. Guarnieri, Folio 26v).

162. Porete, St. John's manuscript, Folio 15v–29r.

163. Porete, St. John's manuscript, Folio 5. The Council of Vienne of 1311–12 condemned this statement in its French version, *congé des vertues* being treated by the council as *licentiat a se virtutes.*

164. Porete, St. John's manuscript, Folio 10r. This statement was also condemned at Vienne.

165. In 1650 in England, Abiezer Coppe, a member of the Ranter sect, promulgated the Brethren's view that ''God dwelt inside them, as an inner light whose authority was above all laws. . . . Sin was thus made to disappear. The consequence was, for some Ranters, sexual license'' (Carey, Foreword, in Nigel Smith, ed., *Ranter Writings,* p. 7). Coppe described the state beyond ''good'' and ''evil'' in terms of ''the mother Eternity, Almightyness, who is universal love'' and to whose child dress and undress, incest and chastity, are alike—he knows no evil (quoted in Cohn, ''Cult of the Free Spirit,'' p. 68).

166. Quoted in Cohn, ''Cult of the Free Spirit, p. 59.

167. Quoted in Cohn, ''Cult of the Free Spirit,'' p. 56.

168. Elizabeth, ''Glass,'' Folio 62v.

169. ''Fay ce que voudras'' (*Gargantua,* ii, chap. 57). The liberty of Rabelais' abbey is often interpreted as mere ''Epicurean'' (i.e., gluttonous) intemperance or desire for a heaven on earth (see, e.g., Kennard, *Friar in Fiction,* p. 58), but it also has a serious libertine aspect. Bakhtin, *Rabelais,* p. 412, compares Thélème to the medieval parody *The Rules of Blessed Libertine.*

170. ''Telles âmes roynes, filles de roy, seurs de roy, et espouses de roy'' (Marguerite Porete, *Miroir,* ed. Guarnieri, Folio 26v).

171. Elizabeth, ''Glass,'' Folio 43r.

172. See the A.D. 1309 charges against Marguerite Porete listed in *Grands Chroniques* 5, 188.

173. English devotional works similarly depict Jesus as a female lover or as both a male lover and a female parent taken together. Among such works are the Middle English translation of Aelred of Rievaulx's *De vita eremitica* (p. 329), Juliana of Norwich's *Revelationes* (pp.58–60), and *Ureisun of Ure Louerde* (Thompson, p. 2). A basic biblical text here, of interest to Marguerite of Navarre and Queen Elizabeth, is St. Paul's Letter to the Galatians. Paul writes that we become "one [*heis*] in Christ, neither man nor woman" (Gal 3:28)—though "one" [*heis*] is here masculine (as in Eph. 2:15, 4:13), not neuter (as in Eph. 2:14). Franciscan theologians often portray maleness and femaleness joined together, not in what we normally assume to be the closest union (marriage), but in roles that would normally be impediments to that union (namely, mother and son, or brother and sister). (Cousins, *Bonaventure*, p. 20.)

174. Porete, St. John's manuscript Folio 13r–13v. Elizabeth, in her "Glass," asks, "For what thing is a man (as for his own strength) before he hath received the gift of faith" (Folio 5r). Bale, in his "Epistle Dedicatory," refers to "the barren doctrine and the works without faith" (Folio 7v). Marguerite of Navarre, in Queen Elizabeth's translation, writes of the "thing a man cannot understand unless he hath a true faith" (Elizabeth, "Glass," Folio 14r); "If we have Him through faith then have we a greater treasure than any man can tell" (Elizabeth, "Glass," Folio 62v).

175. One example of the mockery of "erotic communism" is Jean de Meun's contribution (ca. 1275) to Guillaume de Lorris' *The Romance of the Rose,* esp: "Toutes pour touz e touz pour toutes." Here "the goddess Natura has become the servant of rank promiscuity" (cited in Curtius, *European Literature,* p. 6). Bakhtin discusses *The Rules of Blessed Libertine,* a parody of monastic laws similarly built upon sanctifying that which is forbidden, in *Rabelais,* p. 412.

176. Von Dollinger, *Sektengeschichte* 2, p. 664. See also Schneider, *Libertin,* p. 60.

177. In his *Articles of Visitation,* Melanchthon claims, for example, that Luther's doctrine of the "freedom of a Christian" was interpreted by some Protestant reformers as a charter for moral laxity; Melanchthon argues for the preaching of the Ten Commandments as a guide to the good works that are to follow true faith (Franklin Sherman, Introduction to Luther, "Against the Antinomians"). For a study of the relationship between the Brethren of the Free Spirit and French Protestantism, see Frederichs, "Lutherien français," and for a general discussion of the role of the doctrines of the free spirit in the Reformation, see Guarnieri, "Il movimento del libero spirito dalle origini al secolo xvi" and "Appendici" (in Guarnieri, *Il movimento del libero spirito,* pp. 114–49, 336–58). On the *Miroir* as Protestant in the reformist tradition of Guillaume Briçonnet and Simon du Bois, see Salminen in her edition of the text, pp. 40ff., 62ff., as well as Lefranc, *Les Idées religieuses;* on its Protestant mystical aspects, see Sckommodau, *Die religiösen Dichtungen.*

178. See Cooper, "Incest Prohibitions," p. 2.

179. Elyot, *Image of Governaunce,* 34; Chapman, *Iliad,* 16.501.

180. Wilkinson, *Supplication,* p. 34.

181. Wilkinson, *Supplication,* p. 19.

182. Rogers, *Displaying,* sig. 15r.

183. *The Family of Love,* a comic play probably written by Thomas Middleton and performed by the Children of the Revels, contains a trial scene in which the Family's sexual freedom is institutionalized in law. See Shepherd's introduction to his edition of the *Family,* p. iii; Cherry, *Most Unvaluedst Purchase;* and Halley, "English Family of Love."

184. *Monk* means "alone" (from Greek *monachos*). However, not all monks were *anchorites* living in solitude. Some were *coenobites* (from *koinos,* "common," and *bios,* "life").

Coenobites lived in communities with one or another kind of *arché,* or rule. The *archon* might be a traditional *cenobiarch* in a universally fraternal and/or sororal community; here all members of the community, *including* its ruler, were "Siblings," and called one another Brother and/or Sister. But, on the other hand, the *archon* might rule in a community of a strictly patriarchal or matriarchal sort; here one calls the Superior "Father" or "Mother," and all members of the community are Brothers and/or Sisters, *except* the ruler, who is everyone else's "Parent." There was a parallel struggle in the Elizabethan and Jacobean period as to whether the "monarch"—meaning "alone rule" or "one ruling" (from *monarché*)—is essentially "parentarchal" or "sibling."

185. Matt. 23:19: "Call no man father." *Epistolae ad virgines.* On the pseudo-Clementian *Ad virgines,* see Bugge, *Virginitatis,* p. 72.

186. Bugge, *Virginitatis,* p. 73.

187. Jerome, *Epistulae;* cited in Dumm, *Virginity,* pp. 119, 121.

188. John Chrysostom, *De virginitate;* cited in Dumm, *Virginity,* p. 121.

189. Aristotle, *Politics,* 1.2.5 is generally the Christian ideologists' *locus classicus,* but it is a mistake to assume that this view of the ruler is universal. Indeed, it is generally absent in the nonuniversalist—and to Christendom of the period, politically and ideologically threatening—doctrine of Islam, perhaps, as Lewis remarks in *The Political Language of Islam* (p. 17), on account of its Christian connotation. Only in Turkish does a term meaning "father," *ata,* acquire a political connotation. The image of the ruler as mother, moreover, seems to have been equally abhorrent. If, however, the Moslem state has no parents, it certainly has children, and terms denoting sons and brothers are in common political use. These may be literal, as for example in indicating membership of a tribe or dynasty, or figurative, indicating a relationship to the state. The term "brother," with its various equivalents, is in common use for members of the same group, by allegiance as well as by kinship. And in Turkish usage "elder brother," *aga,* is a common term of respect for holders of authority (see H. Bowen, "Agha," in *Ency. Islam*). The best-known example is the *abna,* "sons," of the 'Abbasid dynasty, an elite group of slaves and freedmen, both soldiers and civilians, who served the state and thus freed it from dependence on Arab tribal support (see K.V. Zettersteen, "Abna," in *Ency. Islam*).

190. Nashe, Introduction to *Menaphon.*

191. For the Elizabethan conjunction of "unrestrained violence and sexual licentiousness," see Saffady, "Fears of Sexual License."

192. Gardiner, *Civil War,* p. 279.

193. Kelly, *Thorn,* p. 3.

194. For the executions: Mutschmann and Wentersdorf, *Shakespeare and Catholicism,* p. 43. For the rebellion: Kaula, *Shakespeare and the Archpriest Controversy,* esp. p. 71.

195. Locke, *Two Treatises,* p. 6.

196. Norman Brown, *Love's Body,* p. 4.

197. Quoted in Staves, *Players' Scepters,* p. 114.

198. "Français, encore un effort si vous voulez être republicains," in *Philosophie,* pp. 179–245; see esp. pp. 221–22.

199. Bale, "Conclusion," Folio 40v.

200. Bale, "Epistle Dedicatory," Folios 3r, 4v, 6v–7r. On this sort of attempt in the sixteenth century to replace the role of blood in the kinship structure with virtue—or "beauty"—Hermann Melville may provide the best commentaries in his *Pierre,* with its ambiguously illegitimate heroine Isabella/Elizabeth. "A beautiful woman is born Queen of men and women both," writes Melville, "as Mary Stuart was born Queen of Scots, whether men or women. All mankind are her Scots; her leal clan are numbered by the nations." According to the narrator, "a plain-faced Queen of Spain [perhaps Queen Isabella I, called

la Católica] dwells not in half the glory of a beautiful milliner's daughter," and people deserve to be excoriated for worshipping "Mary Queen of Heaven" while, at the same time, "for generations refus[ing] cap and knee to many angel Maries, rightful Queens of France"—including many "immortal flowers of the House of Valois" (Melville, *Pierre,* ed. Murray, pp. 46, 47, 56). Among the Marys that Melville may have had in mind: Mary Stuart, daughter of King Francis, who was a Valois/Angoulême; Mary, Queen of France, née Tudor, who was daughter to Henry VII and Elizabeth, sister to Henry VIII, and Queen to Louis XII of France, who was Valois/Capetian; and Mary Tudor, called "Bloody Mary," daughter to Henry VIII and Catherine of Aragon. Melville was apparently an admirer of Robert Melville (1527–1621), who was a representative of Mary Queen of Scots at the English court; among those who begged Elizabeth for Mary's life, Robert Melville later accompanied James VI to England (1603).

201. Bale, "Epistle Dedicatory," Folio 7r, 40r; cf. 40v. "Of the most excellent kind of nobility is he sure (most virtuous and learned Lady) which truly believeth and seeketh to do the will of the eternal father, for thereby is he brought forward and promoted to that heavenly kindred, Io. 1. By that means becommeth he the dear brother, sister, and mother of Christ, Mat. 1.2, a citizen of heaven with the apostles and prophets, Ephe. 2. yea the child of adoption and heir together with Christ in the heavenly inheritance, Rom. 8. No such children left Socrates behind him, neither yet Demosthenes, Plato, nor Cicero with all their pleasant eloquence and wisdom. No such heritage could great Alexander the Macedonian leave his posterity, neither yet Charles, Arthur, nor David" (Elizabeth, "Glass," Folio 6v–7r).

202. Luke 14:26.

203. Bale, "Epistle Dedicatory," Folio 8v.

204. See Bradbrook, "Virtue is True Nobility."

205. In this play Helena, who is of relatively low birth but of virtuous character, and Bertram, who is of relatively high birth but of vicious character, are compared to each other in such way that we might wonder whether he is so far above her station as to make his marriage to her as unbearably painful to him as incestuous marriage would be. Helena is Bertram's sister-by-adoption, whom Bertram's mother the Countess would call "daughter" by transforming her into a daughter-in-law (AWW 1.3.141–156). It requires the King of France to emphasize that nobility resides in the soul, not in blood.

206. Tacitus, *Annals* 12.42. It is worth noting that Agrippina's fourfold kinship arrangement imitates to some extent that of Roman women in general, who were regarded legally "as daughters of their husbands and sisters of their children" (Vico, *New Science,* para. 507). The arrangement was also like that of the old Roman "temple of Ceras, Liber, and Libera, the mother or the spouse, the son, and the daughter, of the Buddhick or Bacchic god" (see Del Mar, *Ancient Britain,* p. 101, citing Livy III.lvi.8).

207. On this politically sensitive form of adoption (the technical term is *arrogatio*), see Hunter, *Exposition of Roman Law,* pp. 202–11. Saint Paul, among others, developed the Roman practice of adrogation of sons by their father into a powerful ideology of adoption by Christ, an adoption whose end was not so much "liberation" in the Roman juridical sense of the word as a homogenizing or leveling of all human beings under God the universal Father. The disappearance of the vestal virgins in the fourth century AD (Emperor Constantine), which was a critical moment in the history of Rome, corresponded to the formal Roman institutionalization of Christianity as the official state religion in the same century. That is why Christian nuns have many of the rituals, powers, and obligations of the vestal virgins. In the Roman ceremony by which a novice becomes a vestal, for example, the novice's hair is shorn and the *pontifex* or head priest calls the novice *Amata* [loved one] and she promises to be chaste (Aulus Gellius 1.12).

208. Butler, introd. to Racine, *Britannicus,* p. 169.

209. Geoffrey of Monmouth, *Historia* 3, 1. On Aeneas, the Trojan War, and the Salic Law, see my "Bale and British Nationhood," in *Elizabeth's Glass.*

210. *Domina* had been an important legal term in Roman times. It meant not only *regina,* or *rex,* but also something like "female head of the household" (*Oxford Latin Dictionary,* p. 570; on the title *domina,* see esp. Onslow, *Empress Maud*). Eventually *domina* came to mean "the [mother] superior of a nunnery," though it always retained some connotation of property ownership, as in Kersey's 1708 definition (in *Dictionarium*) of *domina* as "a Title formerly given to those honorable Women that held a Barony in their own right of inheritance."

211. "All this [throughout Geoffrey's work, about women ruling] is substantially unparalleled in early history, British or otherwise," writes Tatlock. "In Welsh history there is no hint of it, and in France, as no reader of [Shakespeare's] *Henry V* need be told, the Salic law forbad any woman to reign" (*Legendary History,* p. 287).

212. According to tradition, the Salic Law forbad women to inherit in general. Originating with the Salian Franks, it was supposedly invoked in 1316 and 1322 to exclude the daughters of the Capetian Louis X, King of Navarre (1305–16) and King of France (1314–16) and of the Capetian Philip V, King of France (1316–22) from the succession to the throne. Philip VI, 1328–50, who assumed the throne in 1328 after the Capetian Charles IV, King of France (1322–28), died, was the first of the House of the Valois, which ruled France until 1589, when Henry III died and the House of Bourbon assumed power. Marguerite of Navarre was also known as Marguerite de Valois.

213. Cf. Jardine, "Wealth, Inheritance, and the Spectre of Strong Women," "*Still Harping,*" chapter 3.

214. In 1533, Catherine de Médicis (1519–1589) married Henry, second son of Francis I of France, whose sister was Marguerite of Navarre. Catherine became queen of France in 1547, when Elizabeth wrote the "Glass." She had four sons, three of whom became kings of France. In the tradition of queen mothers ("jointresses"), she ruled as regent with great power during the minority of Charles IX (1560–63) and his later reign (1560–74), and was also influential over Henry III (1574–89). Though she planned the Massacre of St. Bartholomew (1572), she did not always side with the Catholics against the (Protestant) Huguenots.

215. His catalogs have been ignored even by Bale scholars: Fairfield, *John Bale,* does not even mention Bale's edition of Elizabeth's *Godly Medytacyon;* cf. McCusker, *John Bale.* Protestant reformers other than Bale, appeasing in later years the Protestant queen Elizabeth, "dwelt more on Elizabeth's exceptional qualities than on the general worth of women" (Travitsky, *Paradise of Women,* p. 92, citing Stenton, *English Women,* pp. 126–27); cf. Levin, "Explication," Jordan, "Feminism and the Humanists," and Knox's distrust of all women in his 1558 *First Blast of the Trumpet.* Lewalski, "Of God and Good Women," p. 213, draws attention to Aemilia Lanyer's contribution to the so-called *querelle des femmes* in Lanyer's "Epistle to the Vertuous Reader," esp. sig. f3v. On catalogs of good women that may have influenced Lanyer, such as Boccaccio's *De claris mulieribus,* Chaucer's *Legend of Good Women,* and Christine de Pisan's *Book of the City of Ladies,* see Lewalski, "Of God and Good Women," note 17. See also Kelso, *Lady of the Renaissance.*

216. See Bale's remarks in Askew, *The first examinacyon,* Folio 11r; and Beilen, "Anne Askew's Self-Portrait," p. 85.

217. In a letter of December 6, 1559 written to five Catholic bishops deprived of their sees for refusing to accept the new Church of England, Elizabeth refers to a tradition of British Christianity separate and independent from the Romish tradition. She mentions "the ancient monument of Gildas" (probably that sixth century Romano-British historian's *De Excidio et Conquestu Britanniae*) and such quasi-historical events as the visit to Britain of Joseph of Arimathea. Long before Augustine came to Britain, she advises the Catholic bishops, there was Christianity in Britain.

218. For remarks on these and other women listed by Bale, see my "Glossary of Proper Names in Bale," in *Elizabeth's Glass.*

219. 1.2.9. In the Hamlet legend Gertrude is, like the Virgin Mary and like Agrippina, related fourfold to the ruler, as we will see in chapter 5.

220. "He (she) understood the custom to be rather more fitting for men than for women." Saxo Grammaticus, *Historiae Danicae,* book 9; cited in Tatlock, *Legendary History,* p. 286.

221. Cited in Tatlock, *Legendary History,* p. 288.

222. Blandina's remarkable martyrdom, which took place under the Roman Emperor Marcus Aurelius in A.D. 177, is discussed by Bale in his preface to Askew's *Exam.,* Folio 7v–9r. Askew was the religious and nationalist model for Bale, who in his autobiographical *Vocacyon of John Bale* "casts himself as a Protestant saint" (Fairfield, *John Bale,* p. 141). See Beilin, "Anne Askew's Self-Portrait," pp. 80, 271 (n. 5).

223. Gwendoline and Martia are called "King" by Bale, "Conclusion," Folio 42v and 43v. Cf. OED 5:704.

224. Bale, "Conclusion," Folio 40v.

225. Elizabeth publicly identified with Saint Elizabeth and other Marian figures, including the Elizabeth who was the niece of Saint Anne (of the Immaculate Conception) and the cousin of Saint Mary (of the Virgin Birth). See Strong, *Cult,* p. 125, and Wells, *Spenser's "Faerie Queene" and the Cult of Elizabeth,* p. 18.

226. Aristotle, *Politics,* 1259a.

227. Aquinas, *Politicorum Arist.,* ed. Spiazzi, p. 47f. On this difference see also my *End of Kinship,* esp. chapter 6.

228. See Eichmann, *Kaiserkrönung* 2, 94ff. By the same token, calling a person both father and son, as in the topos of the *sponsa Christi,* while relatively rare in the early centuries of Christianity, became more common by the twelfth century, thanks partly to the growth of Marian cults. "I am your Father, I am your Son," sang Robert Wace, the Anglo-Norman chronicler of the twelfth century. The material is collected by Mayer, "*Mater et filia*"; Hahn, *Bibliothek der Symbole,* p. 176, 81f; and Wels, *Theologische Streifzüge* 1, 33–51, who considers the father-son antiphrasis a kind of Sabellianism. (See Kantorowicz, *King's Two Bodies,* pp. 99–100.) Some critics have noted that Elizabeth Tudor was sometimes seen as the bride of Canticles and even as a *sponsa Christi*—Edmund Spenser, say some, linked "England's Eliza to the Spouse of Christ" (King, *Tudor Royal Iconography,* pp. 195, 207, 261; cf. Norbrook, *Poetry,* p. 84, L.S. Johnson, "Elizabeth, Bride and Queen," pp. 81–83, and R.H. Wells, *Spenser's "Faerie Queene"*)—but none has remarked on the complete fourfold incestuous aspect of this spousehood.

229. Charles de Grassaille, *Regalium Franciae* 1, ius 20, 217.

230. Coquille, *Oeuvres* 1.3232; cited in Church, *Constitutional Thought,* p. 278, n.16.

231. Kantorowicz, *King's Two Bodies,* p. 212.

232. Godefroy, *Cérémonial,* p. 661.

233. Godefroy, *Cérémonial,* p. 348; for earlier such coronations, see Dewick, ed., *Coronation Book of Charles V,* pp. 33, 83

234. Choppin, *De Domanio Franciae* [1605], III, tit. 5, n. 6, p. 449; cited in Kantorowicz, *King's Two Bodies,* p. 223.

235. Fortescue, *De laudibus legum Angliae,* c.xiii, 30,17. Cf. Cardinal Pole's letter to Henry VIII, in which he says "Your whole reasoning comes to this conclusion that you consider the Church a *corpus politicum*" (Pole, *Ad Henricum VIII,* in Rocaberti, *Bibliotheca,* xviii,204, cited in Kantorowicz, *King's Two Bodies,* p. 229n.

236. The spouse is intragenerational insofar as he/she is, like oneself, the daughter/son of one's parents (in-law), which suggests that marriage is so egalitarian as to be incestuous.

In *The Elementary Structures of Kinship,* Lévi-Strauss writes, "Marriage is an arbitration between two loves, parental and conjugal. [The two] are both forms of love, and the instant the marriage takes place, considered in isolation, the two meet and merge; *love has filled the ocean.* Their meeting is doubtless merely a prelude to their substitution for one another, the performance of a sort of *chassé-croisé.* But to intersect they must at least momentarily be joined, and it is this which in all social thought makes marriage a sacred mystery. At this moment, all marriage verges on incest" (Lévi-Strauss, *Elementary Structures,* p. 489).
By *chassé-croisé,* Lévi-Strauss suggests a chiasmatic situation of reciprocal and simultaneous exchanges having for their end no result.

237. Elizabeth, *Speech to Commons,* 1558/59, in Rice, *Public Speaking of Queen Elizabeth,* p. 117.

238. Elizabeth, "Glass," Folio 19r–v: "Alas, yea, for Thou hast broken the kindred of my old father, calling me daughter of adoption."

239. *Letters,* p. 47. Sir Henry Sidney had married Mary Dudley (sister of Queen Elizabeth's favorite, the Earl of Leicester) and fathered Philip (the poet, statesman, and soldier who later fell out of the Queen's favor when he remonstrated against her proposed marriage with the Duc d'Alençon).

240. Sometimes entitled "On Monsieur's Departure" (in *Poems,* p. 5), this poem was probably written about Francis, Duc d'Alençon in 1582, after marriage negotiations had ended. See Pringle, *Portrait,* p. 44.

241. Elizabeth, *Speech to Commons,* 1558/59, in Rice, *Public Speaking of Queen Elizabeth,* p. 117.

242. It is a commonplace in the current literature that "the Reformation had terminated the *cult of Mary* in England; to a significant extent the 'cult of Elizabeth' replaced it" (Jardine, "*Still Harping,*" p. 177). But the real issue here involves the extent itself and the precise mode of transformation. On the cult of Elizabeth as the virgin queen, though without reference to fourfold kinship, see Montrose, "'Shaping Fantasies,'" and Schleiner, "*Divina virago.*" On visual representations and icons of Elizabeth as the Virgin Mary, see Strong, *Cult of Elizabeth,* esp. p. 66 (on the "Sieve Portrait" of Elizabeth).

243. For the text of John Dowland, the Elizabethan lutenist, see Yates, *Astraea,* p. 78.

244. Chamberlain, *Sayings of Queen Elizabeth,* pp. 61, 68.

245. See Paul Johnson, *Elizabeth,* pp. 94–95.

246. For "earthly paradise," see Crosse, *Lover,* sec. ii.

247. Cited in Paul Johnson, *Elizabeth.* In response to the House of Commons' urging her in 1559 to marry—because "nothing can be more repugnant to the common good, than to see a Princesse . . . leade a single life, like a Vestal Nunne"—Elizabeth said: "I have made choyce of this kinde of life, which is moste free, and agreeable for such humane affaires as may tend to his [God's] service. . . . To conclude, I am already bound unto an Husband, which is the Kingdome of England . . . (And therewithall, stretching out her hand, shee shewed them the Ring with which she was given in marriage, and inaugurated to her Kingdome, in expresse and solemne terms.) And reproach mee so no more, (quoth shee) that I have no children: for every one of you, and as many are English, are my Children" (Camden, *Annales,* bk. 1, 26). John King points out that the association of the ring with Elizabeth's marriage to England was important as late as Elizabeth's approach to death, when the coronation ring had to be cut off her finger; this was interpreted as a sign of "the coming dissolution of her 'marriage with her kingdome'" (John King, "Queen Elizabeth I," pp. 33–34).

248. *Letters,* pp. 27, 188.

249. *Letters,* p. 223. Compare how, on receipt of the news of the death of Anjou, one of her so-called suitors, Elizabeth wrote to his mother, Catherine de Médicis, Queen Mother of

France, that "you will find me the faithfulest daughter and sister that ever Princes had" (letter of July 1584, in *Letters*).

250. Neville Williams, *Elizabeth,* p. 218.

251. Neale, *Elizabeth I and Her Parliaments,* p. 109.

252. Harington, *Letters,* p. 96.

253. With Bale's view compare that of Pope Nicholas I, who argues that "one ought to treat a godparent like a parent, even though the relationship is spiritual and not one of blood. That there cannot be marriage in these relationships is for the same reason that the Roman law does not allow marriage between those one adopts and one's own children" (Nicholas, *Responses to the Questions of the Bulgars,* sec. 2, in Mansi, ed., *Sacrorum* 15, 402).

254. *King Henry the Eighth,* 5.4.9. *Elishabet,* from the name of the priest Aaron's wife (Exod. 6:23) and of the woman in Luke, has been interpreted as "God is fullness," "God of the oath," "God's oath," "God is an oath" (by which one swears), and "consecrated to God." Cf. Gesenius, *Hebrew and English Lexicon,* p. 45, and Bale, "Epistle Dedicatory," Folio 9r: "Elizabeth in the Hebrew is as much to say in the Latin as, *Dei mei requies;* in English, 'The rest of my God.'"

255. *Richard III* 1.3.208.

256. The last scene of *Henry VIII* shows the Duchess of Norfolk and the Marchioness Dorset (the baby princess' two godmothers) substituting for her biological mother (the absent Queen Anne Bullen) who is barely mentioned (5.5.7). And it shows King Henry VIII, presumably the biological father, asking the princess' godfather, Cranmer (5.3.162), "What is [the infant's] name?" (5.5.10). Elizabeth's verifiable gossipred has replaced her consanguineous family; as Bale would have it, Elizabeth is one of "many noble women, not rising of flesh and blood . . . but of that mighty living spirit of His which vanquisheth death, hell, and the devil" (Bale, "Conclusion," Folio 46v).

257. Bale, "Epistle Dedicatory," Folio 9v: "The spirit of the eternal son of God, Jesus Christ, be always to your excellent grace assistant that ye may send forth more such wholesome fruits of soul and become a nourishing mother." In the *Monument of Matrones,* which includes Elizabeth's *Godly Medytacyon,* Bentley praises Elizabeth for styling herself "like a loving mother, and a tender nursse, giving my foster-milke, the foode of thy [God's] word and Gospell aboundantlie to all" (Bentley, *Monument,* 2B4v; King, *Tudor Royal Iconography,* p. 255).

258. "Have I conceived all this people? have I begotten them, that thou shouldest say unto me, Carry them in thy bosom, as a nursing father beareth the sucking child, unto the land which thou swarest unto their fathers?" (Num. 11:12).

259. Jardine would insist, to the contrary, that "Elizabeth I failed to make other than the impact of a *token* woman on the patriarchal attitudes of the early modern period" (*Daughters,* p. 195). I should argue, however, that the parentarchy under Elizabeth did not remain as patriarchal as it had been, that it took on and afterwards kept certain aspects of matriarchy. But the question as to whether Elizabeth's parentarchy was essentially patriarchal or matriarchal is really a red herring. The more important issue here concerns the fate of the old parentarchy itself: Whether male or female, during the Elizabethan period its definite undoing began. The result was the modern English nation state, which laid the basis for an eventual "liberation" of males and females alike.

260. Bennet, "*Measure for Measure* as Royal Entertainment," p. 98.

261. Baker, *Chronicle,* p. 155. For James' homosexuality, see Bennet, "*Measure for Measure* as Royal Entertainment," p. 180, n. 39.

262. James, *Political Works,* pp. 272, 24; cited in Goldberg, *James I,* pp. 141–42; cf. Tennenhouse, "Representing Power," p. 153.

263. See Perrens, *Libertins.*

264. James Otis; quoted in Morison, *American People*, p. 205.

265. Cf. "Therefore shall a man leave his father and his mother" (Gen 2:24). So sang The Animals in San Francisco's spectacular Cow Palace in 1966.

Chapter 5

1. "Tragedy," writes Cavell, "is the result, and the study, of a burden of knowledge, of an attempt to deny the all but undeniable, that a loving daughter loves you, that your imagination has elicited the desire of a beautiful young woman, that however exceptional you may be you are a member of a human society, that your children are yours" (Cavell, "Hamlet's Burden").

2. 2.5.2.

3. 1.2.205–6.

4. Francisco Sánchez, a distant cousin of Montaigne, wrote *Quod Nihil Scitur* [Why Nothing Can Be Known], in 1576 and published it in 1581.

5. For Montaigne, the so-called certainty of consanguinity boils down to mere chance masquerading as natural inclination or likeness. In "Of the Affection of Fathers for Their Children," for example, Montaigne writes about the kinship of fathers, specifically, to their children: "Herodotus relates of a certain district of Libya that intercourse with women is promiscuous, but that the child, when he has the strength to walk, finds to be his father the one toward whom in the crowd natural inclination bears his first step. I believe that this must lead to frequent mistakes [*mescontes*, or false accounts]." The reason we usually give for fathers loving their children—"that we begot them [*pour les avoir engendrez*], wherefore we call them our other selves"—may pass in the case of those "children of the mind [*enfantements de nostre esprit*]" that are essays in fiction or philosophy. But in the case of human children, that reason is insufficient and is based in fictional tales only, or *contes* (Montaigne, *Essais*, in *Oeuvres*, ed. Thibaudet and Rat, bk. 2, chap. 8, pp. 379–80). Montaigne refers here to Herodotus, *Hist.* 4:180: "These people enjoy their women in common. They do not live in couples at all but fuck in the mass, like cattle. When a woman's child is rather older, in three months' space the men come together, and whichever of the men the child most resembles, his the child is regarded as being" (Herodotus, *History*, trans. Grene, p. 347). "Like father, like son" is the measure of legitimacy.

6. 1.2.107–8. Unless otherwise indicated, citations in this chapter are from the Jenkins edition. Abbreviations for other editions I will cite are as follows: *Q1*, the first, or "bad," quarto (1603, printed by Valentine Simmes); *Q2*, the second, or "good" quarto (1604/05), printed by James Roberts; *F*, or first Folio (1623). See J. Dover Wilson, *Manuscript of Shakespeare's "Hamlet"*; and Werstine, "Textual Mystery of *Hamlet*."

7. See, for example, D'Anglure, *Esquimaux*, and cf. the various spiritual kinships endorsed by the Catholic church, including monachism (where the status of Brother supersedes that of brother) and the gossipred. On the gossipred as such, see Gratian, *Decretum*, 30; Rolandus, *Summa*, 144–45; Stephan of Tournai, *Summa*, p. 241; Feije, *De impedimentis*, chap. 16; Council of Trullo (A.D. 692), in *Colección de canones;* and C. E. Smith, *Papal Enforcement*, pp. 48–51. For the anthropological view of gossipred, see Pitt-Rivers, "Spiritual Kinship," p. 55, and "Pseudo-Kinship," p. 408; Gudeman, "Compadrazago"; Mintz and Wolf, "Ritual Co-Godparentage"; and Anderson, "Comparaggio."

8. Montaigne, who was himself put out to nurse with a peasant woman and had sponsors from the same class, is skeptical about the preeminent importance and authority of the consanguineous bond between human parent and human child, i.e., of the "literal" bond between a genitor or genetrix and offspring (see his essay "Of the Affection of Fathers for Their Children" [1578–80]). He refers to instances where nurse mothers even develop for

their nonconsanguineous nurslings "a bastard affection [*affection bastarde*] soon engendered by habit [*s'engendrer bien tost par accoustumance*], more vehement than the natural, and a greater solicitation for the preservation of the borrowed child than for their own [*leurs propres*]." He concludes that the "natural affection to which we give so much authority" of a genetrix for her consanguineous child has "very weak roots" (Montaigne, ed. Thibaudet and Rat, bk. 2, chap. 8, pp. 379–80; trans. Frame, pp. 290–91). On the specific taboo against marriage between milk siblings—people who received the "milk of human kindness" from the same nursemother—, see Crawley, *Mystic Rose* 2, 230. For a discussion of the privileging of collactaneous over blood kinship in the Middle Ages, see Boswell, *Kindness,* p. 359.

9. 1.2.157, 1.5.42, 1.5.83, 3.3.90, 5.2.330. Jenkins discusses how, when Hamlet reproaches Gertrude with her second husband at 3.4.91–92, "the accusation in Q1—'To live in the incestuous pleasure of his bed'—makes him now apply to the Queen what he said of the King (in Q1 as in Q2)" (Jenkins, "Introduction," *Hamlet,* p. 29). Among *Hamlet*'s likely literary sources or analogues are Saxo's Latin *Historiae Danicae,* which was written at the end of the twelfth century and published in 1514, and Belleforest's French *Histoires Tragiques,* whose *Cinquiesme Tome* (1570, 1572, 1576, 1582, etc.) contains a Hamlet story that was probably the basis for the anonymous English *Hystorie of Hamblet* (1608). Saxo says that Feng = Claudius capped "unnatural murder with incest"; Saxo's Amleth accuses his mother of having an "incestuous bosom" and refers to the crime of "fratricide with incest" (Saxo Grammaticus, *Historiae Danicae,* in Gollancz, *Sources,* pp. 101, 115, 139).

10. For the term "sometime" as used in *Hamlet* 1.2.8, see also Bale's reference to "the right virtuous lady Margaret, sister sometime to the French king Francis" (Bale, "Epistle Dedicatory" to Elizabeth Tudor's *Godly Medytacyon* [1548], Folio 9r; in Shell, *Elizabeth's Glass*).

11. Helena in *All's Well That Ends Well* is the daughter by adoption of the Countess. Helena takes great care to determine that she is eligible to marry her legal brother, insisting that she is not the natural daughter of the Countess:

HELENA: Mine honorable mistress.
COUNTESS: Nay, a mother.
Why not a mother? When I said "a mother,"
Methought you saw a serpent. What's in "mother,"
That you start at it? I say I am your mother,
And put you in the catalogue of those
That were enwombed mine. Tis often seen
Adoption strives with nature. . . .
HELENA: Pardon Madam.
The Count Rossillion cannot be my brother. . . .
COUNTESS: Nor I your mother?
HELENA: You are my mother, madam; would you were—
So that my lord, your son, were not my brother—
Indeed not mother! Or were you both our mothers
I care no more for than I do for heaven,
So I were not his sister. Can't no other
But, I your daughter, he must be my brother?
COUNTESS: Yes, Helen, you might be my daughter-in-law.
(1.3.141ff.)

The Countess loves Helena as if she were a natural daughter. "If she had partaken of my flesh and cost me the dearest groans of a mother," says the Countess, "I could not have owed her a more rooted love" (4.5.10–11). Because of that love, the Countess would transform Helena from a daughter by legal adoption into a daughter by nature. But if Helena is to be united with Bertram in chaste marriage, either she must not be the Countess's daughter or Bertram must not be the Countess's son. Helena insists on the first, that she can be the daughter-in-law, but not the daughter, of the Countess.

12. One can no more know whether "fictional" people like Claudius and Gertrude are "literal" brother and sister than know whether "real" people like you and I are related consanguineously. The relationship is fictional whether the people are or not. Cf. Knights, "How Many Children Had Lady Macbeth?"

13. Jenkins complains that, "criticism has made too little of the bond between the two revengers which Hamlet acknowledges with the word *brother*" (Jenkins, ed., *Hamlet*, p. 567). Yet, as we shall see, the play goes well beyond this brotherhood of revenge.

14. A returning son's wondering who his father is, or that father's wondering about the strange young man before him, may be one source for the name *Laertes;* Laertes' "knees were loosed where he stood, and his heart melted," says Homer in the *Odyssey* (24.345–47), "as he knew the sure tokens which [the disguised] Odysseus told him."

15. On the issue of whether bastards could inherit, see chap. 3, n. 23. In Rome, we shall see, filial adoption was virtually a precondition for succession.

16. Textual variants here may be significant. Claudius says to Laertes in one version, "What would you undertake / To show yourself your father's son in deed" (F4); in another, "What would you undertake / To show yourself your father's son indeed" (F); and finally, in a third version, "What would you undertake / To show yourself indeede your father's son" (Q2). Laertes can show himself to be indeed his father's son only in the deed of killing a father's son.

17. *Hamlet* 1.5.42, cf. 46. The term *adulterate* is used by the ghost to describe Claudius, but as "adulteress" it probably describes Gertrude as well. In Marsden's *Antonio's Revenge* the counterpart to Gertrude is clearly unfaithful to her husband. For the view that Gertrude had a guilty relationship with Claudius before her husband's death, see Bradley, *Shakespearian Tragedy*, p. 166, and J. Dover Wilson, *What Happens in 'Hamlet'*, pp. 292–94. Belleforest, *Cinquiesme Tome* (a source for the old Hamlet tale), has it that Geruthe commits incestuous adultery with Feng (Claudius) *before* Feng's murder of Horvendile (Old Hamlet).

18. On the meaning of "too much in the sun" (1.2.67) as "ostracized from home, kindred, and social life," see Blackmore, *Riddles of Hamlet*, p. 103.

19. Electra, who plays the role of Hamlet-like gadfly after the murder of her father Agamemnon in Aeschylus' *Oresteia,* is separated from her exiled brother Orestes, whom she loves just as if he were her brother, father, mother, sister, spouse, and child.

> To call you father is constraint of fact,
> and all the love I could have borne my mother turns
> your way, while she is loathed as she deserves; my love
> for a pitilessly slaughtered sister turns to you.
> And now you were my steadfast brother after all.
>
> —*Libation Bearers*, ll. 239–43

Electra eventually marries a peasant; but she lives out her life like a nun because the peasant fears Orestes' wrath.

20. For the sources: Saxo Grammaticus, *Historiae Danicae*, in Gollancz, *Sources*, pp. 108–9. For collactaneous kinship et al., see pp. 104–5, 109; Hansen, *Saxo*, pp. 130–31, cf. 129; and Jones, *Hamlet*, p. 153. See also Belleforest, *Cinquiesme Tome*, in Gollancz, *Sources*, pp. 202–3. Cf. Belleforest's remark that Ophelia's counterpart was "one that from her infancy loved him" (in Belleforest, *Cinquiesme Tome*, in Gollancz, *Sources*, p. 203).

21. The term *foster* is linked etymologically to a Scandinavian term involving nourish-

ment (as of milk) and recalls such terms as *alumnus*. See Boswell, *Kindness*, pp. 356–60, 388–89.

22. For adoption as an impediment to marriage, and the relevant struggle between secular and religious Rome, see Justinian, *Digest* 1.23, tit. 2, lex 17; Gratian, *Decretum*, caus. 30, q. 3, C. 6; and Ivo Carnotensis, *Decretum, Patrologiae* (Latina) 161, 657. See also G. Oesterlé, "Inceste," in Naz, ed., *Dictionnaire* 5, 1297–314. C.E. Smith, *Papal Enforcement*, p. 6, shows that "adoption has the same effect in precluding marriage as does kinship by blood." Fowler, *Incest Regulations*, p. 40, suggests, however, that this view of adoption has been contested frequently since the fall of Rome. On adoption as an impediment to marriage in Shakespeare, see Bertram's reluctance to marry Helena in *All's Well That Ends Well*.

23. Hansen, *Saxo*, p. 130. The king here is Tamerlaus of England.

24. For these Brothers and Sisters, see the section on "Social Anthropology of Universalist Orders" in chapter 4.

25. Cf. "The Garden of Love," in *Songs of Experience*, in *Poetry and Prose of William Blake*.

26. Perhaps Shakespeare submerged the collactaneously incestuous aspects of the Hamlet-Ophelia story because he believed that consciously desired incest—of the sort we might feel uneasy about in the Claudius-Gertrude story—would not have made for a good tragedy. T. S. Eliot, in his consideration of Elizabethan drama, and Thomas Rymer, in his consideration of the sibling love in *Canace* by Speroni [1546] and in *A King and No King* by Beaumont and Fletcher, would seem to endorse just this belief. They argue on quasi-Aristotelian grounds that a knowingly incestuous hero would be so unlike most people as not to elicit from their audiences the appropriate combination of pity or fear. (Aristotle, *Rhetoric* 1386–87, *Ethics* 1155; Rymer, Critical *Works*, pp. 48, 49; Eliot, *Essays in Elizabethan Drama*, pp. 129, 139; cf. Roper, introduction to Ford, *'Tis Pity*, p. 33, Sherman, introduction to *'Tis Pity*, pp. 45–53, and R. J. Kaufmann, "Ford's Tragic Perspective.") For speculation on whether Ophelia may have consummated a sexual affair with Hamlet (or even Claudius), see West, *Court*, p. 15.

27. Shakespeare, *Pericles* 5.1.197. The basically Protestant view that Hamlet's "nunnery" here means "brothel" has dominated most criticism since Wilson, *What Happens in Hamlet?*, esp. pp. 128–34; for one critique of this interpretation, see Jenkins, ed., *Hamlet*, p. 496, and *Hamlet and Ophelia*. Theobald, in his edition, cleverly reads the twice-repeated term "beautified" (2.2.109–11) as "beatified."

The Sister that Hamlet here wants his quasi-sister Ophelia to become recalls the quasi-nun Olivia in *Twelfth Night* as well as the novice Isabella in *Measure for Measure*. Stephen Greenblatt remarks insightfully of the role of celibacy in *Twelfth Night* that Olivia's vow to become "like a cloistress" in order "to season / A brother's dead love" (*Twelfth Night* 1.1.25–30) "picks up for Shakespeare's Protestant audience associations with life-denying claims of the flesh by entering into holy matrimony" (Greenblatt, *Shakespearean Negotiations*, pp. 69–70, 176). However, few Protestants outside such incest-practicing sects as the Family of Love—and no Catholic—would argue that "life-denying" through celibacy is wrong in those cases where "life-affirming" entails sexual intercourse with one's brother—as is the case for Ophelia's relation to her quasi-collactaneous sibling Hamlet and for the quasi-Sister Isabella's relation to her consanguineous brother Claudio in *Measure for Measure* (see my *End of Kinship*, chap. 4). Where all people are potentially or essentially siblings—as when people either do not know who their kin are or take literally Jesus' catholic words that "All ye are brethren"—the tension between the requirements to reproduce and not to commit incest expresses itself as the conflict between life and death. On the overall question of whether the Shakespearean plays take the Catholic or Protestant side in the great Reformation debates, see Honigmann, *Shakespeare*; Mutschmann and Wentersdorf, *Shakespeare*; and my *End of Kinship*, esp. pp. 49–50, 217.

28. *Pericles,* act 1, chorus, 27–28. For the clever pun on "none"/"nun," see esp. 5.1.128–32. *Much Ado*'s Beatrice touches on the distinctive source of Hamlet's universalist attitude to marriage ("To a nunnery, go") when she says, "I'll none. Adam's sons are my brethren, and truly I hold it a sin to match in my kindred" (*Much Ado,* 2.1.63–64; cf. 4.1.239–42).

29. Despite differences between the nunnery scenes in Q2 and F, this line appears in both. See Werstine, "Text Mystery of *Hamlet,*" p. 24.

30. "Thou dost handle my soul (if so durst I say) as a mother, daughter, sister, and wife, alas, yea, for Thou hast broken the kindred of mine old father, calling me daughter of adoption" (Elizabeth, "Glass of the Sinful Soul," Folio 13r).

31. Belleforest, *Cinquiesme Tome,* in Gollancz, *Sources,* pp. 210–11.

32. Both the profane conflation of kinship positions in the Danish court (where Gertrude and Claudius are sibling-spouses and where Hamlet and Ophelia are sibling-lovers) and their sacred conflation in a Catholic nunnery (where Ophelia would be the Sister and Spouse of Christ) recall the Elizabethan Court of Wards (where the members called one another "cousin" or "brother" and intermarriage was not uncommon). Denmark is, to Hamlet, a "ward" (2.2.246). Edward de Vere (Earl of Oxford), sometimes credited with being the author of the plays we call "Shakespearean," was raised as a ward of the court after the death of his father (and apparent abandonment by his mother). "Ward-son" there to the fatherly William Cecil (Master of the Wards), he called himself Cecil's "most assured and loving brother," formed an alliance with Cecil's consanguineous son Robert, and eventually married Cecil's consanguineous daughter Anne. (The marriage was not of the "hot" type: their first child was probably not Oxford's and there was a five-year period of estrangement. Cf. Bertram reluctantly coming to marry his adoptive sister Helena in *All's Well That Ends Well.*) There is a long critical tradition claiming that Cecil is the historical model for the "steward" Polonius in *Hamlet* (4.5.171). For the argument that Edward de Vere, muzzled by his guardian (Lord Treasurer Burghleigh), wrote under the pseudonym "Shakespeare," see esp. Ogburn, *Mysterious William Shakespeare.*

33. In Belleforest's version, Hamlet regards himself as heir: "I am lawful successor in the kingdom" (Belleforest, *Cinquiesme Tome,* in Gollancz, *Sources,* p. 281).

34. See Hansen, *Saxo,* p. 122. According to historical hypothesis, in the pre-Abramic Hurrian fratriarchy "authority is exercised by the eldest brother and is handed on from brother to brother as something inherited, the inheritance passing to the oldest son of the first brother on the death of the last brother" (de Vaux, *Israel,* p. 197; see also Koschaker, "Fratriarchät," and Gordan, "Fratriarchy"). At the same time, the fratriarch officially adopted spouses into the family as "sisters," either at the time of marriage or earlier. (The former is Speiser's view ["Wife-Sister Motif," pp. 15–18], the latter Seters' [*Abraham,* pp. 72–75].)

35. See Stabler, "Elective Monarchy."

36. Thomson, influenced by Bachofen's views of *Mutterrecht,* comments on similar situations in the *Oresteia,* to the effect that it is as though kinship itself were thought of as having been originally uterine (Thomson, ed., *Aeschylus Oresteia,* p. 385; cf. Engels, *Origin of the Family,* p. 103).

37. Belleforest, *Cinquiesme Tome,* in Gollancz, *Sources,* pp. 182–85. Grebanier, *Heart,* p. 192, says that in Belleforest, "Hamlet's father is rewarded for his services to the crown by being married to the King's daughter."

38. Coke, *Commentary upon Littleton,* L.1, c.5, 36b; see also Clarkson and Warren, *Law of Property,* p. 81. Kittredge, ed., *Hamlet,* p. 139, claims that a "jointress" is "a widow who has jointure, an estate which falls to her on the death of her husband." Clarkson, pp. 83–84, doubts that "jointress" in the play should be taken in this sense. Werder, *Vorlesungen,*

p. 355, discusses Gertrude as an *Erbin* (heiress)—a standard German translation for "jointress" in *Hamlet.* Contrast the legal definition of jointress as "she who had an estate settled on her by the husband, to hold during her life if she survived" (cited in C. K. Davis, *Law in Shakespeare,* p. 255).

39. Brady, *Hamlet's Wounded Name,* p. 15, writes of Claudius and Gertrude that "probably he married her for political advantage as well as love." Jenkins assumes that "what is clear is that Claudius became king before taking her [Gertrude] 'to wife' but consolidated his position by a prudent marriage" (*Hamlet,* p. 434).

40. Saxo Grammaticus, *Historiae Danicae,* in Gollancz, *Sources,* pp. 152–53.

41. Saxo Grammaticus, *Historiae Danicae,* in Gollancz, *Sources,* pp. 152–53.

42. Ency. Brit. (11th ed.) 17:822a.

43. Cf. 4.4.52. On father and mother as one flesh in the sense suggested by Christian doctrine, see Bracton, lib. 5, Tract. 5, cap. 25: "Vir et uxor sunt quasi unica persona, quia caro una, et sanguis unus" (quoted in Rushton, *Legal Maxims,* pp. 21–22).

44. Thomson, ed., Aeschylus, *Oresteia (Eumenides),* l. 741, note.

45. See his paper, "Hamlet's Burden."

46. Strachey comments that "the Queen's want of any clear and distinct views and opinions is in keeping with her whole character" (Strachey, *Shakespeare's 'Hamlet',* p. 173).

47. *Commere,* which means something like "godmother," is Theobald's and Warburton's emendation of *comma.* (*Comma,* which is what modern editors prefer, also makes good sense: it indicates the agency that both separates and links words.) The French term *commerage,* according to Cotgrave's dictionary (1611), refers to "gossiping, the acquaintance or league that grows between women by christening a child together, or one for another." In any case, the spectator can understand the kinship in *Hamlet* in grammatical as well as kinship terms; from one viewpoint, incest is a grammatical violation in which only nomenclature is confused (see Barthes, *Sade, Fourier, Loyola,* pp. 137–38).

48. Gregory I, for example, describes a monastery in which the monks had visits from females which they wished to make seem innocent by having the women appear to be their *commatres* (Letter from Gregory I to Valentinus, A.D. 584, cited in Fowler, *Incest Regulations,* p. 164).

49. One source of the notion that M. Brutus was the illegitimate son of Caesar is Plutarch's report that Caesar's last words were, *Et tu, mi fili* [And you, my son]—which Shakespeare changes to *"Et tu, Brutè"* [And you, Brutus] (*Julius Caesar* 3.1.77); cf. Jones, *Hamlet,* p. 124. For the ancients' belief that Brutus' mother, Servilia, was Caesar's mistress, see *Oxford Classical Dictionary,* p. 980. Shakespeare's Polonius, who played the part of Julius Caesar when he was an actor (3.2.102), is killed by the new Danish "Brutus." For Lucius Brutus, see also Belleforest, *Cinquiesme Tome,* in Gollancz, *Sources,* pp. 193ff.

50. Seneca, *Octavia,* in *Four Tragedies,* p. 258.

51. Claudius was the Roman emperor who undertook the conquest of Britain in A.D. 43, a few years after the philo-Roman British prince Cunobelin (= Shakespeare's Cymbeline) was succeeded by his two sons. In Krantz's *Chronica* (1545), p. 619, Ambletus = Hamlet and the Emperor Claudius are referred to on the same page (*Hamlet,* ed. Jenkins, p. 432). Claudius is cited as the type of bad ruler by Erasmus, *Institutio* (*Hamlet,* ed. Jenkins, p. 163). Skeat, *Shakespeare's Plutarch,* p. xviii, says that the name "Claudius" in *Hamlet* may come from "Clodius" in North's translation of Plutarch's "Antony and Cleopatra," but Montgomerie, "English Seneca," says that the Senecan play *Octavia* was one source for the name "Claudius" (cf. Montgomerie, "More an Antique Roman than a Dane").

Constantine the Great, born in A.D. 288, was the illegitimate son of Constantius I and Flavia Helena, whom Saint Ambrose describes as an innkeeper. "Since Constantine's legal right to the Empire of the West rested on his recognition of Maximus, he now had to seek for a new ground of legitimacy, and found it in the assertion of his descent from Claudius Gothicus, who was represented as the father of Constantius Chlorus. . . . Such is the primary version of the story, implied in the Seventh Panegyric of Eunenius, delivered at Trier in A.D. 310. It would seem that when Christian sentiment was offended by the illegitimate origin ascribed to Constantius, the story was modified and Claudius became his uncle" (*Ency. Brit.* [11th ed.] 6:988).

52. Fowler, "Incest Regulations," esp. pp. 70–80.

53. Tacitus, *Annals* 12.42.

54. Racine, *Britannicus,* 1.2.

55. "Agrippina through a burning desire of continuing her authorite and greatness grew to that shamelesnes that in the midst of the day, when Nero was well tippled and full of good cheer, she offered herselfe to him drunk as he was, trimly decked and readie to commit incest: and the standers by noted her lascivious kisses and other allurements, messengers of her unchast meaning" (Tacitus, *Annals,* trans. Grenewey, 14.2). The *Annals* were translated by R. Grenewey in 1598 (Bullough, ed., *Narrative and Dramatic Sources* 5, 12). For the publication history of Tacitus' work, see Mendell, *Tacitus,* esp. pp. 363–65. Berry ("*Hamlet*") juxtaposes Suetonius' Claudius with Shakespeare's Hamlet: both are kept from power by murderous uncles, both are "wild and whirling" in speech, and both are apparently inactive before their uncles.

56. In Shakespeare's *King John,* the rebels are called "you degenerate, you ingrate revolts / You bloody Neroes ripping up the womb / Of your dear mother England" (5.2.152). *King John* tells the story of John, who occupied the throne in defiance of the right of his nephew Arthur, who was the son of John's elder brother Geoffrey. Of Nero, Lydgate says in his *Falls of Princes* (1494) that "He mysusid his moodir Agripyne," and Marsden, in his *Scourge of Villanie* (1599), describes how "Nero keepes his mother Agrippine" (Lydgate, *Falls,* Bk. 8.I.728; Marsden, *Scourge,* quoted by Montgomerie, "More an Antique Roman," p. 76). Dowden, in his note to the Nero passage in *Hamlet* 3.2, writes that Nero "was the murderer of his mother Agrippina. . . . Perhaps the coincidences are accidental, that Agrippina was the wife of Claudius, was accused of poisoning a husband [Passienus Crispus, her second husband]; and of living in incest with a brother [the Emperor Caligula]."

57. Sophocles' Oedipus says: "Give me a sword, / to find this wife no wife, this mother's womb, / this field of double sowing whence I sprang / and where I sowed my children" (*Oedipus the King,* trans. Grene, 11.1255–58). It is unclear whether Oedipus somehow intends to kill Jocasta. A similar matricidal impulse plays a crucial role in the *Oresteia.*

58. Tacitus, *Annals,* 14.9.

59. Neoptolemus is called Pyrrhus in Cooper's *Achilleus* (1565); cf. Marlowe and Nashe's *Tragedy of Dido* (1594), probably one source of the "Aeneas speech" in *Hamlet* (see Hillebrand's note to his edition of Shakespeare's *Troilus and Cressida,* pp. 451–52).

60. Dusinberre writes that, for Hamlet "to cast out a belief in the indivisibility of man and wife is to justify Gertrude's faithlessness. Hamlet can obey the ghost when Gertrude's death leaves Claudius unprotected, and gives Hamlet a motive for revenge which seems to reunite his parents against the intruder on their marriage" (*Shakespeare and the Nature of Women,* pp. 99–100).

61. Dido does not quite keep her word. She is seduced by Aeneas with such speeches as the one that the Player recites in *Hamlet;* Virgil's *Aeneid* is ambiguous both as to whether Dido and Aeneas are properly married and as to whether Aeneas has a sexual liaison with Dido's sister Anna. (On the love affair between Anna and Aeneas, see Servius' commentary on Varro's remarks about Virgil, *Aeneid* 5.4.)

62. Cf. Tertullian's *Ad uxorem,* in which he advises his own wife to remain unmarried after his death (*Ency. Cath.,* s.v. "Tertullian," 13:1021).

63. For the sources, see Belleforest, *Cinquiesme Tome,* in Gollancz, *Sources,* pp. 186–89. The comparisons between Hamlet and James and between Gertrude and Mary are worth pursuing. King James VI's mother was Mary, initially the hope of the English recusant Catholics, as Carl Schmitt stresses in his book on *Hamlet.* Widow Mary married her cousin Henry Darnley (1565). She probably then conspired with her lover James Bothwell in killing Darnley (1567). She finally married Bothwell in Gertrude-like haste, within three months, thus becoming, like Gertrude, the wife of her husband's murderer and the conferrer on her husband of the title of king.

Her story raises questions about the legitimacy of James VI that match in certain respects those about the legitimacy of Queen Elizabeth. First, although papal dispensation was needed for the cousins Mary and Henry Darnley to marry, the dispensation was never granted. Second, Bothwell may well have been not only James' stepfather but also his adulterous genitor: Bothwell was known as an adventurous libertine and at James' baptism, which Darnley refused to attend, Bothwell stood in for the father. (Incidentally, Bothwell had divorced a previous wife on the dubious grounds of kinship by carnal contagion.) Third, Mary eventually consented to divorce Bothwell only on condition that the divorce not impeach her son James' legitimacy.

64. Cf. Cavell in "Hamlet's Burden": "I do not have to turn aside, as Jenkins, among other critics, is moved to do, the literal possibility that Gertrude is, or that Hamlet thinks that she is, the murderer, or accomplice, in the literal murder, of Hamlet's father." For the arguments on both sides, however, see the anonymously published *Hamlet. An Attempt. . . .*

65. When a *Volk* (*gens,* or kin) is conflated with a species (*genus,* or kind) we have the strange and murderous Neronian consequences which Hamlet seeks to avoid through his doctrine of kind cruelty.

66. The anonymously published *Warning* "recounts how at Lynn in Norfolk a woman was so moved by watching a guilty wife in a tragedy that she confessed to having murdered her late husband" (Jenkins, ed., *Hamlet,* p. 482; citing *Warning,* sig. H2, in Bullough, *Sources,* p. 18). Cf. Heywood's *Apology for Actors* (Gv–G2v).

67. Gifford, in his sixteenth-century treatise, remarks that certain ghosts, "instruments of God's vengeance," "kindle and stir up in [certain reprobate human beings] . . . filthy lusts, and carry them headlong into foul and abominable sins" (Gifford, *Discourse of the Subtle Practices,* quoted in Hoy's edition of *Hamlet,* p. 116). It is in this context, perhaps, that we might ask what were the "foul crimes" of Old Hamlet that they cry out to be purged (1.5.12, cf. 3.3.80–81).

68. The parallel between Claudius and Hamlet as murderous and as both actually or potentially incestuous is patent. In the play-within-the-play, the Player King stands for both Old Hamlet, whom Claudius (his brother) murdered, and Claudius, whom Hamlet (his nephew) threatens to murder; the murder of the Player King by his nephew Lucianus thus not only represents the murder of Old Hamlet by his brother Claudius and the murder of Claudius by his nephew Hamlet, but also conflates the polar opposite topoi of a son who kills his father and a father who kills his son. Shakespeare had depicted such filiacidal and patricidal commonplaces separately in the third part of *Henry VI:* According to the usual stage directions, there enters at one door a "son that hath kill'd his father" (without having known the man to be his father) and then, at another door, "a father that hath kill'd his son" (without having known the man to be his son) (2.5.54–122, cf. Forsythe, "Tacitus"). In *Hamlet,* on the other hand, Shakespeare depicts the father-killing son and the son-killing father in a single "tableau."

69. Belleforest, *Cinquiesme Tome,* in Gollancz, *Sources,* p. 303; cf. p. 301.

70. See the line drawing in Robert, *Oidipus* 1, 55; Gould's note to 1. 36 of his edition of

Sophocles' *Oedipus the King;* and Pindar's reference to the "enigma from the savage jaws of the virgin" (Fr. 177.4). For a relevant psychoanalytic discussion, see Gessain, "Vagina Dentata."

71. Belleforest, *Cinquiesme Tome,* in Gollancz, *Sources,* pp. 287–89. Defeating the Amazon was the greatest of Theseus' Herculean and Stoic tasks. "And when it became known in Greece that [the Greeks] would have peace with the Amazons, never had there been a greater joy, for there was nothing [the Athenians] feared as much as the Amazons" writes Christine de Pisan in *City of Ladies* (p. 47). But what Theseus can do in the comedic Athens of Shakespeare's *A Midsummer Night's Dream,* Hamlet cannot accomplish in tragic Denmark. Hamlet is no Hercules. Hamlet is like Hippolytus, the antisexual son of Theseus and the Amazon Queen Hippolyta: As in the Roman Seneca's and the French Racine's plays on the subject, he is also called upon during his father's adventures in the realm of death to reject the incestuous advances of a mother (for Hippolytus, his stepmother Phaedra, for Hamlet his aunt-mother Gertrude). Queen Elizabeth herself, according to much literature of the period, was the Amazon queen incarnate. Cf. Schleiner, "*Divina virago:* Queen Elizabeth as an Amazon."

72. "Unlike in the *Oedipus* [of Sophocles], where [that fantasy] is brought into the open and realized as it would be in a dream" (Freud, *Interpretation of Dreams,* pp. 264–65). Freud seems to mean that Hamlet both fantasizes wishfully and also represses his fantasies.

73. Suetonius, *Nero,* 52 and 38. Nero claimed that, should he ever be deposed, he would become a great actor in Alexandria. Cf. Hamlet's jest about getting "a fellowship in a cry of players" (3.2.271–72). On Hamlet's singing and dancing like Nero, see Montgomerie, "Folkplay" and Charlesworth, "Nero."

74. See, for example, Gilbert Murray, "Hamlet and Orestes," and Kott, "Hamlet and Orestes." Scholars have noted similarities between the theme of *Hamlet* and both the Orestes theme—as it appears in such works as Aeschylus' *Oresteia* [which Shakespeare probably did not know] and Heywood's *Iron Age* and Seneca's *Agamemnon* [which he probably did]—and the Oedipus theme—as it appears in such works as Sophocles' *Oedipus the King* and Seneca's *Oedipus.* Another Orestes play sometimes considered in this light is Pikerung's *Horestes* (see Prosser, *Revenge,* pp. 41–42). In Seneca's *Agamemnon,* as in *Hamlet,* the ghost of Agamemnon urges his kin to avenge his having been murdered by his brother Atreus.

75. The story of the sibling love between Macareus and Canace was well known from its Roman source in Ovid. On the Canace story in the English tradition, which takes Ovid's *Heroides* as its *locus classicus,* see Gower, *Confessio Amantis,* 3. 142; Chaucer, *Squire's Tale,* 667–69; and Spenser, *Faerie Queen,* 4.3. Cf. Nohrnberg, *Analogy of the Faerie Queene,* pp. 662–63; and Goldberg, *Endlesse Worke,* pp. 114–16.

76. On Pyrrhus' and Hamlet's state of neutrality, see also Levin, "Player's Speech," in *Hamlet,* ed. Hoy, esp. p. 230.

77. As hypocrisy can transform vice to virtue, so it can virtue into vice. Habit is all: "That monster custom . . . is angel yet in this" (3.4.163–64); "For use almost can change the stamp of nature" (3.4.170). In *Pericles, Prince of Tyre* it is said of the incest between Antiochus and his daughter that "custom what they did begin / Was with long use account'd no sin" (Act 1, Chorus, 11. 29–30).

78. *Twelfth Night* 5.5.355. "L'habit ne faict le moyne," is how Rabelais puts the proverb (cited in Kennard, *Friar in Fiction,* p. 45). Cf. Lucio's removal of the Duke's cowl in *Measure for Measure.*

79. Gertrude later says to Hamlet that "these words like daggers enter in my ears" (3.4.95). These are ears like those of Old Hamlet (through which the ghost says that Old Hamlet was killed) and Young Hamlet (through which the ghost speaks words to Hamlet himself). Cf. "The whole ear of Denmark" (1.5.36).

80. Compare Clytemnestra's words, "Bring me quick, somebody, an ax to kill a man" (Aeschylus, *Libation Bearers,* trans. Lattimore, l. 889), which suggest that she plans to kill her son Orestes.

81. Belleforest, *Cinquiesme Tome,* in Gollancz, *Sources,* p. 221.

82. Tacitus, *Annals,* trans. Grenewey, 14.2; emphasis mine.

83. Hamlet may go so far as to compare Polonius, a daughter-sacrificing Jephthah (Judg. 11), with the incestuous Lot (Gen. 19): "As by lot God wot, / And then, you know, / It came to pass, as most like it was" (2.2.411–13).

84. "You Neroes . . ." is from *King John,* 5.2.152. The church is the seat of those guilt feelings that might ordinarily inhibit pursuits of revenge. However, Claudius responds that "revenge should have no bounds"—which implies that an avenger has the right to pursue his victim anywhere at all or nowhere at all.

85. Consider here the fourfold divorce of kinspersons that informs *Coriolanus.* Coriolanus casts off his "father" Menenius Agrippa and claims that he no longer knows "wife, mother, child" (*Cor.* 5.2.71, 83; cf. 5.3.102). Compare Volumnia's remark that Coriolanus is no longer akin to his Roman mother, wife, and child, and other repetitions of the same motif (*Cor.* 5.3.178–80). Coriolanus claims that he has no family at all and allies his rejection of ordinary kinship with the atheist or isotheist hypothesis that "a man [is] author of himself / And [knows] no other kin" (*Cor.* 5.3.36–37).

86. Compare the views of Diogenes of Sinope: "Asked where he came from, [Diogenes of Sinope] said, 'I am a citizen of the world (*kosmopolites*)'" (Diogenes Laertius, *Lives,* VII, 63). Diogenes "preferred liberty to everything else" (Diogenes Laertius, *Lives,* VII, 71). For him this liberty meant overcoming parentarchy through a rejection of particularized consanguinity. ("The only true commonwealth was, [Diogenes of Sinope] said, that which is as wide as the universe. He advocated community of wives, recognizing no other marriage than a union of the man who persuades with the woman who consents. And for this reason he thought sons too should be held in common"; Diogenes Laertius, *Lives,* VI, 72.) Diogenes of Sinope also "assert[ed] that the manner of life he lived was the same as that of Hercules" (Diogenes Laertius, *Lives,* VII, 71; cf. VI, 50), the Stoics' favorite hero and apparently Hamlet's.

87. "Thus the king was drawn one way by his love for his daughter [whom Amleth had married] and his affection for his son-in-law, another way by his regard for his friend, and moreover by his strict oath and sanctity of their mutual declarations, which it was impious to violate. At last he slighted the ties of kinship and sworn faith prevailed." (Saxo Grammaticus, *Historiae Danicae,* in Gollancz, *Sources,* pp. 145–47.)

88. Belleforest, *Cinquiesme Tome,* in Gollancz, *Sources,* pp. 212–13.

89. On the device of the bedtrick in Renaissance and Jacobean drama, see my *End of Kinship,* pp. 145–48.

90. For the general pun on "cozen" and "cousin," see Abraham, "Cosyn and Cosynage."

91. Some time after Ophelia is told by Hamlet "To a nunnery, go," she says to Claudius "God dild you" (4.5.42); Jenkins has "good dild you," a corruption, he says, of "God yield (i.e., requite) you." There are several such allusions in *Hamlet* to artificial members or castration. Thus Hamlet wishes for the melting or resolving of his "too too sullied flesh" (1.2.129), desiring that deliquescence which Paul calls "checking the indulgence of the flesh" (Phil. 1:23–24; cf. Jenkins, note to 1.2.129–30). Compare Rom. 2:29: "Circumcision is that of the heart, in the spirit"; and Matt. 19:12.

92. Polonius would have Reynaldo lay "slight sullies" (2.1.40) on Laertes: "But breathe his faults so quaintly/That they may seem the taints of liberty . . ." (2.1.31–32); "such wanton, wild, and usual slips / As are companions noted and most known / To youth and liberty" (2.1.22–24).

93. The church is the place (topos) where such ordinary offenses as murder and incest, which otherwise seem to cry out for mere retaliation, can be transformed into—even sanctified as—extraordinary wonders such as resurrection and spiritual incest. Claudius' comment that "no place indeed should murder sanctuarize" suggests that no place—not even the church—can sanctuarize, or sanctify, murder.

94. Eisenstadt, "Ritualized Personal Relations."

95. Sollors, *Beyond Ethnicity,* p. 78.

96. Saxo Grammaticus, *Historiae Danicae,* in Gollancz, *Sources,* p. 105. Garber, *Dream,* p. 100, argues that Hamlet, unlike Rosencrantz and Guildenstern, is not one of the "indifferent children of the earth" in that he rejects their "self-interest, policy, and cold-blooded *reasoning.*"

97. Brain, *Friends and Lovers,* p. 30, points out that in the feudal period one sent a boy "to be fostered in the household of one's overlord, where he learned manners and was trained in arms, horsemanship and sports. Two young men thus growing up side by side . . . and competing together in games would become special friends . . . and this intimacy and rivalry continue throughout their lives as warriors."

98. In Saxo Grammaticus' version of the Hamlet story, the counterparts to Rosencrantz and Guildenstern, after their executions in England, are "transformed" into sticks of gold—to Feng = Claudius they will be *wergeld.* Amleth = Hamlet, whom the Danes believe to be dead, takes the sticks with him from England when, exactly a year from his supposed death, he returns to Denmark and presides ghost-like over his own supposed "obsequies." In Denmark, Hamlet presents the sticks: "When the Danes ask Amleth about his missing escorts, he holds forth his two gold-filled sticks and says, 'Here they are, both of them.' The grim jest is familiar from the sagas, in which the aggrieved party sometimes identifies blood money with the dead man" (Saxo Grammaticus, *Historiae Danicae,* in Gollancz, *Sources,* pp. 125–29). Gollancz, *Sources,* pp. 27–31, points out the similarity with the role of *wergeld* in the Brutus tale (cf. Hansen, *Saxo,* pp. 33, 139, on Valerius Maximus, *Memorable Sayings and Deeds,* 7.2).

The presentation of the gold sticks, and the speech accompanying it in which Hamlet points "at the weregild of the slain as though it were themselves," is significantly absent from Shakespeare's revenge-centered play (Saxo in Gollancz, *Sources,* p. 127). Yet here Hamlet manages to wreak vengeance on his incestuous uncle-father only after the latter commands his follower to "set it down" that Hamlet should demand, on behalf of King Claudius, the "neglected tribute" or *Danegeld* [cf. 3.1.171–72] that, in Shakespeare's sources, Hamlet brings back from England in the form of *wergeld.*

99. For the term *adiaphora* [indifference], see Diogenes Laertius, *Lives,* VII.104.

100. On the Blackfriars, see Hillebrand, *Child Actors,* esp. pp. 151–57; Irwin Smith, *Shakespeare's Blackfriars Playhouse,* pp. 177–78; and Furness' note in his edition of *Hamlet,* vol. 1, p. 261. Cf. Melville's discussion of the Blackfriars in his quasi-Elizabethan novel *Pierre* (ed. Murray, forew. Thompson, p. 304).

101. *Hamlet,* 2.2.335–58 is absent from Q. The lines are used by Jenkins as from F.

102. *As You Like It* 2.1.1; see Montrose, "'Place of a Brother.'"

103. The child actors of Shakespeare's day represented on stage the sinful libertine incest to which the Edenic condition of universal fellowship tends in the fallen world. Their production of "Family of Love," for example, contained a trial scene in which the sexual libertinism of the antinomian sect called "Family of Love" was institutionalized in law. This comic play, performed by the "Children of the Revels," was probably written by Thomas Middleton. See Shepherd's introduction to his edition of *Family of Love,* p. iii, and Cherry, *Most Unvaluedst Purchase.* On the sect of the Family of Love generally, see Halley, "Heresy." In *Hamlet* "libertine" is the word with which Ophelia taunts her Paris-loving brother Laertes (1.3.49). Shakespeare may also have known the works of such English Libertines as John

Champneys, who in his *The Harvest is at Hand* (1548) argues on religious grounds that God condones, for his chosen people, such "bodily necessities" as "fornication" and "adultery."

104. Cf. Aristotle, *Ethics*, 115a, 1168b. Diogenes said that "Friends have all things in common" and "Friendship is equality" (Laertius, *Lives*, VIII.10). See Hutter, *Politics as Friendship*, p. 34; cf. pp. 49–51, 126, 455.

105. Barth and Goedeckemeyer, *Stoa*, pp. 25–27; Zeller, *Philosophie der Griechen* 3, 299–303.

106. Plato, *Lysis*, 221e; see Bolotin, Plato's *Dialogue on Friendship*, esp. chap. 8, "Kindred as Friends." As Socrates puts it, "I'll attempt to persuade first the rulers, then the rest of the city, that the rearing and education we gave them were like dreams; they only thought they were undergoing all that was happening to them, while, in truth, at that time they were under the earth within, being fashioned and reared themselves, and their arms and other tools being crafted. When the job had been completely finished, then the earth, which is their mother, sent them up. And now . . . they must think of other citizens as brothers born of the earth" (Plato, *Republic* 414d [trans. Bloom]).

107. Cornford, in notes to his translation of the *Republic* (pp. 161–63), suggests that "Plato did not regard the . . . connections of brothers and sisters as incestuous." See *Republic*, 414d, 460c–d.

108. The Stoic and "cosmopolitan" philosopher Epictetus of Hierapolis was once a slave belonging to a member of Nero's bodyguard. When he became a freedman *(liber)*, he wrote (in the *Discourses*) that a universal siblinghood rules out differences along the lines of social or economic class such as that between Hamlet and Horatio. He cried to a master beating a slave: "Slave, do you not want to help your sibling [*adelphos*], who has Zeus for father, who is born of the same germs as you and is of the same heavenly descent?" (see Zeller, *Philosophie der Griechen*, pp. 299–303, and Barth and Goedeckemeyer, *Stoa*, pp. 25–27).

109. In the *apatouria* ("feast of men of the same fathers"), all *phrateres* became at once brothers and sons of the same fathers. On the *apatouria*, see Jeanmaire, *Couroi et Courites*, pp. 133–44, 379–83; Thomson, *Aeschylus and Athens*, p. 28; and Kretschmer, "Benennung des Brüders," in his *Einleitung* 2, 2–10.

110. Diogenes Laertius, *Lives*, VII.188, reports: "In his *Republic* [Diogenes of Sinope] permits marriage with mothers and daughters and sons. He says the same in his work *On Things for their own Sake Desirable*, right at the outset." Cf. Diogenes Laertius, *Lives* VI.72; and, on Zeno, *Lives* VII.131.

111. Diogenes of Sinope, among others, wrote tragedies, now lost (Sayre, *Greek Cynics*).

112. H. Von Arnim, *Stoicorum veterum fragmenta* 1, 59–60 (citing Pearson, *Exposition* pp. 210ff). Cf. Copleston, *History* 1, 140.

113. Thomas Aquinas argues that incest is natural but prohibited.

114. 2 Cor. 3:17.

115. "Fay ce que voudras" (Rabelais, *Gargantua*, chapter 57). See the interview with Trouillogan (*Gargantua*, chapter 36): "Do what you want!" is the only piece of positive advice that Trouillogan gives.

116. See, for this argument, among others, Allo, *Saint Paul*, and Conzelmann, *Corinthians*, p. 97. On the Family, see Wilkinson, *Supplication*, p. 91. Similarly, the English seventeenth-century Ranters wrote "I belong to the liberty of nature, and all that my nature desires I satisfy. I am a natural man" (quoted in Cohn, "Cult of the Free Spirit," p. 59).

117. *Munera Pulveris*, p. 126n.

118. Montesquieu, *Greatness of the Romans*, chap. 12; cited in *Ency. Phil.* 8:43; emphasis mine.

119. See Fly, "Accommodating Death."

120. The Stoic doctrine was that one should confront Death *(la mort)* as did Hercules,

who met Hades and worsted him (Homer, *Iliad,* 5.395ff). Hamlet admires Hercules, who killed the "Nemean lion," and he seems to have contempt for Claudius, who is "no more like my father / Than I to Hercules" (1.4.83; 1.2.152–53).

121. Belleforest, *Cinquiesme Tome,* in Gollancz, *Sources,* pp. 192–95. On this Brutus see especially Livy I.lvi.8.

122. Livy I.lvi.10–12. On parallels between the Brutus and Hamlet stories, see Detter, "Hamletsage." Cf. the tale of Brutus as treated by Machiavelli: "It is very wise to pretend madness at the right time"; "In order to maintain newly gained liberty, Brutus' sons must be killed" (*Discourses,* 3.1.2–3).

123. Brown, *Love's Body,* p. 33, refers to Brutus as liberator from the tyranny of Tarquin.

124. Livy II.v.8.

125. *Digest,* 11, 7, 35; Post, "Two Notes," p. 287; cf. Shakespeare, *Henry VI,* pt. 3, 2.5.54–122. Durandus wrote, "Nam pro defensione patriae licitum est patrem interficere" (*Speculum iuris,* IV, part iii, sec. 2, n. 32; 3:321).

126. Salutati, *Epistolario,* 1, 10; trans. Kantorowicz, *King's Two Bodies,* p. 245.

127. *Codex Justiniani* 10, 70, 4, n.7, p. 345; cf. Kantorowicz, *King's Two Bodies,* pp. 245, 248.

128. *Apokolokyntosis* 12, 2. In "English Seneca," Montgomerie considers the possible influence on Shakespeare of this work, only sometimes said to have been written by Seneca.

Chapter 6

1. Butler, introd. to Racine's *Britannicus,* p. 12.

2. Mère Angélique de St. Jean Arnauld, daughter of Antoine Arnauld (1560–1619), secured the abbess' chair in 1599 when she was eight years old and started to reform her convent in the direction of its original rule in 1608. Her brother was the great Antoine Arnauld (1612–1694), the most famous of the Jansenist theologians. She is to be distinguished from her niece, Angélique de St. Jean Arnauld d'Andilly (1624–1684), who herself eventually became abbess and produced important writings.

3. Delcroix, *Le Sacré dans les tragédies,* esp. pp. 329–94; Mauron, *L'Inconscient dans l'oeuvre,* p. 202; my translation.

4. Picard, in Racine, *Oeuvres,* p. 35.

5. Racine, *Abrégé,* in *Oeuvres,* ed. Clarac, p. 346.

6. Racine was sent in 1653 to study at the Port-Royalist grammar school with such masters as Nicole and Le Maître.

7. Racine went to live with Father Sconin, vicar-general in Uzès (Languedoc), in November 1661.

8. Sainte-Beuve suggests that Racine had a sister Marie who was Oblate at Port-Royal. (Sainte-Beuve, *Port-Royal,* 3:538).

9. Vuillart, *Lettres* (April 30, 1699). For the conventual grill, see Sainte-Beuve, *Port-Royal,* 3:555.

10. For Hamon: Butler, introd. to Racine's *Britannicus,* p. 13; cf. Gazier, *Ces Messieurs.* For Antoine Le Maître: see his letter to Racine of March 21, 1656: ". . . aimez toujours vôtre papa." Cf. Louis Racine's *Mémoires* (in *Oeuvres,* ed. Clarac, pp. 17–66) and Vaunois, *Racine,* p. 97.

11. See my *End of Kinship,* esp. pp. 102, 209, 227.

12. Mauron interprets Angélique's divorce of parents only from the perspective of a genitor or genetrix who has lost a daughter to a Being who is both Paternal and Spousal (Christ) and regards himself or herself as betrayed. "Aggression against the real father is consum-

mated by a sort of social suicide at the profit of the mystical spouse" *(L'Inconscient dans l'oeuvre*, p. 202, my translation).

13. The psychiatric tradition concerning hysteria and religious celibates here includes Charcot's "Hystériques" (1878) and *Leçons du mardi* (1892) as well as earlier works by Richer, Ferran, Rathéry, and even by Briquet (1859). For Freud and Breuer in the 1890s, "hysterical deliria often turn out to be the very circle of ideas which the patient in his normal state has rejected, inhibited, and suppressed with all his might" ("Mechanism of Hysterical Phenomena: Lecture," *Standard Edition* 3:38). With Charcot, Freud presumes to explain why "the hysterical deliria of nuns revel in blasphemies and erotic pictures" ("Footnotes to Charcot," SE 1:138; cf. "A Case of Successful Treatment," SE 1:126; and Freud and Breuer, "On the Psychical Mechanism of Hysterical Phenomena," SE 2:10–11). For other traditional psychoanalytical articles on religious celibacy, see Steffen, "Zölibat"; Levi-Bianchini, "La neurosi antifallica"; and Gilberg, "Ecumenical Movement."

14. Comment by Freud, quoted in Hitschmann, "Über Nerven- und Geisteskrankheiten," p. 271.

15. "Je te fais aimer comme ma fille et mon espouse. Voilà la leçon que je te donne, sur laquelle tu dois souvent faire réflexion, anéantissant toujours tes pensées en ma présence . . ." (*Autobiographie*, ed. Charcot, p. 46). Jeanne des Anges was seduced by the priest Grandier, who had written a well known book on spiritual sexuality, the *Célibat des prêtres* (see Jeanne des Anges, *Autobiographie*, pref. Charcot, p. 10).

16. Sainte-Beuve, *Port-Royal*, 3:538, 549.

17. "Denn was wäre schon diese Revolution / ohne eine allgemeine Kopulation" [And what's the point of this revolution without general copulation] (Weiss, *Marat/Sade*, p. 122). Cf. Sade's "Français, encore un effort si vous voulez être républicains," in *Philosophie*, esp. pp. 221–22.

18. Josephus, *Jewish War* 2.2, in Racine, *Oeuvres*, ed. Clarac, p. 599. On Racine's "Des Esséniens," written between 1655 and 1658, see Vaunois, *Racine*, p. 152. The view that Jesus resided with the Essenes is probably a romanticism promulgated by Christian biblical scholars who have been fond of seeing this sect as *the* direct link between pre-rabbinic Judaism and Christianity.

19. Josephus, *De vita contemplativa*, in Racine, *Oeuvres*, ed. Clarac, p. 605.

20. Josephus, *De vita contemplativa*, in Racine, *Oeuvres*, ed. Clarac, p. 605.

21. For the legal definition of "spiritual incest" as a Brother or Sister's having sexual intercourse with anyone at all, see chap. 1, n. 69.

22. Josephus's description of the Essenes' aversion for marriage is relevant: "[It] comes not from a desire to abolish the succession of children from fathers . . . but from their belief in the incontinence of women, who, in their opinion, almost never remain faithful to their husbands" (Josephus, *Jewish War* 2.2, in Racine, *Oeuvres*, ed. Clarac, p. 599).

23. This was in October, when he was nearly eighteen; he entered the Collège d'Harcourt, where he boarded with his second cousin Nicolas Vitart, steward of the Duke of Luynes. His Jansenist surroundings continued at the Collège, since the Duke of Luynes was a severe Port-Royalist.

24. The baptismal certificate of Jeanne-Thérèse Olivier is in the registries of Notre-Dame, the Auteuil parish (in Racine, *Oeuvres*, ed. Mesnard, 1,187–8, and ed. Picard, p. 32).

25. When Du Parc conceived again, she was poisoned by the notorious Catherine Voisin. Racine was nearly charged with murder. Some literary historians compare "la Voisin" with Locuste in *Britannicus*.

26. *Racine*, pp. 42–43 Though the word "incest" appears infrequently in Racine's works (J. G. Cahen, *Vocabulaire*), the theme is still pervasive.

27. *La Thébaïde,* 4.1. The love between the siblings has existed since "infancy" (*Bajazet,* 5.5, cf. 1.4, 5.6). They have loved "since almost forever" (*Mithridate,* 1.1, 1.2, 3.5).

28. Josephus, *Jewish War* 2.2, in Racine, *Oeuvres,* ed. Clarac, p. 602. Josephus refers to the "unmarried Essenes" and to the "marrying Essenes" (*Jewish War,* 2.160 and 2.1 61). On the Essenes' ideas about preserving "mankind" or the human "race," see Hippolytus, *Philosophumena,* 11, 28 (*Patrologiae [Graeaca]* 16, pt. 3) and Vermes, "Essenes," p. 101. Cf. J. Massingberd Ford, *Wisdom and Celibacy,* pp. 28–29.

29. Racine, Letter to Mme. de Maintenon of March 4, 1698, in *Oeuvres,* ed. Mesnard, 7:228–29. On historical difficulties concerning this claim of Racine, cf. Picard, *Racine,* pp. 304–08.

30. For a fuller discussion, see chapter 4.

31. For Racine's translations of ecclesiastical writings (1655–1658), see Vaunois, *Racine,* p. 155.

32. On Lancelot (the scholar of Greek), see Cognet, *Claude Lancelot,* and R.C. Knight, *Racine et la Grèce.* Cf. Mauron, *L'Inconscient dans l'oeuvre,* p. 200.

33. Whether Racine knew Pascal personally is considered by Vaunois, *Racine,* pp. 149–51.

34. Racine, *Abrégé,* in *Oeuvres,* ed. Clarac, p. 332.

35. Cited in Woodgate, *Jacqueline Pascal,* p. 82.

36. "Leurs coeurs n'étaient qu'un coeur." Cited in Mauriac, *Pascal,* pp. 5, 43.

37. Cousin, *Jacqueline Pascal,* p. 163.

38. "Vous savez assez que c'est de [Dieu] seul que procède tout l'amour" (Cousin, *Jacqueline Pascal,* p. 167).

39. "Si vous n'avez pas le force de me suivre, au moins ne me retenez pas" (Cousin, *Jacqueline Pascal,* p. 170).

40. "Non seulement il n'avoit point d'attache pour les autres, mais il ne voulait pas du tout que les autres eussent pour lui" (Cousin, *Jacqueline Pascal,* p. 338). Cf. the discussion of siblings in Giraud, *Soeurs de grands hommes.*

41. Cousin, *Études sur Pascal.*

42. Cousin, *Jacqueline Pascal,* p. 399; *Études sur Pascal,* p. 452.

43. Chateaubriand, *René,* trans. Putter, p. 103.

44. Chateaubriand, *René,* trans. Putter, p. 108.

45. Chateaubriand, *René,* trans. Putter, pp. 108, 111. Soon afterward, Sister Amelia tells her brother René that "for the most violent love, religion substitutes a sort of burning chastity in which the lover and the loved are one," develops a burning fever, and dies (Chateaubriand, *René,* p. 142).

46. On Chateaubriand's devotion to Lucile, see Aubrée, *Lucile et René;* on his entering the orders, see Chateaubriand, *Mémoires* 1, 78, and *Oeuvres romanesques,* pp. 121–22; and on his libertinism, see Barbéris, "René," pp. 51, 249–50.

47. Chateaubriand, *René,* trans. Putter, p. 106.

48. Homer, *Iliad* 6:429–30.

49. *Libation Bearers,* trans. Lattimore, 11. 239–43.

50. Cf. the remark of Coriolanus' mother Volumnia that her son is no longer akin to his Roman mother, wife, and child (5.3.178–80; 5.3.101–3). One parallel in Shakespeare's work to the Roman Catholic profession in which a woman takes leave of her earthly family ("dies to the world") and enters the heavenly family ("is reborn") by becoming the wife, sister, daughter, and mother of God is the scene in which Coriolanus, banished by the Romans, his "brothers," and "servanted to others" (the Volscians—5.2.84), claims that he no longer knows "wife, mother, child" (5.2.83).

51. Dante, *Paradiso,* 33: "Vergine Madre, figlia del tuo Figlio."

52. *Die Geburt der Tragödie,* sec. 9, in *Werke* 1:58.

53. Butler, introd. to Racine's *Britannicus,* mentions Racine's weeping at the funeral of Thérèse Du Parc.

54. Butler, introd. to Racine's *Britannicus,* p. 12; Sainte-Beuve, *Port-Royal,* 3:596.

55. Louis Racine, *Mémoires,* with reference to the profession of Sister Lalie. Cited in Mauron, *L'Inconscient dans l'oeuvre,* p. 217; emphasis mine.

56. Cf. Mauron, *L'Inconscient dans l'oeuvre,* p. 216.

57. Lacretelle, *La Vie privée de Racine,* cited in Mauron, *L'Inconscient dans l'oeuvre,* p. 217. Marie eventually married in 1699, the year of Racine's death. The youngest children, Françoise and Madeleine, remained single, contemplated taking the veil, but did not enter the cloister.

58. Aristotle, *Rhetoric* 1387a and *Ethics* 1155a.

59. Racine, letter to Louis Racine, November 10, 1698; cf. Racine's letter to Agnès, written November 9, 1698 (*Oeuvres,* ed. Picard, 2:643–44; 2:641–43).

60. Quesnel, letter to M. Willard, February 14, 1697.

61. Seneca, *Apokolokyntosis,* 14.1.

62. Narcissus/Pallas ensured Messalina's death; *Britannicus,* 1123. Tacitus suggests that Pallas was Agrippina's lover.

63. This is according to the Twelve Tables (see Watson, *Rome of the XII Tables,* esp. pp. 52–70).

64. Pavel argues that in *Britannicus* "the main character [Britannicus] has a stronger hereditary claim to the throne of Rome than his elder step brother Néron" (Pavel, "Racine and Stendhal," p. 274).

65. Caligula consciously imitated the incestuous despots of Egypt like Cambyses and the Ptolemys, claimed (according to Suetonius) that Augustus had had incestuous relations with his daughter Julia, and announced in A.D. 38 that he would marry his sister Drusilla and make her Empress. Suetonius reports that Caligula "had violated Drusillia during their adolescence." Caligula had lived in habitual incest with all his sisters (Santiago, *Children,* p. 58). After the death of Drusilla, Caligula exiled his two other sisters because they were "uncooperative."

66. With this double meaning compare Latin *sacer.* For other such terms, see Freud, *Totem,* p. 18.

67. C. E. Smith, *Papal Enforcement,* p. 6, shows that "adoption has the same effect in precluding marriage as does kinship by blood"; Fowler, "Incest Regulations," p. 40, says, however, that this view has been contested. On marriage between adoptive children, see relevant passages in Pope Nicholas I, "Responses to the Questions of the Bulgars," sec. 2, in Mansi, ed., 15:402. Nicholas "states that one ought to treat a godparent like a parent, even though the relationship is spiritual and not one of blood. There cannot be marriage in these relationships for the same reason that the Roman law disallows marriage between those one adopts and one's own children" (Fowler, "Incest Regulations"). For adoption in Roman Christianity as an impediment to marriage, see Justinian, *Digest,* 1.23, tit. 2, lex 17; Gratian, *Decretum,* caus. 30, q. 3, c. 6; Ivo Carnotensis, *Decretum, Patrologiae (Latina)* 161:657; and Oesterlé, "Inceste," in Naz, ed., *Dictionnaire* 5, 1297–314.

68. *Britannicus,* 480; Tacitus, *Annals,* 1.6.

69. *Britannicus,* 876; 1450.

70. On Roman women being barred from inheritance, see Watson, *Rome of the XII Tables,* esp. pp. 134–50 (on *mancipatio*), and pp. 150–57 (on *usus*). Barthes (*On Racine,* pp. 9, 38n) claims Agrippina is the patriarch of Racine's *Britannicus.* I think he is mistaken: *If*

there is a single patriarchal figure in the play it is not Agrippina but Augustus Caesar. Agrippina does have "male" qualities (like Clytemnestra's in the *Oresteia*), just as Nero has "female" qualities (like the ones he loved to portray on stage); see Montgomerie, "More an Antique Roman than a Dane."

71. Del Mar, *Augustus Caesar.*

72. Tacitus, *Annals,* 12.1–3.

73. *Britannicus,* 1157.

74. Tacitus, *Annals,* 12.37; in Racine, *Britannicus,* ed. Butler, Appendix 1, p. 191.

75. Tacitus, *Annals,* 12.42; emphasis mine; in Racine, *Britannicus,* ed. Butler, Appendix 1, p. 191.

76. *Britannicus,* 156.

77. Butler, notes to *Britannicus,* p. 153.

78. Gazier, *Étude sur Racine,* pp. 114–20; Picard, *Racine,* p. 306n; citing Chateaubriand, *Génie du Christianisme.*

79. *Britannicus,* 1415.

80. Barthes, *On Racine,* p. 84.

81. *Britannicus,* 254. In the sources, the woman with whom Nero fell in love was not Junia, but Acte. Racine says in his preface that the "Junia" with whom the Roman historians claim Nero had an affair is not the same person he depicts in *Britannicus:* The Junia of the historians, he reminds us, was accused of incest with her brother. Venesoen, *Racine,* p. 139n, suggests the playwright "historise avec coquetterie" the relations between Agrippina and Nero.

82. *Apokolokyntosis,* 14.1. Racine's Burrhus in *Britannicus* (805–6) calls attention to Seneca's absence. See Levitan, "Seneca in Racine," pp. 185–86.

83. Suetonius, *Nero,* in *Lives,* p. 8. The republican Julius Caesar had similarly refused the title.

84. *Hamlet,* ed. Jenkins, 5.2.402.

85. As in seventeenth-century France, the word *parricide*—literally "one who kills a father"—was taken to include "every attack against authority, including Father, Sovereign, State, and the gods" (Barthes, *On Racine,* p. 39).

86. Tacitus, *Annals,* 14.9.

87. *Britannicus,* 644.

88. Mozart's opera *La Clemenza di Tito,* with libretto by Metastasio, tells the story of a man who, like Hamlet, refuses to become cruel like Nero. The opera is set just after the affair of the incestuous Berenice and the Roman emperor Titus.

89. Already in 1670, Edmé Boursault argued that this twists historical facts and does not allow the spectator to pity Junia in the proper Aristotelian way (*Artémise et Poliante,* in *Britannicus,* ed. Caput, p. 127).

90. On Nero's reaction when he learns that Junia has successfully entered the vestals' sanctuary, see Butler, notes to Racine's *Britannicus,* 1. 1756.

91. In the story of Phaedra as told by Euripides, Hippolytus, son of the Amazon Antiope—"Hippolyte" is another name for "Antiope"—tries to devote his life to chaste Artemis (Euripides' *Hippolytus,* 81ff). But his attempt to be chaste angers the unchaste Venus (*Phèdre,* 257–58). Lemaître, *Racine,* stresses Phèdre's consciousness of the importance of the chastity she would violate, and he compares her to "some religious woman or nun consumed in her cloister by an incurable and mysterious passion" (cited in Racine, *Phèdre,* pp. 121, 123).

92. The profane aspect of *Britannicus* concerns the struggle between Nero and Agrippina (Hubert, *Essai,* and Edwards, *Tragédie*). The sacred aspect concerns "the conflict between

[novice-vestal] Junia, whom Racine treats as a tragic figure, and the world" (Goldmann, *Dieu caché*, p. 367). Cloonan, *Racine's Theatre*, p. 46, dismisses the idea that there is any real struggle: "Néron shares center stage with no one."

93. The kinship arrangement of the institution of the vestal virgins was like that of Roman society generally. See chap. 4, n. 206.

94. Boursault comments on the ending of *Britannicus* that Junia "becomes a nun in the order of Vesta."

95. See Angélique Arnauld's description of the Port-Royalist profession; for the ceremony, see Aulus Gellius 1.12.

96. Marbeck, *Book of Notes*, p. 15. On Roman adrogation, see also Watson, *Rome of the XII Tables*, pp. 41–42.

97. Blaise Pascal had opposed Jacqueline's intentions to "die to the world," citing propertal considerations. See Cousin, *Pascal*, p. 170.

98. Butler, notes to Racine's *Britannicus*, p. 169. Butler adds that the convent "was used . . . as a means of enforcing obedience upon a recalcitrant daughter."

99. See Brody's consideration of "the voice of blood" in Racine ("Racine's *Thébaïde*") and Barthes' suggestion that in the "Racinian metaphysics" of blood, "what is in question is not a biological reality but a form" (*On Racine*, p. 390).

100. One of her brothers, Silanus, had committed suicide on the day of Agrippina's triumphant marriage to Claudius, and Agrippina poisoned the other (*Britannicus*, 66, 226, 1141, 612). Silanus, who had been betrothed to Octavia (as Racine notes from Tacitus), was expelled from Rome as the result of the charge made by anonymous enemies (including perhaps Agrippina) that Silanus had committed incest with his sister.

101. Matt. 22:17–21. Cf. Luke 20:21–25 and Mark 12:14–17. Snowden, *Coins*, and G. Williamson, *Money*, pp. 69–70, suggest that the apostles refer to a silver denarius of Tiberius Caesar.

102. *Dieu caché*, pp. 368–71.

103. These words of Isaiah speaking to God (Isa. 45:15), central to Goldmann's analysis, are emphasized by Saint John of the Cross, *Spiritual Canticle*.

104. "I learn that, to save her child from death, Andromache tricked the clever Ulysses, while another child, torn from her arms, was led to death under the name of her son" (Racine, *Andromaque*, 1.1).

105. *Pace* Barthes, it is not so much infanticide as supposititiousness of parenthood that counts in Racine (Levitan, "Seneca in Racine," pp. 205–6; Barthes, *On Racine*, p. 39).

106. Thayer, *Preliminary Treatise*, pp. 346–47.

107. Fuller, *Legal Fictions*, p. 300.

108. Dewey, "Legal Personality," pp. 566–67.

109. See Saint-Victor, *Deux Masques*, pp. 3, 99, cited in Compagnon, "Proust and Racine," p. 41; cf. Lemaître, *Racine*, p. 283.

110. Gazier, *Racine*; and Orcibal, cited by LeRoy in his Notes to Sainte-Beuve, *Port-Royal*, 3:936.

111. *Esther*, 47, 48, 97.

112. *Esther*, 54.

113. *Esther*, 1033, 1037.

114. *Esther*, 485, 170. Certain critics, including Mauron, aver that Esther's "crime" is that she is a Jew.

115. Some of the Spanish and Portuguese Marranos, or *anusim*—the "raped ones": the Jews who were "compelled" to convert to Christianity during the sixteenth century but who also secretly remained Jews—prayed to Saint Esther as to a patron saint. The Marranos, some say, worshipped Esther (Yovel, *Spinoza*, p. 21). But was it really a secret that Esther

was a Jew? Although the Book of Esther (2.10; cf. 7.3ff) insists it was, it is nevertheless the case that she was taken from Mordecai's house—that of her uncle and foster-father—, who was well known to be a Jew (3.4; cf. 6.13). One Sabbath, in the (old) Spanish and Portuguese synagogue in Montréal as a child, I wondered if Esther, blessed as the savior of her people, could actually have converted—had been compelled to convert—to the Persian creed. Could it be, in fact, that Esther was one of the *anusim*? Perhaps her mother was not Jewish. Her being both crypto-Jewish and Persian would explain Esther's status as a Marrano heroine.

116. Racine, preface to *Esther*. On the presentation of these plays at Port-Royal, see Vossler, *Jean Racine*, p. 90, and Sainte-Beuve, *Port-Royal*, pp. 581–602.

117. *Esther*, 952, 996.

118. See King, *Tudor Royal Iconography*, esp. fig. 80.

119. Vaunois, *Racine*, p. 127. See Racine's letter to Vitart of May 30, 1662, in *Oeuvres*, ed. Picard, 2:437–39.

120. Boswell estimates that during the first three centuries of the Roman Empire, 20 to 40 percent of urban children were foundlings (*Kindness*, p. 135).

121. Matt. 27:46.

122. See Boswell, *Kindness*, p. 151. Cf. Psalm 22:1; the biblical Hebrew term *a-zav*—"abandon," "forsake"—is also used for a man's leaving his parents ("Therefore a man leaves his father"; Gen. 2:24), a wife's forsaking her husband ("You will be saved from the loose woman . . . who forsakes the companion of her youth"; Prov. 2:17), and an animal's abandoning its young ("Even the hind in the field forsakes her newborn calf because there is no grass"; Jer. 14:5). See Gesenius, *Hebrew Lexicon*, pp. 736–37.

123. Gen. 21:18.

124. *Tolle, tollere, subtuli.*

125. *Oxford Dictionary of Latin*, p. 1974A. *Aufhebung*, the German translation of the Roman *sublatio*, similarly refers to a "higher" recognition of something as one's own.

126. See my discussion of *Aufhebung* and *modus tollens* in my *Money, Language, and Thought*, pp. 139–42. Mozart, *Waisenhaus-Messe*, K. 139, was performed in Rennweg on December 7, 1768; the Empress and her children were in attendance. The passage from the Catholic Mass is here quoted from *Missel Quotidien*, p. 1130. It recalls John 1:29.

127. See Jesus' use of the terms for "adoption of abandoned children," *giothesia* and *adoptio* (Rom. 8:15 and 23, 9:4; Gal. 4:5; Eph. 1:5).

128. 1 Sam. 1–2.

129. For the argument that oblation is a form of abandonment, see Boswell, *Kindness*, p. 228. Cf. Boswell, "*Expositio* and *Oblatio*."

130. *Offero, offerre, obtuli.*

131. Cf. Daube, "Lex Talionis."

132. OED 7:18.

133. Boswell, *Kindness*, p. 74. In the greater oblation, or offering, physical bread and wine are consecrated in the Eucharist as the mysterious body and blood of Christ; in ordinary oblation consanguineous kinship is entirely sublated by kinship in Christ.

134. Bernard's cousin Robert, an oblate from Cluny who had his profession at Clairvaux, wanted to return to Cluny; Bernard said he couldn't and the Popes said he could. See Bernard, letters 1, 324, 325 (in *Patrologiae* [*Latina*] 182), and Boswell, *Kindness*, pp. 312–13.

135. In exceptional historical circumstances—the expulsion of the Jews and Muslims from Roman Catholic Spain in 1492 and the attempted genocide of Jews by Nazi Germany in the twentieth century—Jews and Moslems were willing to abandon their children, or at least permitted child abandonment.

136. *Ency. Judaica* 11:840–41.

137. Maimonides, *Book of Holiness,* pp. 21–23; cf. Boswell, *Kindness,* p. 351.

138. Boswell, *Kindness,* pp. 157–160. Vico condemns propagation outside the bond of the traditional "cyclopean" family as nonhuman, insofar as the children of such unions will be abandoned and devoured by dogs or raised as if they were animals, reverting to incest ("sons with mothers and fathers with daughters"—*New Science,* sec. 336; citing Xenophon, *Memorabilia* 4.4.19–23).

139. Some historians say that nearly 150,000 children were shipped out between the years 1853 and 1917 (100,000 placed by the Children's Aid Society founded in 1853 by Charles Loring Bruce and 50,000 placed by other institutions); see McOllough, "Orphan Train," p. 145. For the view of the Church, see Fry, "Children's Migration," p. 79.

140. Boswell, *Kindness,* p. 415.

141. In Paris in 1721, 9 percent of births were foundlings; by 1790, it had risen to 29 percent. Between 1713 and 1722, 8.6 percent of the births were bastards, whereas between 1785 and 1794, it was 19.8 percent (Meyer, "Illegitimates and Foundlings," p. 252; cf. the lower rates given in Yves Blayo, "Illegitimate Births in France"). In 1793 the revolution leveled the legal distinction between marital and extramarital children: even children of adulterous unions were given the same right as others to inherit. See Law of 12 Brumaire, Year 2 (1793). The term *enfants hors mariage* was used. In 1803 a new law entitled "Paternity Affiliation in the Civil Code" returned France to its more traditional concerns with illegitimacy.

142. Rousseau writes: "Five children resulted from my liaison with the poor girl who lived with me, all of whom were put out as foundlings. I have not even kept a note of their dates of birth, so little did I expect to see them again" (letter to the Maréchale de Luxembourg [née Madeleine-Angélique de Neufville de Villeroy], June 12, 1761 [Rousseau, *Correspondance,* 6:146]; my translation). Elsewhere he explains that "by destining my children to become workers and peasants instead of adventurers and fortune-hunters I thought I was acting as a citizen and father, and looked upon myself as a member in Plato's republic" (*Confessions,* p. 333). Jacobus, "Incorruptible Milk," comments aptly that "the *Republic* is imagined as Platonic parent, both the *enfant trouvé* and the culpable parent disappear from the record." Rousseau's "common law wife" was Thérèse Levasseur.

Rousseau's assignment of his consanguineous children to "oblivion" has encouraged moralists to condemn him as an unloving parent and even as a child-killer (since conditions in the foundling hospitals were often unhealthy). Rousseau's abandonment of his children is discussed by Blum, *Rousseau and the Republic of Virtue,* esp. pp. 74–92. For Rousseau's account in the *Confessions* of placing out his children, see his *Oeuvres,* 1:356–58.

143. For Sade, see his "Français, encore un effort si vous voulez être républicains," in *Philosophie,* esp. pp. 221-22.

144. For Rousseau, see his *Confessions,* in *Oeuvres* 1:357.

145. In 1717 the infant Jean le Rond d'Alembert was left on the doorstep of a church by Tencin, who denied maternity throughout her life. (Mme. de Tencin broke off her vows before 1714, when she received technical permission for her "secularization" from the Pope himself.) The illegitimate Jean was raised in foundling and foster homes. Thanks to the family of his natural father—the Chevalier Le Camus Destouches—d'Alembert was eventually educated at the Jansenist Collège, where he wrote a *Commentary on St. Paul.* See Hankinds' introduction in d'Alembert, *Traité de Dynamique,* p. xiii.

146. Le Rebours, *Avis aux mères,* p. 57.

147. Garden, *Lyon.*

148. See the frontispiece to the 1787 French edition of *Letters from an American Farmer* by Crèvecoeur, a French immigrant to America, entitled *Ubi panis, et libertas, ibi Patria.* It depicts "American sucklings feed[ing] . . . on their Indian Ceres" (p. 80; reproduced in Sol-

lors, *Beyond Ethnicity*, p. 77). On the idea of the American alma mater generally, see chapter 1.

149. Vovelle, *Revolution française* 4, 142. (For the Bastille, see 4, 143.)

150. Michelet, *Histoire* 1:6, 8, 9; 8:193, 194.

151. Revolutionary thinkers generally attacked the Catholic orders themselves, including the Cistercian houses, finding transcendent spiritual incest by participation in the Holy Family repressive and teleologically genocidal. They charged that even sincere monks and nuns, including those in the Jansenist tradition, were unable to sublimate their desires, and that most religious celibates were, in any case, insincere. (See Diderot's *La Religieuse* and other texts discussed in Ponton, *La Religieuse*.) They confiscated the property of the orders and executed their members. Francis Poulenc reminds us of the result, however, in his opera *Dialogues of the Carmelites:* The central scene of the *Dialogues* involves a sister's decision to become a Sister against the will of her brother; the last scene depicts a series of triumphantly sacrificial decapitations of the Carmelite Sisters, with the last to lose her head being the sister. (On monachism and the French Revolution generally, see Estève, "'Le Théâtre 'Monacal.'") The sister in the *Dialogues* is twice victimized, first in the name of the Christian family, when she dies "to the world," and then in the name of the nation, when she is decapitated.

152. De Quincey, *Autobiographical Sketch,* in *Collected Writings,* v, chap. i.

153. Coleridge, *Notebooks,* no. 1637. See Coleridge's interpretation of *Romeo and Juliet,* in *Lectures and Notes,* ed. Ashe, pp. 110–13.

154. *Contarini Fleming,* pt. 1, chap. 1; cf. pt. 1, chap. 7. Disraeli himself was very close to his sister Sarah.

155. Coleridge, fearful of incest, dropped his utopian ideals and excepted his sister from the group of all human beings. See letters of September 20 and October 14 (in Robert Southey, *Life and Correspondence* 1:219, 227). Cf. Coleridge, *Notebooks,* no. 1637 (November 1803), on William Paley's approach to incest *(Principles,* bk. 3, chap. 5).

156. Byron, *Selected . . . Works,* p. 35. Shelley's "Revolt of Islam" is, in part, an attempt to revive these ideas in the sphere of poetry.

157. As Chinese couples are compelled to have no more than one child and so were pressured into giving up or even selling any extras, so Romanian couples under the tyrannical Ceauçescu were compelled to have so many children that they had to abandon some and donate others to national hospices where they were raised, if at all, as equal brothers and sisters. Perhaps, if these children had been raised in decent living conditions—under the watchful eyes of consanguineous parents, as on Israeli *kibbutzim,* for example—such hospices might have encouraged the sort of liberal equality and communalist fraternity that Rousseau pretended to have in mind instead of abetting the death by AIDS that now faces so many Romanian children.

158. See Speiser, "Wife-Sister Motif," pp. 15–18, and Seters, *Abraham,* pp. 72–75.

159. See Arthur Wolf, "Adopt a Daughter-in-law."

Chapter 7

1. The average cost of keeping a pet suggests the importance pets have to their owners. In 1986 a ten-pound cat cost about $3957 during its lifetime. (This figure does not include the cost of "extras" such as licensing and grooming.) An eighty-pound dog cost $8353.

2. Each of these benefits has been studied separately. For protection and security, for example, see Sebkova, "Anxiety Levels." For companionship, see Beck and Katcher, *Between Pets and People;* consider also the advertising literature distributed by marketers of

"companion pets" to old age homes such as the Bide-a-Wee Association and the Pet-a-Pet Program.

3. See Erika Friedmann et al., "Pet Ownership and Coronary Heart Disease." For a general overview of the positive health consequences of pet ownership, see Erika Friedmann et al., "Health Consequences of Pet Ownership." Several recent essays in *The American Journal of Public Health* suggest that the health benefits of pet ownership have been much overstated.

4. On the use of pets to facilitate interactions among residents of homes for the aged, see Corson et al., "Socializing Role of Pet Animals."

5. Gomperz adds that one can also learn constancy in conjugal affections from some species of fowl (*Moral Inquiries*, pp. 20, 21).

6. The death of a pet is generally the American child's first experience of death, and widows and widowers often make pets their surrogate spouses. See esp. Kay et al., *Pet Loss*.

7. Some psychotherapists suggest that pet ownership may be particularly useful in cases of disturbed children. See Levinson, *Pet-Oriented Child Psychotherapy*, Link, "Helping Emotionally Disturbed Children" and "Pets and Personality Development." Cf. Mugford and M'Comiskey, "Value of Cage Birds With Old People," Arkow, *Pet Therapy*, and Rynearson, "Humans and Pets and Attachment." In the academic press, the "positive" benefits of pet ownership are stressed by various North American academic and veterinary institutions; see, for example, Canadian Veterinary Medical Association, *Proceedings*.

8. *Time* ("The Great American Petmania") reminds us that pets, if they are not considered "as cherished companions worthy of love and protection," can be considered only as "representing the more trivial part of our consumer-oriented society" (cf. Beck, "Population Aspects," p. 47, in Kay et al., *Pet Loss*). In a similar vein, Aillaud writes that "the practice of keeping pets . . . is part of that universal but personal withdrawal into the private small family unit, decorated or furnished with mementos from the outside world, which is such a distinguishing feature of consumer societies" ("Looking at Animals," p. 12). Tuan, in *The Making of Pets*, says that "the making of and maintenance of pets is, after all, a relatively innocuous occupation" yet focuses generally on the largely unacknowledged cruelty to animals that accompanies the institution of pethood.

9. Smart, "Of Jeoffry, His Cat," *Collected Poems*, pp. 118–20.

10. *Talking to Animals*, p. 200; cf. Woodhouse's movie, *Love Me, Love My Dog*, discussed in *Talking to Animals*, p. 190. The same theme runs current in Ruth Silverman, ed., *Dog Observed*.

11. Rynearson, "Pets as Family Members."

12. OED 7:745.

13. P. 246. Cf. Woodhouse's description of her relation to her new puppy Juno: "I called my new puppy Juno, and all the love for dogs I possessed now went to Juno, who from about ten weeks old became almost a human being to me—just as if the spirit of Jyntee [her recently deceased dog] had passed into her" (*Talking to Animals*, p. 164). Vicki Hearne, "How to Say 'Fetch!,'" p. 12, writes of her relationship to her dog Salty that "love, of course, is getting into things," but does not define the precise quality of her love as trainer; see too Hearne's "Moral Transformation of the Dog," also written from the trainer's viewpoint.

14. The ordinary definition of pet as animal assumes not only that the pet is a nonhuman animal but also that the pet owner is a human being. In some cases, however, the pet owner may think of himself as a nonhuman animal or may actually be a nonhuman animal. Consider here the example of the gorilla "Koko" and her pet kitten "All Ball." The psychologist Francine Patterson (*Koko's Kitten*) says that Koko "asked for" and received a pet kitten subsequently named "All Ball." (Patterson claims that Koko used sign language to make this request and to indicate such humanoid emotions as love and grief.) The human being Pat-

terson assumed a super-special kinship between herself and the gorilla Koko; by the same token, Koko supposedly assumed a kinship between herself and the kitten All Ball.

15. See Beck and Katcher, *Between Pets and People,* pp. 47, 49.

16. See Horn and Meer, "Pleasure of Their Company."

17. See Meer, "Pet Theories." See also "Undertaker for Pets" (anon.). On the ancient çustom of burying pets as though they were human kin, see Pollard, *Birds in Greek Life,* p. 136.

18. *Odyssey* 10,432–35: "Why are you so enamoured of these woes, as to go down to the house of Circe, who will change us all to swine, or wolves, or lions, so that we may guard her great house perforce?"

19. OED 7:745, The relationship between man and dog has been of special interest for the photographer—see, for example, the famous work of William Wegman and Man Ray.

20. *Pet-Love,* p. 16.

21. According to Johnson's *Dictionary* (1775) a "pet" is archetypally "a lamb, or a kid, taken into the house and brought up by hand, a *cade* lamb." Synonymous terms are "cosset," "sock," "tiddle," and perhaps also "Anthony pig." There are few, if any, translational equivalents into other languages.

22. On pets as children, see Cain, "Study of Pets in the Family System," and Beck and Katcher, *Between Pets and People,* esp. p. 73. On pets as grandchildren, see D. Taylor, "Grandchildren Versus Other Semi-Domesticated Animals." On pets as idealized mothers, see the anthropologist Constance Perin (discussed in Beck and Katcher, *Between Pets and People,* pp. 84–85). And on the general function of pets as surrogate relatives, see Wessels, "Family Psychotherapy Methodology," Keddie, "Pathological Mourning," and Rynearson, "Humans and Pets and Attachment."

23. In a recent survey, a broad spectrum of Americans were "asked to rate several aspects of their lives in order of how important they were." "Five out of six respondents naturally named their immediate families as number one. But so many put pets second and third that, combining the top three ratings, pets ranked right behind friends and relatives, and ahead of the job" (Horn and Meer, "*PT* Survey," p. 54).

24. The example of the dog's relationship to humankind may be instructive. *Canis familiaris* (literally "familiar dog") is Linnaeus' "scientific" name for the group of dogs suitable for human domestication (i.e., able somehow to join a family or household or, if you want, able to domesticate man). Linnaeus' choice of name (in *Animal Kingdom*) suggests that he was classifying animals by their human family or by their familiarity to humans. He thus distinguished the familiar, faithful dog *(Canis familiaris)* from the wolf and the wild dog, and subdivided the category of the faithful dog into varieties such as the sheepdog *(Canis domesticus)* and the turnspit.

25. Beck and Katcher (*Between Pets and People,* p. 73) write that "since the pet has the status of a favored child in the family, sexual exploitation of pets is a kind of incest," and they claim that "zoophilia can be a kind of incest" (p. 77). Yet Beck and Katcher seek ultimately to distinguish the one taboo from the other, insisting, for example, that the taboo on bestiality is more "effective" than that on incest and ignoring the logical connections between species and family boundaries that make both taboos parts of an ideological or political whole.

26. Concerning pets as transitional objects, consider the view of Rappaport, "Zoophily and Zooerasty," 565–66: "The assumption that animal pets . . . allow children to maintain a healthy skepticism in separateness from the universe [an assumption made by Searles in *Non-Human Environment*] and prepares them for future interpersonal relationships is controversial, unless the animal serves only as a transitional object which automatically and in

time becomes decathected before a too intensive identification with the animal has been established." (On the transitional object in the sense in which Rappaport here uses the term, see Winnicott, "Transitional Object and Transitional Phenomena.") The view that animals can play an important role in the transference of affection is not a specifically twentieth-century one; thus Richard Steele, in one of his essays (1710; no. 266), writes of a woman's "transfer[ring] the amorous Passions of her first Years to the Love of Cronies, Petts, and Favourites [a dog, monkey, squirrel, and parrot]."

27. In *Totem and Taboo* (1912–13), Freud "observed the similarities between the reactions of children and of primitive men to animals. Among both there is no trace of the common 'adult' arrogance towards animals; the child regards the animal as its *equal* and feels more *akin* to the animal in its uninhibited awareness of its needs" (Heiman, "Relationship Between Man and Dog," pp. 582–83). Heiman writes that "the dog may be considered a descendant from a totem animal used by man in his development and useful to him in the process of civilization. . . . The domesticated animal, in particular the dog, is for civilized man what the totem animal was for the primitive" (p. 584).

28. OED 8:1610.

29. For an instance of this meaning, see Caruthers, *Kentuckians in New York* 1,175: "Oh! it is nothing more than puppy love." And for "puppy" as an asexual human plaything, see Besant and Rice, *Chaplain of the Fleet* 1,10: "I was once the pet and plaything of ladies, a sort of lapdog."

30. OED 8:1610; emphasis mine. According to the dictionary, toy dogs are small dogs "of little value or importance" (OED), "value" being used here in the sense in which we say that nonworking animals lack it. The breeding of toy dogs marks a new beginning in the history of the interspecies relationship between man and dog.

31. For the links between *puppy, poupée,* and *puppet,* see OED 8:1610.

32. An accurate counterpart to *puppy lover,* which suggests a relationship that is at once bestial and not bestial, is *kissin' cousin,* which suggests the possibility of a relationship that is at once both incestuous and nonincestuous.

33. Horn and Meer, "*PT* Survey," p. 58. Kinsey et al., *Sexual Behavior in the Human Male,* report that between 40 and 50 percent of farm boys surveyed acknowledged having sexual activity with animals.

34. "Who will know a generation hence that a snugglepup is a young man who attends petting parties, and that a petting party is a party devoted to [mere] hugging?" (Krapp, *English Language in America* 1,117).

35. *Playboy*'s word choice may be especially appropriate: An older word for "rabbit" is the English *coney,* from the Latin *cuniculus.* Leach ("Animal Category and Verbal Abuse," p. 50) writes that "the eighteenth-century rabbit was a *cunny,* awkwardly close to *cunt,* and he draws a parallel between a Playboy Bunny Club and a London eighteenth-century cunny house. Cf. the *National Lampoon* parody magazine, *Pethouse,* which includes sexy animal photographs.

36. Such doubt is one gist of an article in the *Chicago Tribune,* February 24, 1948, III.1/2.

37. OED 7:745.

38. Cf. Chapman, ed., *New Dictionary of American Slang,* s.v. "Heavy petting," p. 203.

39. The double-entendre becomes more explicit, perhaps, in the case of the lap-dog. The breeding of lap-dogs (cf. the German *Schosshundchen*) might make for a special chapter in sexual and cultural history. "The smaller dogs they be," writes John Caius, "the more pleasure they provoke, as more meet playfellows for mincing mistresses to bear in their bosoms, to keep them company withal in their chambers, to succour with sleep in bed, and nourish with meat at board, to lay in their laps and lick their lips as they ride in wagons"

(Caius, *Of Englishe Dogges*, p. 21; a translation of Caius' sixteenth-century *De canibus britannicis*). On women who train dogs to lap their genitalia, see Leigh, "Psychology of the Pet Owner."

40. The *Athenaeum* of April 27, 1889 (no. 534) contains a fine instance of this use of the term: "His fatherly affection for his children . . . takes the form of unreasonable petting."

41. Friedmann et al., "Health Consequences," p. 26; cf. Katcher et al., "Men, Women, and Dogs."

42. For Freud, see his "Analysis of a Phobia in a Five Year Old Boy" (1909) and "From the History of an Infantile Neurosis" (1918). For Deutsch, see her "A Case of Hen Phobia." For Ferenczi, see his "Little Chanticleer," in *Sex in Psychoanalysis.*

43. Other relevant psychological literature includes Schneck, "Zooerasty and Incest Fantasy," and Beryl, "A Patient and Her Cats."

44. French writes of Cobbe, the pet-loving leader of the anti-vivisectionist movement in the 1870s, thus: "Her dog and her cat are a great deal to her; and it is the idea of *their* suffering which excites her. . . . She is not defending a right inherent in sentient things as such; she is doing a *special* pleading for some of them for which she has a *special* liking" (French, *Antivivisection*, p. 375; for Cobbe, see Cobbe, *Italics*, pp. 443–44 [emphasis mine]).

45. The title of a book by Boone that is popular among pet lovers. Concerning the kinship with all life that some pet owners feel, see also M.W. Fox, *Soul of the Wolf.*

46. *Canterbury Tales*, General Prologue, ll. 144–49.

47. Smart, *Collected Poems* 1,227.

48. The Persian Artaxerxes shared his throne with a real bitch (Schochet, *Animal Life*, p. 141). For an early discussion and condemnation of marriage contracts between human beings and animals, see Genesis Rabba 26:5.

49. Perrault's influential collection of folk tales (1697) lacks a "Beauty and the Beast." That tale did not appear in something like its nineteenth-century version until its publication in the collection of Madame Leprince de Beaumont (1756). But Perrault did include "Peau d'Âne," with its similar concern with liberty and incest; see Dournes, "L'Inceste préferentiel." The first volume of the Grimm brothers' collection was published in 1812.

50. For an example of this rational or Enlightenment view of the story, see the interpretation of Robert Graves (cited in Opie, *Classic Fairy Tales*, pp. 182–95).

51. See, for example, Bettelheim, *The Uses of Enchantment*, and Barchilon, "Beauty and the Beast."

52. See Mintz, "Meaning of the Rose."

53. In this context, "Beauty and the Beast" is a defining fulfillment of the literary form called the "animal groom story." The term *groom* both names and skirts the quandary involving the all-important distinction in the tale between legitimate and illegitimate sexual intercourse. *Groom*, as *bridegroom*, means a man both just before his marriage and just after it. However, we are less concerned here with whether the union of the groom with his bride is premarital or marital than with whether, as one who attends on horses, his union with his counterpart is bestial, or intraspecies.

The genre includes movie fantasies such as *King Kong* (the ape loves Fay Wray), *The Fly* (who has a wife), and Paul Anka's "Puppy Love." In the folktale, Beast is both in tune with Beauty's family and species and not in tune with them; he is, in this sense, a "'toon," at once human and animal, as in the recent feature-length part-cartoon movie *Who Framed Roger Rabbit?*, in which Roger the talking rabbit is married interspecially not to a human woman, but to a 'toon of woman. (For a more genre-oriented treatment of the topos of the animal groom in folk tales, see Lutz Röhrich, *Märchen und Wirklichkeit.*)

In most "beast fables" other than "Beauty and the Beast" the talking animal is a thinly disguised human being and the "moral" is moralistic, or, as Rousseau would have it, immoral. Authors of "beast fables," which originated with Aesop during the Age of the Tyrants, disguise their men as animals because they fear the political persecution that would ensue from their speaking outright (it is safer for a slave such as Aesop to argue for animal than for human liberation) or because they find it easier to depict a fox that is really a sly man in bestial disguise than to depict a sly man. (See my analysis of Rousseau's and Locke's interpretation of Aesop's "The Fox and the Crow" and "The Fox and the Hedgehog" in *The Economy of Literature,* pp. 113–28.) "Beauty and the Beast" goes beyond the simple characterization of human traits by means of exaggerated and disfiguring cartoon-like masks, to the point where it might demarcate those species and family boundaries between beast and human that themselves ground and help explain the perennial popularity and aesthetic force of the beast fable, the animal caricature, and the twentieth-century cinematic animal cartoon.

54. The version of the story told by Madame Leprince de Beaumont includes three brothers as well as a father and sisters. The 1761 English translation of de Beaumont's version is included in Opie, *Classic Fairy Tales,* pp. 182–95. In Cocteau's movie, Beauty has an apparently exogamous suitor (Avenant) who dies and/or is transformed into a dead beast at the moment that Beast himself is transformed into a handsome man.

55. *History of Human Marriage* 2.37–47.

56. So described by Cocteau in *Diary,* October 2 (p. 52).

57. "I'll *never* leave you," says Beauty to her father in Cocteau's film. And insofar as her father is Beast she never does leave him.

58. Psyche's jealous sisters tell her that, though she did not realize it, she was spending her nights with a great serpent (cf. Opie, *Classic Fairy Tales,* pp. 180–81; Rouse, preface to Apuleius' *Cupid and Psyche,* pp. xxiv–xxvii; and Neumann, *Amor and Psyche,* p. 11). In "Cupid and Psyche" the oracle's awful pronouncement is: "Nec speres generum mortali stirpe creatum / Sed saevum atque ferum vipereumque malum."

59. Yet there are hints of incest in Apuleius' tale. In his "Cupid and Psyche," the sea mew speaks thus to Venus/Psyche: "And so there has been no pleasure, no joy, no merriment anywhere, but all things lie in rude unkempt neglect; wedlock and true friendship and parents' love for their children have vanished from the earth; there is one vast disorder, one hateful loathing and foul disregard of all bonds of love" (cited in Neumann, *Amor and Psyche,* p. 31).

60. On the two realms of the amphibian groom as land and water, see Heuschner, *Psychiatric Study of Myths and Fairy Tales,* p. 213. For centuries many people believed that tadpoles and frogs were two separate animal species, the individuals of one species somehow transforming themselves into those of the other (cf. Bettelheim, *Uses of Enchantment,* p. 290).

61. See 1 Pet. 5:14; for Paul, see Rom. 1:16, 1 Cor. 16:20, and 1 Thess. 5:26. See also Crawley, *Mystic Rose,* 344ff., and Perella, *Kiss,* esp. pp. 12–50.

62. The doctrine that "All ye are brethren" is acted out by Christians at Christmas festivities, where celebrants are masked, no one can tell who is a consanguineous kin and who is not, and everyone is enjoined to kiss everyone else equally, much as in the "original" Saturnalia of libertine Rome on which Christmas is partly based. In his essay on the Roman carnival, or Saturnalia, Goethe claims that "though it postponed the festival of the Saturnalia with its liberties for a few weeks, the birth of Christ (Christmas) did not succeed in abolishing it" ("Roman Carnival," in *Italian Journey,* p. 446). Not only can a son pass for his father or a father for his son—resulting in liberty of the kind Goethe witnessed in the celebration of the Saturnalia at Rome—but one's sister can pass for a woman who is not one's kin, resulting in incest. "In the world of the carnival," writes Bakhtin, "all hierarchies are canceled. All castes and ages are canceled. During the fire festival a young boy blows out his father's candle, crying out . . . 'Death to your father, sir!'" (*Rabelais,* p. 251). For details of the festival as it pertains to animals, see Frazer (*Golden Bough* 11, 291n).

63. Bettelheim, *Uses of Enchantment*, p. 199, writes that Beauty, the maiden, "transfers her attachment from father to lover." Bettelheim argues in the same way that "only marriage made sex permissible, changed it from something animal-like into a bond sanctified by the sacrament of marriage" (*Uses of Enchantment*, p. 283).

64. Opie, ed., p. 195.

65. The blend of human with beast does not require a male "animal"; female counterparts to the animal father/animal groom are common in the literature. So too are animal nursemaids, or wet nurses, who transmit kinship (and species kind) through their "milk" just as an animal mother or father transmits kinship through the "blood." Milk kinship as such, according to the regulations of many societies, resulted in the same diriment impediments to marriage as blood kinship. In the eighteenth century, the social institution of the nursemaid was widespread; there were many historical and literary accounts of foundling and orphan children fostered by animals in the forest. What did this mean for the child fostered by bears or wolves? It made sexual intercourse between them incestuous (they and their lupine foster siblings were of the same milk) just as it made intercourse with other human beings bestial (they were now human wolves, or werewolves).

66. The Cathars, renounced by the Church for their Albigensian heresy, refused to recognize the distinction between kin and nonkin, and so renounced procreation, just as they refused to recognize any difference between men and animals, and so became vegetarians. See Lea, *Inquisition* 1,97.

67. On the specific taboo against marriage between milk siblings—people who received the "milk of human kindness" from the same nurse—see Crawley, *Mystic Rose* 2,230. On the Virgin Mary as nurse mother, see Warner's "The Milk of Paradise" in her *Alone of All Her Sex*. On Queen Elizabeth and Moses as nurse parents, see chap. 4, nn. 257–58, in the present volume.

68. Montaigne, *Essais*, in *Oeuvres*, ed. Thibaudet and Rat, bk. 2, chap. 8, pp. 379–80; trans. Frame, pp. 290–91.

69. For Feuerbach's aphorism, see Barth's introductory essay in Feuerbach, *Essence of Christianity*, p. xiv; and Feuerbach, *Die Philosophie der Zukunft*, pp. 89–90.

70. "But let mothers deign to nurse their children, morals will reform themselves" (Rousseau, *Emile*, trans. Bloom, p. 46). On the general history, see Fildes, *Wet Nursing: A History;* and Le Rebours, *Avis aux mères*, p. 57; cited in Jacobus, "Incorruptible Milk."

71. Garden, *Lyon*.

72. When England and Scandinavia adopted the widespread use of cow's milk, many people were similarly concerned about long-term "animalizing" effects on the culture. For further references, see Meyer, "Illegitimates and foundlings in pre-industrial France," in Laslett et al., eds., *Bastardy*, pp. 249–63.

73. In 1802 William Paley wrote that "the experiment of transfusion proves that the blood of one animal will serve for another" (*Natural Theology*, p. 484). The history of blood transfusion proper begins in the latter half of the seventeenth century, when scientists conducted successful experiments transfusing the blood of one animal "into another [animal] of the same or a different species" (*Philosophical Transactions of the Royal Society* [1666], 353).

74. The term *vaccination* was taken from the name of the eruptive disease of cattle called *vaccinia*, or cowpox.

75. In 1796 he wrote, "Will you try to look out for a fit servant for us . . . scientific in vaccimulgence? That last word is a new one" (*Biographica Literaria*, cited in OED 7:5).

76. 1.2.62; 1.2.30.

77. 1.21–24.

78. Cf. the comparison in *Merchant of Venice* between Laban's breeding of lambs (which Christians call natural) and Shylock's breeding of money (which Christians call unnatural or perversely artful). On the relationship between the offspring of sexual and monetary gen-

eration—both are indicated by the Greek term *tokos*—see my *Money, Language, and Thought,* pp. 48–55.

79. 4.4.82,98; cf. OED 1:695.

80. Cf. p. 33: "We practice . . . all Conclusions of Grafting and Inoculating, as well of Wilde-Trees as Fruit Trees."

81. Definitions from OED, 5:317.

82. Inoculation was already controversial in 1722, when Nettleton (*Philosophical Transactions of the Royal Society* 32, 214) wrote about people who had died after having been inoculated with the smallpox. See OED 5:317.

83. "Inoculation with the virus of syphilis, as a means of cure or prevention" (OED 10:389).

84. Stern, *Should We Be Vaccinated?,* p. 14.

85. Cited in Stern, *Social Factors,* p. 61.

86. Moseley, *Medical Tracts,* pp. 182–83.

87. Rowley, *Cow Pox Inoculation No Security.*

88. Squirrel, *Observations.*

89. Stern, *Factors,* pp. 56–57.

90. Stern, *Should We Be Vaccinated?,* p. 22.

91. The Minotaur had the body of a man and the head of a bull. His, or its, half-sister was the incestuous Phaedra, who married the minotaur-fighting Theseus—as Racine reminds us in his *Phèdre* (cf. Euripides, *Hippolytus,* esp. 1.337). Seneca's *Phaedra* compares such incestuous desires as Phaedra's for her (step-)son Hippolytus with such bestiality as Pasiphaë's; Theseus says that "even the beast abhors forbidden union, instinct teaches proper respect for laws of generation" (*Four Tragedies,* p. 134). See chapter 2 for a discussion of the historical and conceptual connections of the Spanish bullfight with the Minotaur.

92. Moseley, *Medical Tracts,* pp. 182–83.

93. Stern, *Factors,* p. 59.

94. Stern, *Should We Be Vaccinated?,* p. 82. Cow maniacs argued with some reason that most objections to vaccination came from the merely literary or philosophically inclined. (There were also economic interests at work: many of the cow-phobic doctors had lucrative practices to protect. And certainly the cow-maniac scientists' methodology and cleanliness were not always above reproach.) Dr. Jenner, who first popularized vaccination in the last years of the eighteenth century, recognized this "philosophical" tack of his detractors when he criticized the well known German Dr. John Ingenhousz. (Dr. Ingenhousz had tried inoculation in 1768, when he had vaccinated some members of the imperial family at Vienna, but by 1799 he was arguing vociferously against it; Stern, *Should We Be Vaccinated?,* p. 10.) "This very man Ingenhousz," wrote Dr. Jenner, "knows no more the real nature of the cow-pox than Master Selwyn does of Greek. Yet he is among the philosophers what Johnson was among the literati" (Stern, *Should We Be Vaccinated?,* p. 11, citing John Baron, *Jenner,* p. 296). And in trying to defend himself against the cow-phobic Dr. John Sims, President of the London Medical Society, Jenner wrote of him as "the philosophic and medical critic" (Jenner, *Medical and Physical Journal,* vol. 1, p. 11; cited in Stern, *Should We Be Vaccinated?,* p. 12).

95. Adams, *Answers,* p. 29.

96. Minutes of the Third Festival of the Royal Jennerian Society in 1805.

97. Dr. Stromeyer of Hanover stated in 1800 that Herz's position was the norm: "Most of our physicians here exclaim against the vaccine inoculation" (Stromeyer, letter to the *Medical and Physical Journal,* vol. 3, p. 471). Among a few physicians the reception was more generous. Dr. I. Faust of Buckenburg, Professors Juncker and Sprengler of Halle, and Dr. Hufeland of Jena soon adopted Jenner's views, and his book was soon translated. For the German background, see Ring, *Treatise on Cow Pox.*

98. De Quincey, *Last Days*. For this and some of the following references, I am indebted to Feuer, *Lawless Sensations*.

99. Cf. W. Wallace, *Kant*.

100. See *Jewish Ency.* 6:368. Dr. De Carro, of Vienna, a chief proponent of vaccination, mentions Herz as an important opponent. Throughout the nineteenth century, there continued to be so much "hysterical" German opposition to vaccination—especially to its being compulsory, as the German Emperor wanted it to be—that Bismarck convened the Imperial Vaccination Commission in 1884.

101. We might observe here the linkage between biological and social tolerance. Biological tolerance involves an organism's willingness or ability to survive *despite* infection with a parasite or otherwise discomforting organism, much as religious or social toleration involves an individual's or a group's putting up with the discomfort of having strangers around. Jenner's description of immunological vaccination involves an organism's surviving infection *because* of infection: his patients survived smallpox thanks to the tolerance they gained from cowpox. The dangerous analogy here between medical terminology's "pathogenic organism" and sociological terminology's "stranger among us" is common since at least sixteenth-century Spain.

102. In some tribes "the contrast between man and not-man provides an analogy for the contrast between society and the outsider" (Douglas, *Implicit Meanings*, p. 289; cf. Needham, *Primordial Characters*, p. 5).

103. I am not concerned here with working out all implications of this historically controversial model—that Christianity has itself idealized and accepted the doctrine is enough for us. However, the central fact about the Christian approach to kinship—that it substitutes an apparently extraordinary kinship for an ordinary one—still needs clarification, insofar as social anthropologists have generally failed to note the primary role it plays. They ignore, for instance, the Christian debate about kinship in the New Testament, and neglect to consider the kinship practices of Christian society as suggested by the Catholic orders (Brothers and Sisters all), the subgroup comprising the clergy and laity (Parents and Children), and even the bond of kinship between Jesus and both his female progenitor (Mary, the sister, wife, child, and mother of God) and his male Progenitor (the father who is also the son).

104. This representation may be inaccurate in two ways at least. First, Christianity, however much it idealizes itself as universalist ("All men are my brothers"), becomes emphatically particularist in practice ("Only my brothers are men"). Indeed, the extremism of the claim to universalism would seem to ensure particularism in practice. Second, the connection between universalism and particularism in Judaism is more complex than the ordinary Christian view of Judaism would allow. On the one hand, Judaism certainly does enjoin several particularist legal doctrines, the most famous such enjoining Jews to distinguish between brothers and others when making monetary loans: "Thou shalt not lend upon usury to thy brother. . . . Unto a stranger thou mayest lend upon usury" (Deut. 23:19–20, cf. Deut. 28:12 and Lev. 25:35–37, and for further discussion, see my discussion of the connection between monetary and sexual generation in *Money, Language, and Thought*, chap. 3). But, on the other hand, Judaism also has a powerful universalist tendency in both doctrine and practice (on which see chap. 8, esp. nn. 92–95, in the present volume).

105. White, "Historical Roots of Our Ecologic Crisis."

106. See Thomas, *Man and the Natural World*, p. 137, on the Protestant reaction to such legends. Montaigne says of men and animals: "we are neither above nor below the rest" ("Apology for Raymond Sebond," *Essais*, bk. 2, chap. 12; trans. Frame, p. 336). See Howard Williams, *Ethics of Diet*, pp. 113–22, 128–50.

107. For the carol, see "Carol of the Beasts," sometimes also called "The Friendly Beasts," in Paola, *The Friendly Beasts*.

108. Although "stable" or "manger" is probably an inaccurate translation of the Greek

word in the New Testament, Christians throughout the world persist in depicting the place and singing about it as such (fabulous *crèche* scenes are good examples).

109. The quotations are taken from the well-known Christmas carol "Away in a Manger."

110. Woodhouse, *Talking to Animals*, p. 11.

111. In the Christian story of God's earthly birth, the human Mary is filled by the spirit of God and subsequently delivers the godman. The tale of her literal "enthusiasm" was often compared to similar tales about interspecies generation by human mothers. One example would be the well-known and often credited eighteenth-century report of Mary Toft's giving birth to bunnies. *The British Journal* commented, "A fine Story—Credat Judaeus Apella" (December 3, 1726), and *A Letter from a Male-Physician in the Country to the Author of the Female Physician at London* argued that "it is impossible for women to generate rabbits or other animals" (cited by Paulson, *Hogarth* 1, 169). William Hogarth, mocking the Methodist John Wesley and the English "Enthusiasts" who generally believed in the productive ability of interspecies generation, represented the report in his engravings entitled "Cunicularii" (1726), "Enthusiasm Delineated" (1760), and "A Medley—Credulity, Superstition, and Fanaticism" (1762). See Paulson, *Hogarth* 2, 170, 299, 354; and *The Journal of the History of Medicine* 28, 3 (July 1973): 282–83.

112. "The animal was sometimes dressed in clothes and tied in a sitting position during the trial. In 1386, a pig in Falaise, Normandy, that had torn the face and arm of a small child was dressed in clothes and sentenced to be maimed in the same manner as the child. In 1685, a wolf in Austria that had killed several people was dressed in clothes, wig, and beard. His snout was cut off and a human mask tied over it during the trial. He was sentenced to be hanged." Sometimes the punishment was kinder: "An Austrian dog that bit a man was sentenced to only a year in jail in 1712." In many courts all animals—including insects—were allowed to have lawyers. (Mannix, *Torture*, cited by Schochet, *Animal Life*, pp. 141–42.)

113. This point is developed by Thomas, *Man and the Natural World*, pp. 17–24. There are, of course, factors other than religious doctrine to help explain the particular kind of exploitation of animals that one finds in Christendom.

114. Luke 8:33. "They were about two thousand," writes Saint Mark (Mark 5:13).

115. Augustine, *Catholic and Manichaean Ways of Life*, p. 102; emphasis mine; cited in Schochet, *Animal Life*, p. 274. The ellipses here represent references to the Parable of the Withered Fig Tree (Luke 13).

116. Singer, *Animal Liberation*, pp. 213–14. Pius was Pope from 1846 to 1878.

117. See God's statement to Noah (Gen. 9:8–10): "Behold I establish My covenant with you . . . and with every living creature that is with you, the fowl and the cattle and every beast of the earth with you."

118. Schopenhauer wrote that the "revolting, gross, and barbarous, view, peculiar to the West . . . that our conduct in regard to animals has nothing to do with morals or that we have no duties towards *animals* . . . has its roots in Judaism" (quoted in Unna, *Tierschutz*, p. 6). He was wrong about Judaism, but since he was attributing to Judaism what he did not like about Christianity, his argument was influential among millennialist "pro-animal" and "anti-Semitic" propagandists such as Richard Wagner, Adolf Hitler, and Heinrich Himmler. Himmler linked the vegetarian world to come with the extermination of the Jews. The millennium would soon arrive, he said, promising a definitive defeat of Great Britain and the United States. And with the coming of that millennium, insisted the great vegetarian—even as he oversaw the shipping of meat to members of the Gestapo and the slaughter of millions of human beings in the death camps—all animal-killing would be punished as a capital crime for the very reason that it is really homicide. Felix Kersten reports that during the war Himmler told him: "After the war I will issue the most rigorous laws for the protection of animals. In schools the children will be systematically taught to love animals, and I will give special

police authority to the societies for the protection of animals" (quoted in Syberberg, *Hitler,* p. 207).

119. Exod. 23:5; Deut. 22:4.

120. Exod. 23:12.

121. Deut. 25:4.

122. Prov. 12:10. Other relevant biblical texts are Num. 22:28, Deut. 22:7, Isa. 1:11, and Jon. 4:11. The passage from the Book of Jonah reads: "And should I [God] not pity Nineveh, that great city, in which there are more than a hundred and twenty thousand persons who do not know their right hand from their left, and also much cattle?"

123. For examples, see Thomas, *Man and the Natural World,* p. 358, n. 7. On pets in general in the ancient and early Christian world, see Clutton-Brock, *Domesticated Animals from Early Times.*

124. "For it is written in the law of Moses, Thou shalt not muzzle the mouth of the ox that treadeth out the corn. Doth God care for the oxen? Or saith he it altogether for our sakes? For our sakes, no doubt" (1 Cor. 9:9–10).

125. Gen. 1:26 is a key biblical text in this area: "Then God said, 'Let us make man in our image, after our likeness; and let them have dominion over the fish of the sea, and over the birds of the air, and over the cattle, and over all the earth, and over every creeping thing that creeps upon the earth." See also Thomas, *Man and the Natural World,* pp. 17–24.

126. Pococke, *Commentary on the Prophecy of Hosea,* pp. 95, 97.

127. Thomas Aquinas, *Summa contra gentiles* 3, 113. Cf. the different context of the argument of Maimonides that we should be kind to animals so that cruelty will not become a habit (*Moreh Nebukhim* 3,17, 3,48). "Since nature makes nothing purposeless or in vain it is undeniably true that she has made all animals for the sake of man."

128. Calvin, *Commentaries on the Last Four Books of Moses* 3, 56–57.

129. In Judaism special emphasis is given to the *tzaar* and to the argument of Moses Nachmanides (1194–1270) and others concerned with the disappearance of a whole species (Nachmanides, Commentary to Deut. 22:6; *Sefer haHinnukh,* Mitzvah 294, 545; and the *Kol Bo* [late thirteenth century], chap 3). Cf. Schochet, *Animal Life,* p. 216.

130. There are some interesting exceptions: The Franciscan document *Dives and Pauper* (1410), for example, allows flesh-eating, yet outlaws cruelty, much as the Jewish law codes do (see Priscilla Barnum, ed., *Treatise on the Ten Commandments*).

131. Thomas, *Man and the Natural World,* writes that "the opponents of animal cruelty drew primarily on the doctrine, which they found to be latent in the Old Testament, of man's stewardship over creation" (p. 154).

132. Thomas *(Man and the Natural World),* whose historical scholarship is otherwise quite excellent, is one of those who adopt, unquestioningly, the sentimentalist Christian view of the Jewish understanding of kindness and cruelty to animals.

133. Consider here the Fruitarian movement, whose members claim that humans should eat only fruit without seeds. The commercially successful fad for buying and caring for stuffed animals and pet rocks (pet rocks were sold in the hundreds of thousands during the mid-1970s) suggests how the institution of pethood might enable us to toe the line not only between human and animal beings but also between animate and nonanimate beings.

134. Fox, *Returning to Eden: Animal Rights and Human Responsibility,* presents a typical sentimentalist and idealist attack on "the usual motivational and cognitive approach of scientific inquiry" into the question of animal welfare; Fox criticizes that inquiry for being "based upon Baconian utopianism, Cartesianism, dominionism, and hubris" (M. W. Fox, "Pet Animals and Human Well-being").

135. On slaughtering mothers apart from children and vice versa, Nachmanides, Commentary to Deut. 22:6; on bird's nest law, Schochet, *Animal Life,* 179–86, and Lev. 22:28.

136. One rule that interpreters sometimes use to illustrate the connection of kindness

(or regard) to animals with Kashrut is that one must not boil a kid in its mother's milk (Exod. 23:19 and 34:26; Deut. 14:21). To do so would be "unseemly." Such unseemliness is said to be one basis of the Jewish prohibition against mixing milk with meat.

137. I do not mean that all people who are Christians are omnivorous, only that the Christianity they profess allows them to be so. Thus a Christian Englishman, when he refuses to eat dog meat, does not do so because of a Christian injunction; a Jewish Englishman, however, when he refuses to eat pork, does so because of a religious injunction. The Christian might say that the flesh he refuses to eat is not food, so he will not eat it; the Jew might say that the flesh he refuses to eat is food, but he will not eat it. (Cf. Leach, "Animal Category and Verbal Abuse," p. 32.)

138. Fox, "Pet Animals" (in Kay et al., eds., *Pet Loss,* p. 16), writes that "for the British, eating dog is akin perhaps to cannibalism." Lyttelton, *Dialogues of the Dead* (1795) (cited in Thomas, *Man and the Natural World,* p. 55), writes that "Montstuart Elphinstone, in the 1840s, reacted with horror to the Italian habit of cooking robins [which were kept as pets in England]. 'What! Robins! Our household birds! *I would as soon eat a child*'" (emphasis mine). "Yet in the Elizabethan age 'robin red-breasts' had been 'esteemed a light and good meat'" (cited in Thomas, *Man and the Natural World,* p. 116). Cf. the punishment in Kashmir of vaccicide (cow-killing) as if it were homicide (human-killing). For a general discussion of symbolic aspects of "spiritual cannibalism," see my *Money, Language, and Thought,* ch. 2, and *Art & Money,* ch. 2.

139. Among the sources here are the regulations set down for the religious celibates of the medieval era. On the ban on pets in the monasteries and nunneries, see Power, *Nunneries,* pp. 305–7; and Platt, *Southampton,* p. 104.

140. In some French societies it is still taboo to give human names to dogs (Lévi-Strauss, "Religion, langue, et histoire"). We English-speaking humans are sometimes offended when our fellows give "Christian" names to animals; see, for example, Taylor, *Wit and Mirth* (1630), in Carew, ed. *Shakespeare's Jest Book,* p. 35.

141. Thoreau, "Higher Laws," in Thoreau, *Walden,* ed. Krutch, p. 265. Goldsmith, *Citizen of the World,* in *Complete Works,* ed. Friedman, vol. 2, p. 60 (cited in Schochet, *Animal Life,* p. 280).

142. See Mary Howitt's poem, "The Sale of the Pet Lamb," in *Ballads.*

143. Leach, "Animal Category and Verbal Abuse," pp. 42–43.

144. By the same token, as I have argued, there are people for whom all intercourse with human beings is condemned outright as a kind of incest, as in the regulations of the Catholic orders.

145. Similarly, for some people, all meat-eating is forbidden as cannibalism.

146. Leach, "Animal Category and Verbal Abuse," p. 45.

147. See Teppe, *Chamfort,* p. 53; see also chap. 8, n. 122, in the present volume.

148. Rogers, *Horrible Secte . . . the Familie of Love* (1578), sig. I.vii.

149. Cf. the thesis of Des Pres, *Survivor,* that the Germans tried to turn the Jews into animals in order to make killing them morally acceptable.

150. By "to pet" I here mean "to treat a human being as an animal." On the controversial view that Nazism is a kind of millennialist Christian sect, see Cohn, *Pursuit of the Millennium,* esp. pp. 285–86.

151. One of the first measures that Heinrich Himmler often passed upon conquering neighboring countries was to ban *shehitah* as inhumane. (The Nazi invasion of Belgium is an example; see Steinberg, *Question juive.*) In this way he won praise from the SPCA for ending the butchering of animals in a supposedly inhumane way. In the same millennialist spirit, he frequently announced that, when the war ended, anyone who killed any animal would be prosecuted for murder. Cf. Hitler's saying that "the Jews are undoubtedly a race, but they

are not human," which is quoted as Spiegelman's epigraph to his comic book *Maus,* where the Nazis are cats and the Jews are mice.

152. See Gomperz, *Moral Inquiries on the Situation of Man and of Brutes.* It is worth noting here the opinion that it was a Jewish scholar and teacher, Sherira ben Hananiah, Gaon of Pumbedita (c. A.D.906–1006), who wrote the first "defense of animal rights" (Rosenberg, ed., introd., *Jewish Cat Book,* p. iii).

153. *Complete Poetical Works* 1, 74–75.

154. *Complete Poetical Works* 1, 68–69.

155. *Complete Poetical Works* 1, 74.

156. Num. 22:21–30.

157. In his poem "On the Prospect of Establishing a Pantisocracy in America," Coleridge looks forward to dwelling as an absolute equal with "kindred minds." Although here these "collegial" minds are those of human brothers and sisters, in "To a Young Ass" they are presumed to be animal.

158. Despite such "spiritualization," the incestuous and bestial implication of Francis's thought helps to explain why the Franciscan Spirituals, who were the most literal-minded followers of Saint Francis, were massacred under the direction of several Roman Catholic popes as "heretics": Though the official church liked the idea that we are all siblings ("All ye are brethren"), it disliked the corollary sexual and familial implications of this idea. These implications have been virtually forgotten today, with the result that it was really meaningless, if pleasant, for the Pope in 1980 to have named Francis the patron saint of "ecology."

159. *Complete Poetical Works* 1, 75.

160. Note in Coleridge, *Complete Poetical Works* 1, 75.

161. Shakespeare, *Timon of Athens,* 4.3.

162. See the Protestant hymn with the refrain, "All God's creatures / Have a place in the Choir, / Some sing lower, / Some sing higher."

163. Thomas, *Man and the Natural World,* p. 117.

164. Some researchers might argue in this context that we should delve into the neurotic sexual condition not only of pet owners but also of the pets themselves. See, for example, Leigh, "Psychology of the Pet Owner," p. 518; and for an overall view of animal neuroses, see Brion and Ey, eds., *Psychiatrie Animale.*

165. Just as one can apparently flee incestuous petting or the fear of it through pet-love, so one can flee cannibalism or the fear of it by eating the family pet. Just as fondling the familiar (animal) is a means to transcend the incest taboo, so eating the familiar (animal) is a means of transcending the cannibalism taboo.

166. Those who share this ideology dislike the institution of *Playboy* "Bunnies" and *Penthouse* "Pets" not so much because it reduces human beings (both male and female) and animals to "mere" things—which would be the radical position—but because it reduces female human beings to "mere" animals. See, for example, Cantor, "The Club, the Yoke, and the Leash," *Ms. Magazine.*

167. This essay is not about animals. It is about pethood—a human and social institution in which animals happen to play a part. Therefore one does not need to be an expert in animal physiology, psychology, and sociology to begin to consider the role that a particular group of animals (or even things)—the ones we call "pets"—play for us humans in our needful attempt to define or express the familial and species boundaries of the world in which we live. Even the question about what animals are "in themselves"—a question whose answer clearly requires expertise in animal physiology, psychology, and sociology—pertains to human institutions and language.

168. "Living things" is Aristotle's term in his *Politics* for slaves.

169. The incorporation of a chimpanzee into the human scientist's family as part of a

scientific experiment is one subject of Temerlin's *Lucy: Growing Up Human*. Desmond, *Ape's Reflection*, calls Lucy "a chimpanzee who lives literally as one of the family. She is analyzed according to Freud, but not treated according to Darwin, by which I mean that whenever possible she is interpreted as a primal human." In the book a photograph of Lucy shows her and her human "foster mother" looking at an issue of *Playgirl* magazine with expressions of enjoyment. Cf. Sperling, *Animal Liberators*, p. 171. Humans joining troops of apes—as opposed to animals joining families of humans—enjoy a special prominence in the popular imagination, as in the scientific experience of Jane Goodall and fictional tales about Tarzan. In Hugh Hudson's film version of the Burroughs novel, *Greystoke: The Legend of Tarzan, Lord of the Apes*, Tarzan grieves in Edwardian England over the dead body of the old ape who raised him as a son in the jungle ("He was my father"). Then he decides to leave his biological father's civilized land and return to the jungle. Jane, a little like Beauty, loves the half-Beastly Tarzan well enough, but she is cowed by the taboo on bestiality and does not follow Tarzan to the Land of the Apes. Much less would she kiss a gorilla. Jane would marry one of her own kind—or is it kin?—, preferring the apparent other to the real brother.

170. William Blackstone (*Commentaries on the Laws of England*, bk. 4, chap. 23), says that the Forest and Game Laws were founded on the "unreasonable" notion of permanent property in wild creatures. On the other hand, Adam Smith (*Lectures on Jurisprudence*, p. 15) says that nonwild living things—crops and herds—were the *earliest* form of private property. There is a longstanding debate about whether animals that are neither wild nor useful can be property (Thomas, *Man and the Natural World*, p. 112), that is, whether pets can be private property.

Chapter 8

1. P. 238. Compare Freud's discussion of Saint Francis and the ideal of universal sibling love (*Civilization and Its Discontents*, pp. 49, 56, cf. p. 59). "Love on the hippie scene," writes Yablonsky in his *Hippie Trip*, p. 309, "tends to be egocentric and onanistic in practice—even though vast feelings of love are felt in a general way. In the psychedelic drug reverie the individual is loaded with oceanic feelings of love and compassion; but in action, aside from a casual embrace or a sexual act, little concretely is done. . . . Very little action is unselfishly taken for another or others. . . . To feel love for everyone and everything is to love nothing."

2. Aristotle, *Poetics*, 2.1.12–13.

3. In the fifth book of the *Republic* one Platonic doctrine concerning enemies and foes arises in considering the connection between the doctrine of the national autochthonously generated family, whose existence is the gist of the "noble lie," with the practice of incest. Wondering when it is proper for a man to be kissed by a person he likes—whether male or female and whether kin or nonkin—Socrates remarks that "when Greeks fight with barbarians and barbarians with Greeks we'll assert they are at war and are enemies by nature and this hatred must be called war. While when Greeks do any such thing to Greeks we will say they are by nature friends, but in this case Greece is sick and factious and this kind of hatred must be called faction" (*Republic* 470c, trans. Bloom).

4. Matt. 5:44; Luke 6:27.

5. Cf. Schwab, "Enemy oder Foe," and Schmitt, *Concept of the Political*, p. 10, n. 20.

6. See Heidegger's ambiguous letter to Schmitt (April 22, 1933). Heidegger, who had once entered a Jesuit novitiate and would seem to have intended to become a Catholic Brother, wrote at least two essays praising Brother Abraham a Sancta Clara, the barefooted Augustinian of the second half of the seventeenth century who claimed Messkirch as his native ground (like Heidegger) and was a well-known German nationalist and influential anti-Semite: Brother Abraham wrote propaganda that was both anti-Jewish (Jews are the

enemy *within*) and also anti-Turkish (Moslems are the enemy *without*). Heidegger's two essays on Brother Abraham were written half a century apart, "Abraham a Sancta Clara: Zur Enthüllung seines Denkmals" in 1910 (*Gesamtausgabe* XIII) and *Über Abraham a Sancta Clara* in 1964. Essays specifically about Heidegger and Abraham a Sancta Clara include Cancelo, "Reflexiones de Martin Heidegger," Capánaga, "Martin Heidegger," and IJsseling, "Martin Heidegger."

Heidegger's "Self-Determination of the German University"—his Rector's Address of May 27, 1933—follows the Nazi line in grounding national spirit in blood and earth. "The spiritual world of a people . . . is the force of the deepest preservation of its powers of earth and blood, the power of the innermost excitement and most profound shock *(Erschütterung)* of its existence. A spiritual world alone bestows greatness on a nation, for it forces it to the ultimate decision as to whether the will to greatness or a tolerance for decline will become the law for our nation's future history" (Heidegger, *Selbstbehauptung der deutschen Universität,* p. 5). A couple of weeks earlier Heidegger might have read in the local right-wing student newspapers that "the synthesis of blood and land . . . is decisive for the fate of a nation" and that the fight against the Jews will allow for "the gigantic spiritual revolution that National Socialism has set in motion" (*Freiburger Studentenzeitung,* May 16, 1933, p. 2; trans. Heidegger, *German Existentialism,* pp. 52–53).

The specifically Catholic background to the development of National Socialism in Germany is the subject of considerable debate. On relevant Catholic German xenophobia of the late nineteenth and early twentieth centuries, see Weinzierl, "Der österreichisch-ungarische Raum," and Pulzer, *Enstehung des politischen Antisemitismus.* A relevant work for the study of Heidegger in particular is the essay by his one-time associate Gröber ("Altkatholicismus in Messkirch"). Cf. Farías's quasi-biographical *Heidegger,* which Lacoue-Labarthe, (*Heidegger, Art, and Politics*) severely criticized for its inattention to textual analysis; for an overview of literature of relevant seventeenth-century German anti-Semitism, see Frankl's 1905 work *Der Jude in den deutschen Dichtungen.*

7. Pp. 28, 29.

8. Stratagems attributed by the Old Testament to Jehu (2 Kings 10:18–20) were permissible to Christians in their ferreting out of heretics. In Marlowe's *The Jew of Malta,* Barabas remarks that Christians "hold it a principle, / Faith is not to be held with heretics" (*Jew of Malta,* 2.3.311–12; cf. Marlowe, *Tamburlaine the Great* [pt. 2], 2.1.33–63). The debate within Christianity about whether the church condones the breaking of such promises often centers about the treatment accorded to the so-called heretic John Hus, as in Molanus' *De Fide Haereticis Servanda.* Lupton, in his *Persuasion* (p. 47), writes: "For it is a maxime and a rule with the Pope and his partakers that *Fides non est seruanda haereticus,* Faith (or promise) is not to be kept with Heretickes." For the contradiction of the view that faith need not be kept with heretics, see the *Noli* in the *Decretum* of Gratian (Nelson, *Idea of Usury,* p. 27n).

9. See Vico's consideration in his *New Science . . . concerning the Nature of the Nations,* sec. 638, of the problem of the alien or enemy *(hostis)* in relation to the idea of the nation as expressed in the Twelve Tables of the Romans (3.7 and 2.2; in Bruns, *Fontes iuris Romani antiqui*).

10. Balthasar-Portia, who pretends to uphold the Christian ideal of universal brotherhood, uses a particularist law that discriminates between citizens and aliens.

> If it be proved against an alien
> That by direct or indirect attempts
> He seeks the life of any citizen,
> The party 'gainst the which he doth
> contrive
> Shall seize one half his goods; the other half

Comes to the privy coffer of the state;
And the offender's life lies in the mercy
Of the Duke. . . .

(*The Merchant of Venice,* 4.1.347–54)

Portia turns the attention of the Venetian court to "the first and most obvious division of the people," as William Blackstone says, the division "into aliens and natural-born subjects" (Blackstone, *Commentaries* 1, 354). This turn contradicts a universalist ethic ("All men are my brothers, none are 'aliens' ").

11. Schmitt also passes over how the Platonic doctrine concerning enemies and foes arises in connection with sometimes discomforting problems involving incest and the doctrine of the nation as family.

12. Mitchel, "Advice to the First-Time Butcher," who writes in practical terms of how it might be possible to conceive of a shepherd as "good" at the moment of slaughtering his lambs.

13. *Hamlet,* 4.3.19.

14. Ps. 23:5.

15. Nationalist liberals often seek to distance themselves from "mere racialists" by claiming that whereas the racialist wants to violate the incest taboo, presumably in order to keep the blood pure, nationalist liberals want to be chaste (literally, nonincestuous). The liberals are correct about the racialist's tendency toward the endorsement of incest. For example, Gobineau criticizes French attempts to ratify in law prejudices against consanguineous marriages and praises the Seleucids, Ptolemies, and Incas, who he supposes married their sisters as a means to keep their tribes pure (foreword to *Essay on the Inequality of the Human Races,* 2nd ed., in Gobineau, *Selected Political Writings* [hereafter SPW], p. 234); he lauds, in his play *The Renaissance,* the supposed incest of Lucrezia Borgia (pp. 199–200); he draws parallels, after befriending Richard Wagner in 1876, between race and species, endorsing incest in the same way he condemns miscegenation; he encourages incest among the pure *Volk;* and he attacks that "liberalism" which, in its dislike of "parochial exclusiveness . . ., celebrates the union of Negro and white man as much as possible—hence the mulatto" (foreword to *Human Races,* 2nd ed., in Gobineau, SPW, p. 234). But the nationalist liberals, by the same token, are incorrect about liberalism. Even universalist liberal ideology, which would enlarge the particular siblinghood to include all humankind, compels all people either to marry within the same siblinghood or not to marry. The liberal maxim "All men are brothers" requires a lifting of the incest taboo in much the same way as the racialist rule "Marry only your brother."

16. Arendt, "Race-thinking Before Racism," p. 5.

17. OED 7:30; emphasis mine.

18. Gobineau, foreword to *Human Races,* 2nd ed., in Gobineau, SPW, p. 232; see also Gobineau, *Human Races,* in Gobineau, SPW, p. 59.

19. Gobineau, *Histoire d'Ottar-Jarl.*

20. Gobineau, *Human Races,* in Gobineau, SPW, p. 71.

21. Gobineau, *Third French Republic,* in Gobineau, SPW, p. 213.

22. See Cassirer, *Myth of the State,* p. 240, on Gobineau's *Human Races.*

23. Gobineau, *Human Races,* in Gobineau, SPW, p. 71: "They end up one day by summing up their views in the words which, like the bag of Aeolus, contain so many storms—'All men are brothers.' "

24. The reunion of all persons living within the boundaries of German lands east and west into a single national siblinghood is expressed in hand-held placards reading "Wir sind ein Volk" [We are one people], which in the first days of German "reunification" celebrated Chancellor Kohl's idealistic promise of one boundaried nation. But at violent rallies in March

1991, there were already signs reading "Wir sind (k)ein Volk!?" [We are one people—or none]. (See Regis Bossu-Sygma's photograph.)

25. *Ency Phil.* 5:443.

26. Renan, *Oeuvres complètes* 1:904. Prime Minister Trudeau, in a well-known speech to the Canadian House of Commons in 1980 about whether Québec might legally disassociate from Canada as the result of a one-time referendum proposed by the nationalist Parti Québécois, gave special prominence to Renan's view (Trudeau, "Un plébiscite de tous les jours," p. 400).

27. Leibniz adds that "nature has seen fit to keep these at a distance from us so that there will be no challenge to our superiority on our own globe" (*New Essays*, pp. 472–73).

28. Leibniz, *New Essays*, p. 314.

29. *New Essays*, p. 314.

30. See the criticism of this view in Godwyn, *The Negro and Indians Advocate*, pref. and ch. 1 (cited in Popkin, "Medicine, Racism, Anti-Semitism," p. 411).

31. See Matt. 1.

32. Chateaubriand, *Les Martyrs*, in *Oeuvres* 3:641.

33. Chateaubriand, *Le génie du Christianisme*, in *Oeuvres* 3:37. For some of the following references I am indebted to Crowe, *Extraterrestrial Life Debate;* here p. 181.

34. Edward Young, *Night Thoughts*, 9:1777–79.

35. Or perhaps Oliver Goldsmith—the attribution is uncertain.

36. Citation is from Engdahl, *Planet-Girded Suns*, p. 67.

37. Rittenhouse, *Oration*, pp. 19–20.

38. Dostoevsky, "Dream of a Ridiculous Man," pp. 314, 316. Describing Alyosha gazing at the stars after the death of Father Zosimov, Dostoevsky writes: "There seemed to be threads from all those innumerable worlds of God, linking his soul to them"; that soul was trembling all over "in contact with other worlds" (*The Brothers Karamazov*, p. 340).

39. "Two things fill the mind with ever new and increasing admiration and awe, the oftener and more steadily they are reflected on: the starry heavens above me and the moral law within me. . . . The former . . . broadens the connection in which I stand into an unbounded magnitude of worlds beyond worlds and systems of systems. . . . The former view of a countless multiplicity of worlds annihilates, as it were, my importance as an animal creature, which must give back to the planet (a mere speck of the universe) the matter from which it came" (*Practical Reason*, p. 569).

40. See Paneth, *Chemistry and Beyond*, pp. 110–111.

41. Kant, *Theory*, p. 182. I refer to the concluding third part of the 1755 edition. In the 1791 edition, much of Kant's discussion of extraterrestrial life has been omitted.

42. *Theory*, p. 190.

43. Paul, *Tibetan Symbolic World*.

44. Noyes' American Perfectionists held that the godhead was both male and female. Ann Lee ("Mother Ann"), founder of the Shaker society in America in the 1770s, regarded herself as "the female element that supplemented Jesus and thus completed the revelation to the world of a father-and-mother God" (Andrews, *Shakers*, pp. 96–97; cf. Benjamin Young, *Testimony*). Eldress Anna, a Shaker, says that her community "is the only society in the world, so far as we know, where women have absolutely the same freedom and power as men in every respect" (Evans, *Autobiography*, p. 268).

45. Gal. 3:26–28. Concerning the millennialist works: In the *Gospel according to the Egyptians* it is written: "When Salome inquired when the things concerning which she asked should be known, the Lord said: 'When ye have trampled on the garment of shame, and when the two become one and the male with the female neither male nor female.'" Another version of the Egyptian Gospel goes: "For the Lord himself being asked by someone when

his kingdom should come said: 'When the two should be one, and the outside (that which is without) as the inside (that which is within) and the male with the female neither male nor female'" (James, ed., *Apocryphal New Testament,* p. 11; cf. G. Wilson Knight, *Christian Renaissance,* p. 11, and Heilbrun, *Androgyny,* pp. 20, 179).

46. See 1 Pet. 5:14; for Paul, see Rom. 1:16, 1 Cor. 16:20, and 1 Thess. 5:26. See Perella, *Kiss,* esp. pp. 12–50.

47. See the Middle English translation of Aelred of Rievaulx's *De vita eremitica,* 329; Juliana of Norwich, *Revelations,* 58–6; and Thompson, ed., "Ureisun of Ure Louerde," p. 2.

48. Cousins, *Bonaventure,* p. 20.

49. Condorcet, "Admission," pp. 99–100, 102. Condorcet's essays appeared in the *Journal of the Society of 1789.* For this and some of the following references I am indebted to Landes, *Public Sphere;* here pp. 115–16.

50. Condorcet, "Admission," p. 100.

51. Landes, *Public Sphere,* p. 122.

52. Landes, *Public Sphere,* p. 139.

53. Palm, *Appel,* p. 123.

54. Hippel, *Status of Women,* pp. 120, 121.

55. Wollstonecraft, "Vindication," p. 53.

56. Wollstonecraft, "Vindication," p. 5.

57. Cf. Harding, *Coleridge.* In Coleridge the ideal of marriage is expressed mathematically: "1 and $1 = 2$; but I cannot be multiplied $1{:}1 \times 1 = 1$" (Coleridge, Letter to H. C. Robinson, 1811). On the connection between spousal and sibling love in Coleridge, see below.

58. See the French law of September 20, 1792; cf. Landes, *Women and the Public Sphere,* p. 122.

59. Godwin, *Political Justice,* p. 763.

60. Sade, *Philosophie,* pp. 179–245, esp. pp. 221–22. On incest and the revolutionary ideals of Mirabeau and Shelley, see Jones, *Hamlet,* pp. 89–102. Shelley's "Revolt of Islam" is in part an attempt to revive in poetry some of these ideas about incest.

61. Cited in Kramnick, introd. to Godwin, *Political Justice,* p. 14.

62. Landes, *Public Sphere,* p. 148. Compare MacCannell's question in her *Regime of the Brother,* p. 27: "We [women] accepted (far too long) the patriarchal power to define, shape, and twist woman's identity, woman's desire, at will, and to go without fulfilling her claims to identity, her demands for love, or even meeting her minimal material needs. But what about now, during—and after—the Regime of the Brother?"

63. On the Catholic Sisters' participation in the politics of 1789, see Landes, *Public Sphere,* p. 107.

64. Landes, *Public Sphere,* p. 175.

65. For "Sister in Humanity," see the dedication to the 1842 edition of Tristan's *London Journal.*

66. Tristan, *Workers' Union,* p. 88.

67. See Tristan, *Workers' Union,* introd., pp. xv–xvi; Tristan, *La Tour,* pp. 102, 139. On the life of Saint Teresa, see chap. 4, above.

68. Origen, *De oratione,* 15.4.

69. Basil, *Regulae brevius tractatae, Patrologiae (Graeaca)* 31:1153; Gregory of Nazianzus, *Epistolae,* 238; and Macarius, *Homiliae,* in *Patrologiae (Graeaca)* 34:468. When Jerome said, "Te universa salutat," he meant that only his own community of Eremites are brothers to him (Jerome, *Epistulae* 134).

70. See Cyprian, *Epistolae,* 53, in *Corpus scriptorium ecclesiasticorum latinorum* 3:620.

71. Optatus, *Contra Parmenianum,* in *Corpus scriptorium . . .,* p. 5.

72. Gal. 3:26–28.

73. Among them: Ignatius of Antioch, *To the Ephesians,* p. 653; Clement of Alexandria, *Stromata,* p. 61; Justinian, *Dialogue with Tryphon,* 96, p. 704; and Tertullian, *Apologeticus,* 39, pp. 8–9.

74. Matt. 23:9.

75. 3.2.11–12. On *Measure for Measure,* see my *End of Kinship.*

76. For the Ten Commandments: Exod. 20:12—"Honor thy father and thy mother." For the name "father" in the mystery cults, see Preisendanz, *Papyri graecae magicae* 4, 115ff; and, for additional evidence, see Henzen et al., *Inscriptiones,* nos. 406, 727, 2233.

77. See Cyprian, *Epistolae,* 53, in *Corpus scriptorium . . .,* 3:620.

78. F. Dolger, "Brüderlichkeit," in *Reallexikon* 2:641–46.

79. In Christianity political liberty is possible, if at all, only in the death through celibacy universalized of the body politic. Paul, seeming to retreat from the Christian ideal of a free association of *liberi,* or free siblings, writes: "For, brethren, ye have been called unto liberty; only use not liberty for an occasion to the flesh, but by love serve one another. For all the law is fulfilled in one word, even in this: Thou shalt love thy Neighbour as thyself" (Gal. 5:13–14). We are enjoined to love all people equally (as free children, or *liberi,* of the Father), while we are enjoined not to disobey the old rule against loving all people equally. (The old rule enjoined a different kind of love for siblings and kinsmen than for others, and also a different kind of love for spouses—who were others before marriage—than for siblings, kinsmen, or others.) Paul thus refuses both to transcend and to accept the old rule: he ejects the incestuous Corinthians from the Christian community and may even have killed them.

80. That is the crux of Shakespeare's *Timon of Athens.* Timon who has no particular family of his own, slides easily from thinking that he loves every member of humankind to thinking that he hates every one. *Timon of Athens,* more Hellenist in spirit than either Greek or Roman, demonstrates the slide from the Stoics' and Christians' doctrine of the universal love of mankind to the universal hatred of it. From admired brother to despised other is but the journey from "What a piece of work is man!" to "Man delights not me!" (*Hamlet* 2.2.300–305).

81. Gilson (*Dante,* p. 166) argues that in his *Monarchia* (esp. I, 16, 23 and III, 10, 44), Dante "borrowed from the Church its ideal of universal Christendom and secularized it" by substituting "mankind" for "Christendom."

82. Douglas, *Implicit Meanings,* p. 289.

83. Gal. 5:14.

84. Matt. 5:43. Stade, *Geschichte des Volkes Israel,* considers the view that Jews cannot understand universal love.

85. For Islam, see chapter 2. It is worth mentioning here that the Koran includes Sabianism as well as Christianity and Judaism as religions of the Book worthy of protection. And since at different times varied groups such as Zoroastrians in Persia and Hindus in India were said to be Sabians (see Carra de Vaux, "*al-Sabi'a,*" p. 20), the distinction between a non-Muslim people of the Book and a pagan people was not always clear. Islamic doctrine also included the doctrine of universalism based on common descent: "Mankind is from Adam and Eve, and all of you are alike in your descent from them. On the Day of Judgment, God will not ask you about your noble descent or your lineage; rather the most honoured of you before God on that Day will be the most righteous of you" (Ibn Kathir [A.1]; commentary on Sura 49:13; cited in Chapra, "Economic System of Islam"; cf. Bernard Lewis, *Political Language of Islam,* p. 17). Goitein ("Concept of Mankind in Islam") discusses aspects of this Muslim universalist humanism.

86. For profession: see chapter 6. For John of the Cross: *Collected Works,* pp. 656–57.

87. Freedman, note to Sanhedrin 58a, in *Babylonian Talmud,* ed. I. Epstein.

88. The *rikkub* principle in the logical system of a group of Jewish Karaites, the *Ba'ale ha-Rikkub,* includes the rule of "nominalism," whereby, for example, a stepsister has the status of a sister if she is called one. See L. Epstein, *Marriage Laws,* pp. 266–67; on the *rikkub* principle, in general, see Jeshua ben Judah, *Sefer ha-Yashar,* cited in Nemoy, ed., *Karaite Anthology,* esp. pp. 127–32. For a related consideration of the wife-sister motif in the Bible, see also Sanhedrin 58a–b, in *Babylonian Talmud,* ed. I. Epstein.

89. Isa. 56:7 (cf. Bickermann, *Ezra to Maccabees,* p. 19). Zech. 8:23 (cf. Bickermann, *Ezra to Maccabees,* pp. 19–20).

90. For the Stoics, see chapter 5. For Alexander and the Jews, see Torrey, *Second Isaiah,* p. 126, and Finkelstein, *Pharisees,* 2, 566.

91. Genesis Rabba, 39.

92. Mal. 2:10. This verse might seem to extend fraternity from all human beings to all animals, including both domestic and wild ones. Kalonymos ben Kalonymos' *Book of Animals and Men* (A.D. 1316), composed by the "Society of Pure Brethren"—whose remarkable membership included Jews, Christians, and Muslims—is a philosophical dialogue whose central question is, "By what right, if any, do men justify the enslavement of animals?" The dialogue include's the Rooster's Lament: "At midnight I rise to pray . . . / But the sleeping ones lay hold of me . . . / They slaughter me and eat me. / Have we not all one father? / Has not one God created us all?" (trans. Schochet, *Animal Life,* p. 256). Cf. Eccles. 3:21: "Who knoweth the spirit of man whether it goeth upwards, and the spirit of the beast whether it goeth downward to the earth?"

93. On loving neighbors: Lev. 19:18. On strangers who dwell within the community: Lev. 19:34.

94. Hillel, in *Pirque Aboth* 1.12 (cf. Hillel, Shabbath, 31a); Meir, in *Pirque Aboth* 6.1.

95. Aaron ibn Hayyim, *Korban Aharon.* In accordance with the tradition of the imitation of God—"as he is merciful so be you merciful" (Shabbath, 33b)—mercy transcends familial bonds to encompass the entire range of human relationships. Cf. Eccles. 18:13 and Genesis Rabba, 33. For the doctrine of universal human love in modern Judaism, see Hirsch, *Horeb;* Hermann Cohen, *Religion der Vernunft;* Buber, *I and Thou; Ency. Judaica* 11:530; and the ruling of the synod at Leipzig held in 1869 and the German-Israelite Union of Congregations in 1885. On the injunction to love even outlaws and criminals, see Sanhedrin 45a. Compare Judaism's potentially universalist notion of pedagogic kinship. "One who teaches another's child Torah is regarded by the tradition as one who gave birth to the child" (Sanhedrin 19b).

96. See chapter 2 on how Spinoza's way of opposing Judaism's real particularism to Christianity's ideal universalism might amount to a "Machiavellian" strategy . . . to lay out a practical groundwork for a real universalism based on the ancient Hebrew constitution and on a catholicism where Christians and Jews might live together tolerably well (Strauss, *Spinoza,* pp. 16, 18).

97. Deut. 23:19–20. Cf. Deut. 28:12 and Lev. 25:35–37. On usury and brotherhood, see my *Money, Language, and Thought,* esp. chap. 2.

98. Saint Jerome, *Comment. in Ezechielem* 6:18, in *Patrologiae (Latina)* 25:176. In this vein Henry Smith wrote: "Of a stranger, saith God thou mayest take usury; but thou takest usury of thy brother therefore this condemneth thee, because thou useth thy brother like a stranger" ("Examination of Usury: The First Sermon," pp. 97–98).

99. Aristotle, objecting to the identification of monetary offspring with natural offspring, argues that "currency came into existence merely as a means of exchange; usury tries to make it increase [as though it were an end in itself]. This is the reason why usury is called by the word we commonly use [*tokos*]; for as the offspring resembles its parents, so the interest bred by money is like the principal which breeds it, and [as a son is styled by his father's

name, so] it may be called 'currency the son of currency.' Hence we can understand why, of all modes of acquisition, usury is the most unnatural" (Aristotle, *Politics,* 1258). As Francis Bacon puts it, "it is against nature for money to beget money ("Of Usury," *Works* 12, 218). Luther says, that "money is the sterile thing" [*pecunia est res sterilis*] (*Tischreden* 5, 5429). On the identification of monetary offspring with natural offspring in Shakespeare, see also my *End of Kinship,* esp. pp. 29–30, 126.

100. Clement of Alexandria followed the Hellenic Jew Philo in interpreting the Old Testament term *brother* to mean merely "a child of the same parents," for example. Philo said that *brother* meant "not merely a child of the same parents, but anyone who is a fellow-townsman and fellow tribesman (Philo, *De virtutibus* 14.82; discussed in Wolfson, *Philo,* 2:365). Clement of Alexandria said that *brother* might refer to "whoever is of the same tribe, [namely] of the same faith, and who participates in the same *logos*" (Clement of Alexandria, *Stromata, Patrologiae (Graeca)* 1023–24; cited by Nelson, *Idea of Usury,* p. 3).

101. See the argument connecting Calvin with these sects in Bouyon, *Réfutation,* esp. pp 186–93. *Patarenes,* or *Patarelli,* was the name first used in the eleventh century to denote the extreme opponents of clerical marriages. It was appropriated by the Cathars in the thirteenth century. Like the Anabaptists, Patarenes are closely linked with the Bogomils, who held that marriage was not a sacrament (*Ency. Brit.* 5:119). See too Nelson, *Idea of Usury,* p. 73. On the relationship between the "invention" of purgatory and the gradual acceptance of the practice of usury, see Jacques Le Goff, *Your Money or Your Life,* p. 93.

102. John of the Cross, *Collected Works,* pp. 656–57. On Saint Teresa's "judaizing" grandfather, see Serís, "Nueva genealogía."

103. Seneca, *De Clementia,* bk. 1, chap. 26; my translation.

104. Maine, *Village-Communities,* pp. 225–27; cited in Nelson, *Idea of Usury,* p. xvi.

105. Bing, in *Lettres,* pp. 8–9. For this and many of the following references I am indebted to Berkovitz, *Shaping of Jewish Identity.*

106. For other implications of the biblical passages here referred to, see Nelson, *Idea of Usury,* and Rosenthal, "Interest from the Non-Jew."

107. Tama, *Transactions,* p. 198.

108. Bing, in *Lettres,* pp. 54–55.

109. Durham, *Report on the Affairs of British North America,* 2:16.

110. Re: Clermont-Tonnerre, quoted by Berkovitz, *Shaping of Jewish Identity,* p. 71. The relevant benediction, "Bless the Revolution which will make us all brothers," was heard in the General Assembly in 1790 (*La Révolution française et L'emancipation des juifs* 7:34, cited by Trigano, "French Revolution and the Jews," p. 171).

111. Napoleon, *Selected Letters,* trans. Thompson, pp. 155–59.

112. Tama, *Transactions,* pp. 133–34.

113. Tama, *Transactions,* pp. 76–80.

114. On the offical French Commissioners' dissatisfaction with this answer, see Schwarzfuchs, *Napoleon, the Jews, and the Sanhedrin,* p. 206n. On the controversy surrounding such rulings, see Katz, *Exclusiveness and Tolerance,* esp. pp. 182–93.

115. Tama, *Transactions,* pp. 199–200.

116. Cahen, *Précis,* pp. 12, 25–26.

117. Halévy, *Instruction religieuse,* pp. 99–100.

118. Bloch, "Sur l'esprit," p. 73.

119. Cited in *Ency. Brit.* (11th ed.), 3:563.

120. See Mendelssohn, *Jerusalem,* trans. Jospe, esp. pp. 106–8; and chap. 2.

121. Nelson, *Idea of Usury,* p. 113.

122. Chamfort's 1793 remark about the potentially deadly quality of the Jacobin definition of liberty as an association of sons (*liberi*) led to his arrest (Teppe, *Chamfort,* p. 53).

The epigram confuses the familiar and unfamiliar forms of pronouns. For example, the speaker's command that his interlocutor should become a family member implies that he is not already familiar. (The speaker says "Soyez mon frère" instead of "Sois mon frère.") By the same inversion, the speaker's threat to kill his interlocutor implies that he is already familiar. (The speaker says "Je te tuerai" instead of "Je vous tuerai.") Compare the American Patrick Henry's revolutionary slogan, "Liberty or Death!"; in the 1990s, automobile license plates in the American state of New Hampshire still read "Live free or die."

123. Quoted by Berlin, *Crooked Timber of Humanity.*

124. Emerson, "Ability," in *Essays and Lectures,* p. 810; cf. *Journal* LM (1848), 10:310.

125. Chamfort said: "The fraternity of such people is the fraternity of Cain and Abel" (quoted in Teppe, *Chamfort,* p. 53). Connolly (*Unquiet Grave,* p. 78) suggests that "the complexity of Chamfort's character would seem to be due to his temperament as a love-child; he transmuted his passionate love for his mother into a general desire for affection."

Conclusion

1. *Bartlett's Familiar Quotations,* p. 720, quoting Wells, *Outline of World History* (1920), chap. 41.

2. See Wells' comment on the extermination of the Tasmanians in his *War of the Worlds,* p. 113; quoted as an epigraph to the section "War of the Worlds" in chapter 8. In the wake of a devastation he called "the war to end wars," Wells, his science-fiction writing now decades behind him and Hiroshima still ahead, wondered where might lead "the 'legitimate claim' . . . of every nation to manage its own affairs . . . regardless of any other nation"? (Wells, *Outline,* p. 782). Cf. his view of the idea that "men form one universal brotherhood" (*Outline,* pp. 426–27).

3. In *From Generation to Generation* Eisenstadt discusses "nonkinship, universalistic principles"—that is, principles according to which persons "act towards other persons without regard to familial kinship, lineage, ethnic, or hierarchal properties of those individuals in relation to their own." He remarks that "it should, of course, be emphasized that no society can be entirely and wholly universalistic" (*From Generation to Generation,* pp. 116, 117). Eisenstadt does not discuss the incestuous implications of nonkinship or universal brotherhood.

4. Shakespeare, *Measure for Measure,* 3.1.138; see my discussion of incest as the antonomasia of unchastity in *End of Kinship,* pp. 103–4.

5. For the use of the term "distant-close" [*loingpres*], see Dagens, "Le 'Miroir des simples ames' et Marguerite de Navarre."

6. See Révah, "La controverse sur les statuts de pureté de sang," p. 265.

7. Barthes, *Sade, Fourier, Loyola,* pp. 137–38.

8. Wells, *Outline,* p. 780.

BIBLIOGRAPHY

Aaron [ben Abraham ben Samuel] ibn Hayyim. *Korban Aharon* [Aaron's Offering]. 1st ed., Venice, 1609. 2nd ed., Dessau, 1742.

Abercrombie, Thomas J. "When the Moors Ruled Spain." *National Geographic* 174 (1988): 86–119.

Abraham, David. "Cosyn and Cosynage: Pun and Structure in the Shipman's Tale." *The Chaucer Review* 11 (1977): 319–27.

Abravanel, Isaac. *Ma'ayene ha-yeshu'ah*. Amsterdam, 1647.

Acta sanctorum. Editio novissima. Ed. and pub. Society of the Bollandists. 67 vols. Brussels, 1863.

Adams, Jos. *Answers to All the Objections Hitherto Made Against Cow-pox*. London, 1805.

Aelred of Rievaulx. *De vita eremitica*. Ed. Carl Horstmann. *Englische Studien* 7 (1884): 304–44.

Aeschylus. *The Oresteia of Aeschylus*. Ed. with introd. and commentary by Walter Headlam and George Thomson. Rev. and enl. ed. Amsterdam, 1966.

———. *Oresteia*. Trans. Richmond Lattimore. Chicago, 1969.

Aillaud, Gilles. "Why Look at Animals." In *About Looking*, ed. John Berger. New York, 1981, pp. 1–26.

d'Alembert, Jean. *Traité de Dynamique*. Ed. with an introd. by Thomas L. Hankinds. New York and London, 1968.

Allo, E.-B. *Saint Paul: Première épitre aux Corinthiens*. Paris, 1956.

Alonso de Cartagena (Bishop). *Defensorium unitatis christianae* [1449–50]. *Tratado en favor de los judíos conversos*. Ed. P. Manuel Alonso. Madrid, 1943.

Alta California. Untitled article. July 24, 1864: 3.

Ames, Percy W., ed. *The Mirror of the Sinful Soul, by Princess (afterwards Queen) Elizabeth*. Facsimile edition with introd. Royal Society of Literature of the United Kingdom. London, 1897.

Anderson, Gallatin. "Il Comparaggio: The Italian Godparenthood Complex." *Southwestern Journal of Anthropology* 13 (1957): 32–53.

Andrews, Edward D. *The People Called Shakers*. New York, 1953.

Animals, The [rock and roll group]. "We gotta get out of this place." Single 45 rpm. MGM # 13382. 1965.

Antisthenes of Athens. *Fragmenta*. Ed. Fernanda Decleva Caizzi. Milan, 1966.

Apuleius. *"Cupid and Psyche" and Other Tales From the "Golden Ass" of Apuleius*. Ed. W. H. D. Rouse. London and Boston, 1907.

Aquin, Hubert. "La Fatigue culturelle de Canada français." *Liberté* 4 (mai 1962): 299–325.
Aquinas, Thomas. *In Libros Politicorum Aristotelis expositio*. Ed. R. M. Spiazzi. Turin and Rome, 1951.
———. *Summa contra gentiles*. Trans. English Dominican Fathers. 2 vols. New York, 1924.
———. *Summa Theologica*. Trans. English Dominican Fathers. 3 vols. New York, 1947–48.
———. *Tractatus de regimine principum*. In Aquinas, *Selected Political Writings*, pp. 3–84. Ed. A. P. Dentrèves, in Latin (ed. Mathis). Trans. J. G. Dawson. Oxford, 1948.
Archivio Histórico Nacional. Sección Inquisición.
"Are the Human Race All of One Blood?" *American Ladies Magazine* 6 (1833): 359–62.
Arendt, Hannah. "Race-thinking Before Racism." *The Review of Politics* 6 (1944): 36–73.
Aristophanes. *Assembly of Women*. Ed. and trans. by Robert Glenn Ussher. Oxford, 1973.
Aristotle. *Aristotle's Theory of Poetry and Fine Art*. Trans. S. H. Butcher with an introd. by J. Gassnero. 4th ed. New York, 1951.
———. *On the Generation of Animals*. Trans. A. L. Peck. Cambridge, Mass., 1943.
———. *Nicomachean Ethics*. Trans. H. Rackham. Cambridge, Mass., 1934.
———. *Politics*. Trans. H. Rackham. Cambridge, Mass., and London, 1967.
———. *Politics*. Trans. Carnes Lord. Chicago, 1984.
———. *Rhetoric*. Trans. John Henry Freese. London and New York, 1926.
Arkow, Phil. *Pet Therapy: A Study of the Use of Companion Animals in Selected Therapies*. Colorado Springs, Colo., 1982.
Ascoli, G. *La Grande-Bretagne devant l'opinion française*. Paris, 1927.
Askew, Anne. *The first examinacyon of Anne Askewe . . . with the elucydacyon of Johan Bale*. Wesel, 1546.
———. *The lattre examinacion of Anne Askewe with the elucydacyon of Johan Bale*. Wesel, 1547.
Athanasius, Saint (?). *Vita Antonii*. In *Vitae Patrum*, ed. H. Rosweyd. PL 73:117–91.
Aubrée, E. *Lucile et René de Chateaubriand chez les soeurs à Fougères*. Paris, 1929.
Augustine. *The Catholic and Manichaean Ways of Life*. Boston, 1966.
"Les autochtones et nous: vivre ensemble." Extracted from *La situation des autochtones au Québec: Comité d'appui aux nations autochtones de la Ligue des droits et liberté*, 1980. *See* in Boismenu et al., *Le Québec en textes*, pp. 557–67.
L'Avenir de Français au Québec. Actes du colloque tenu à Montréal les 2 et 3 mars 1987 par l'Union des écrivains québécois. Montreal, 1987.
Babylonian Talmud. Ed. I. Epstein. 18 vols. London, 1935–48.
Bachofen, Johann Jakob. *Das Mutterrecht: Eine Untersuchung über die Gynaikokratie der alten Welt nach ihrer religiösen und rechtlichen Natur* [1861]. Vols 2–3 of Bachofen, *Gesammelte Werke*. Ed. Karl Meuli. 10 vols. Basel, 1943–67.
Bacon, Francis. *The Works of Francis Bacon*. Ed. J. Spedding, R. L. Ellis, and D. D. Heath. London, 1857–74.
———. *The New Atlantis*. London, 1631.
Bainton, Roland. *Women of the Reformation in France and England*. Boston, 1975.
Baker, Richard. *Chronicle of the Kings of England*. London, 1643.
Bakhtin, Mikhail. *Rabelais and His World*. Trans. Helene Iswolsky. Cambridge, Mass., 1968.
Balbus, Bernardus [Bishop of Faenza and Pavia]. *Summa decretalium*. Ed. E. A. T. Laspeyres. Ratisbon, 1860.
———."Elucydaycon." *See* in Askew, *The first examinacyon*.

Bale, John. "Epistle Dedicatory" & "Conclusion." In Elizabeth I, *A Godly Medytacyon of the Christen Sowle* (1548). *See* in Shell, *Elizabeth's Glass.*

———. *Illustrium majoris Britanniae scriptorum.* Wesel, 1548–49.

———. *The Image of Both Churches.* Ed. Henry Christmas. In Bale, *Select Works.*

———. *King Johan.* In *Four Morality Plays,* ed. with an introd. and notes by Peter Happé. Harmondsworth, Middlesex, Eng., 1979.

———. *Kynge Johan.* Ed. J. P. Collier. London, 1838.

———. *Select Works* (1849). Ed. Henry Christmas. Repr. New York, 1968.

———. *The Three Laws of Nature.* London, 1908.

Balzac, Honoré de. *Le Vicaire des Ardennes.* Paris, 1882.

Bandinelli, Orlando [Alexander III]. *Summa.* Ed. Friedrich Thaner. Innsbruck, 1874.

Barbaud, Philippe. "Parlerons-nous Cajun?" *See* in *L'Avenir de Français au Québec,* pp. 61–69.

Barchilon, Jacques. "Beauty and the Beast: From Myth to Fairy Tale." *Psychoanalysis and the Psychoanalytic Review* 46 (1959): 19–29.

Barnum, Priscilla Heath, ed. *A Treatise on the Ten Commandments* (1410). Early English Text Society, 275. London, 1980.

Baroja, Julio Caro. *Los Judíos en la España moderna y contemporanea.* 3 vols. Madrid, 1962.

Baron, John. *The Life of Edward Jenner.* 2 vols. London, 1827, 1838.

Baron, Salo W. *A Social and Religious History of the Jews.* 2nd ed. 12 vols. New York, 1952.

Bart, Benjamin F., and Robert Francis Cook. *The Legendary Sources of Flaubert's "Saint Julian."* Toronto, 1977.

Barth, Paul, and Albert Goedeckemeyer. *Die Stoa.* Stuttgart, 1941.

Barthes, Roland. *On Racine.* Trans. Richard Howard. New York, 1964.

———. *Sade, Fourier, Loyola.* Trans. Richard Miller. New York, 1976.

Bartlett, John. *Familiar Quotations: A Collection of Passages, Phrases and Proverbs Traced to Their Sources in Ancient and Modern Literature.* 15th edition. Ed. Emily Morison Beck. Boston, 1980.

Basil. *Regulae brevius tractatae. Patrologiae* (Graeca) 31: 1051–1321.

Basnage, Henri sieur de Beauval. *Tolérance des religions.* Rotterdam, 1684.

Bayle, Pierre. *Dictionnaire Historique et Critique.* Rev. ed., with notes by Chaufepié et al. 16 vols. Paris, 1820–24.

Beauchemin, Yves. "Le Français au Québec: Combat des les broussailles." *See* in *L'Avenir de Français au Québec,* pp. 135–63.

Beaumont, Francis, and John Fletcher. *A King and No King.* Ed. Robert K. Turner. Lincoln, 1963.

Beaumont, Madame Leprince de. *Magasin des enfans, ou dialogues entre une sage Gouvernante et plusieurs de ses Élèves.* London, 1756.

Beck, Alan M. "Population Aspects of Animal Mortality." *See* in Kay et al., eds., *Pet Loss,* pp. 42–48.

Beck, Alan M., and Aaron H. Katcher. *Between Pets and People: The Importance of Animal Companionship.* New York, 1983.

Bédier, Joseph. "La Tradition manuscrite du *Lai de l'Ombre.*" *Romania* 54 (1928): 161–96, 321–56.

Behiels, Michael D. "The Commission des écoles catholiques de Montréal and the Néo-Canadian Question: 1947–63." *Canadian Ethnic Studies* 13 (1986): 38–64.

Beilen, Elaine V. "Anne Askew's Self-Portrait in the Examinations." In Hannay, ed., *Silent But for the Word,* pp. 77–91.

Belleforest, François de. *Le Cinquiesme Tome des histoires tragiques.* Paris, 1582. English trans. (1608). *See* in Gollancz, ed. *Sources of "Hamlet,"* pp. 94–165.

Benardete, Seth. "A Reading of Sophocles' *Antigone*." *Interpretation* 4.3 (1975): 148–96; 5.1 (1975): 1–56; 5.2 (1975): 148–84.

Benito Ruano, Eloy. "El memorial contra los conversos, del Bachiller Marcos García de Mora (Marquillos de Mazarambroz)." *Sefarad* 17 (1957): 314–51.

Benjamin, Walter. *Illuminations*. Trans. Harry Zohn. Ed. with an introd. by Hannah Arendt. New York, 1968.

Bennet, Josephine Waters. *"Measure for Measure" as Royal Entertainment*. New York, 1966.

Benrath, Karl. *Bernardino Ochino von Siena: Ein Beitrag zur Geschichte der Reformation* (1892). Rpt. Nieuwkoop, 1968.

Bentley, Thomas. *Monument of Matrones*. London, 1582.

Bercovitch, Sacvan. "Emerson, Individualism, and the Ambiguities of Dissent." Unpublished manuscript. 1991.

Berg, William J., Michel Grimmaud, and George Moskos. *Saint Oedipus: Psychocritical Approaches to Flaubert's Art*. Ithaca, New York, 1982.

Berkovitz, Jay R. *The Shaping of Jewish Identity in Nineteenth Century France*. Detroit, 1989.

Berlin, Isaiah. *The Crooked Timber of Humanity: Chapters in the History of Ideas*. Ed. Henry Hardy. New York, 1991.

Bernal, Martin. *Black Athena: The Afroasiatic Roots of Classical Civilization*. Volume 1: *The Fabrication of Ancient Greece 1785–1985*. New Brunswick, New Jersey, 1987.

Bernard of Clairvaux, Saint. *On the Song of Songs*. Trans. Kilian Walsh with an introd. by M. Corneille Halflants. Cistercian Fathers series, nos. 4, 7, 31, 40. Vols. 2–5 of *The Works of Bernard of Clairvaux*. Spencer, Mass., 1970.

Bernardine of Siena. *Opera Omnia*. Florence, 1950–56.

Bernardo, Reta Mohney. "The Problem of Perspective in the *Miroir de l'âme pecheresse*, the *Prisons*, and the *Heptameron* of Marguerite de Navarre." Ph.D. diss., State University of New York at Binghamton, 1979.

Berry, E. G. "*Hamlet* and Suetonius." *Phoenix* 2 (1948): 73–80.

Beryl, Sandford. "A Patient and Her Cats." *Psychoanalytic Forum* (1966): 170–176.

Besant, Walter, and James Rice. *The Chaplain of the Fleet*. 3 vols. London, 1881.

Bettelheim, Bruno. *The Uses of Enchantment: The Meaning and Importance of Fairy Tales*. New York, 1976.

Beyle, Marie-Henri de [Stendhal]. *The Shorter Novels of Stendhal*. Trans. C. K. Scott Moncrieff. New York, 1946.

Bible. Authorized King James Version (1611). London, 1957

The Geneva Bible (1560). Facsimile ed. with an introd. by Lloyd E. Berry. Madison, Wis., 1969.

Novum Testamentum Graecae et Latine. Ed. August Merk. 6th ed. Rome, 1948.

Revised Standard Version. New York, 1962.

The Twenty-Four Books of the Old Testament. Hebrew text and English Version. Trans. rev. Alexander Harkavy. 4 vols. New York, 1928.

Bible Communism. A Compilation from the Annual Reports, and Other Publications of the Oneida Association and Its Branches; Presenting, in Connection with Their History a Summary View of Their Religious and Social Theories. Brooklyn, 1853.

Bickermann, Elias J. *From Ezra to the Last of the Maccabees: Foundation of a Postbiblical Judaism*. New York, 1978.

Bing, Isaiah Berr. "Le Cri du citoyen contre les Juifs" (1787). Reprinted in *La Révolution française et l'émancipation des Juifs*, vol. 8: *Lettres, mémoires et publications diverses*, 1787–1806. Paris, 1968.

Black, William Henry, ed. Robert of Gloucester. *The Life and Martyrdom of Thos. Beket*. Vol. 19. Percy Society. London, 1845.

Blackmore, Simon Augustine. "Hamlet's Right to the Crown." In *The Riddles of Hamlet and The Newest Answers*. Boston, 1917.
Blackstone, William. *Commentaries on the Laws of England*. New ed. 4 vols. London, 1813.
Blake, William. *The Complete Writings*. Ed. G. Keynes. London, 1957.
———. *Poetry and Prose of William Blake*. Ed. David V. Erdman with a Comm. by Harold Bloom. New York, 1968.
Blayo, Yves. "Illegitimate Births in France from 1740 to 1829." *See* in Laslett et al., eds. *Bastardy*, pp. 278–83.
Bloch, Simon. "Sur l'esprit et la tendance de ce recueil." *La Régénération* 1 (1836): 65–67.
Blum, Carol. *Rousseau and the Republic of Virtue: The Language of Politics in the French Revolution*. Ithaca, 1986.
Bodin, Jean. *Les six livres de la république*. Paris, 1583.
Boerhaave, Herman. *Traité des maladies de l'enfant*. Avignon, 1759.
Boismenu, Gérard, Laurent Mailhot, and Jacques Rouillard. *Le Québec en textes: Anthologie 1940-1986*. Nouvelle édition. Montreal, 1986.
Bollain, L. *El Toreo*. Seville, 1968.
Bonaventure. *The Soul's Journey into God, the Tree of Life, and the Life of Saint Francis*. Trans. Ewert Cousins. New York, 1978.
Boone, Allen J. *Kinship With All Life*. New York, 1954.
Bossu-Sygma, Regis. Photograph of placard at nationalist rally in Leipzig. *Newsweek* (April 1, 1989), 28.
Boswell, John. "*Expositio* and *Oblatio*: The Abandonment of Children and the Ancient and Medieval Family." *American Historical Review* 89 (February 1984): 10–33.
———. *The Kindness of Strangers: The Abandonment of Children in Western Literature from Late Antiquity to the Renaissance*. New York, 1988.
———. *The Royal Treasure: Muslim Communities under the Crown of Aragon in the Fourteenth Century*. New Haven, 1977
Bouissac, P. *Circus and Culture: A Semiotic Approach*, Bloomington, 1976.
Boursault, Edmé. *Atrémise et Poliante*. Paris, 1739.
Bouthillier, Guy, and Jean Meynaud, eds. *Le Choc des langues au Québec (1960–1970)*. Montréal, 1972
Bouyon (Abbé). *Réfutation des systèmes de M. l'abbé Baronnat et de Mgr. de la Luzerne sur la question de l'usure*. Clermont-Ferrand, 1824.
Boyesen, Hjalmar Hjorth. "Trilby" (in"The World of Arts and Letters"). *Cosmopolitan* 18 (December 1894): 246–47.
Bradbrook, Muriel C. "Virtue is the True Nobility." *Review of English Studies* 1 (1950): 289–301.
Bradley, Andrew Cecil. *Shakespearean Tragedy*. London, 1904.
Brady, Haldeen. *Hamlet's Wounded Name*. El Paso, 1964.
Brain, Robert. *Friends and Lovers*. New York, 1976.
Brântome, Pierre de Bourdeille. *Les Dames galantes*. Ed. Maurice Rat. Paris, 1960.
Brion, A., and Henri Ey, eds. *Psychiatrie Animale*. Paris, 1964.
Briquet, Pierre. *Traité clinique et thérapeutique de l'hystérie*. Paris, 1859.
British Association for the Advancement of Science. Reports. 1831–.
Brodsky, Claudia. *The Imposition of Form: Studies in Narrative Representation and Knowledge*. Princeton, 1987.
Brody, Jules. "Racine's *Thébaïde*. An Analysis." *French Studies* 13 (1959): 199–213.
Brokmeyer, Henry C. *Hegel's Logic*. A translation from the German. Manuscript. Missouri Historical Society.
Brown, Charles Brockden. *Wieland; or, The Transformation*. New York, 1798.

Brown, Norman O. *Love's Body*. New York, 1966.

Brown, William Hill. *The Power of Sympathy; or, The Triumph of Nature*. Ed. Herbert Brown. Boston, 1961.

Brown, William Wells. *Clotel; or, The President's Daughter* (1853). New York, 1969.

Browne, Sir Thomas. *Pseudodoxia Epidemica, or Enquiries into Very Many Received Tenets*. London, 1646.

Bruns, Carl Georg. *Fontes iuris Romani antiqui*. Leipzig, 1893.

Bryant, Jacob. *A New System, or Analysis of Ancient Mythology*. 3 vols. London, 1774–76.

Buber, Martin. *I and Thou*. Trans. Ronald Gregor Smith. Edinburgh, 1937.

Bugge, John. *Virginitas: An Essay in the History of a Medieval Ideal*. The Hague, 1975.

Bullough, Geoffrey. *Narrative and Dramatic Sources of Shakespeare*. 8 vols. New York, 1957–75.

Burns, Robert I. "Christian-Islamic Confrontation in the West: The Thirteenth Century Dream of Conversion." *American Historical Review* 76 (1971): 1386–1434.

Byron, George Gordon, Lord. *Selected Verse and Prose Works*. Ed. Peter Quennell. London, 1959.

Cagigas, Isidro de las. *Los Mudéjares*, in *Minorías étnico-religiosas de la Edad Media española*. 4 vols. Madrid, 1947–49.

Cahen, J.-G. *Le Vocabulaire de Racine*. Geneva, 1946.

Cahen, Samuel. *Précis élémentaire d'instruction religieuse et morale pour les jeunes français israélites*. Paris, 1820.

Cain, Ann. "A Study of Pets in the Family System." Lecture presented at the Georgetown Family Symposium, Washington, D.C., on October 27, 1978.

Caius, John. *De canibus britannicis*. London, 1570. Also in Christian Franz Paullini. *Cynographia curiosa seu canis descriptio*. Nuremberg, 1685.

————. *Of Englishe Dogges, the Diversities, the Names, the Natures, and the Properties*. Trans. Abraham Fleming. [n.p., 1880].

Calendars of Letters, Despatches, and State papers relating to the negotiations between England and Spain. Ed. G. A. Bergenroth et al. London, 1862–1954.

Calvin, John. *Calvin's Sermons of the Epistles to Timothy and Titus*. Trans. Laurence Tomson. London, 1579.

————. *Commentaries on the Last Four Books of Moses*, ed. Charles Bingham. 4 vols. Edinburgh, 1852–1954.

————. *Opera omnia*. Eds. J. W. Baum, E. Cunitz, E. Reuss, P. Lobstein, and A. Erichson. 59 vols. Brunswick/Berlin, 1863–1900.

Camden, William. *Annales: The True and Royall History of the Famous Empress Elizabeth, Queen of England, France, and Ireland, etc*. London, 1625.

————. *Tomus Alter, & Idem: Or the Historie of the Life and Reigne of that Famous Princesse Elizabeth*. London, 1629.

Campbell, P. G. C. "Christine de Pisan en Angleterre." *Revue de littérature comparée* 5 (1925): 659–70.

Campbell-Jones, Suzanne. *In Habit: A Study of Working Nuns*. New York, 1978.

Campos de España, R. "España y los toros." In Carlos Orellana, ed., *Los Toros en España*, 3 vols., Madrid, 1969–.

Cancelo, José Luis. "Reflexiones de Martin Heidegger sobre un agustino Abraham a Santa Clara." *Revista Agustiniana de Espirtualidad* (Calahorra) 13 (1972): 33–61.

Cantor, Aviva. "The Club, the Yoke, and the Leash: What We Can Learn From the Way a Culture Treats Animals." *Ms. Magazine* 12 (August 1983): 27–30.

Capánaga, Victorino. "Martin Heidegger y el P. Abraham de Santa Clara." *Crisis* (Madrid) 21 (1974): 89–97.

Cardoso, Isaac [Fernando]. *Las excelencias de los Hebreos.* Amsterdam, 1679.

———. *Philosophia libera.* Venice, 1673.

Carlyle, Thomas. *Oliver Cromwell's Letters and Speeches.* London, 1845.

Carra de Vaux, B. "al-Sabi'a." In *Encyclopedia of Islam.* Luzac, 1913.

Carroll, Lewis. *Alice's Adventures in Wonderland; Through the Looking Glass; The Hunting of the Snark.* Ed. Donald J. Gray with backgrounds and essays in criticism. New York, 1971.

Cartwright, Samuel. *The Prognathous Species of Mankind* (1857). *See* in McKitrick, ed. *Slavery Defended,* pp. 139–47.

Caruthers, W. A. *The Kentuckians in New York; or, The Adventures of Three Southerners.* 2 vols. New York, 1834.

Casgrain, H. R. (Abbé). *Un Pèlerinage au pays d'Évangeline.* 1888.

Cassirer, Ernst. *The Myth of the State.* New Haven, 1946.

Castro, Américo. *The Structure of Spanish History.* Trans. Edmund L. King. Princeton, 1954.

———. *Cervantes y los casticismos.* Madrid, 1966.

Cavell, Stanley. "Hamlet's Burden of Proof." In Cavell, *Disowning Knowledge in Six Plays of Shakespeare,* pp. 179–91. Cambridge, Mass., 1987.

———. *Pursuits of Happiness: The Hollywood Comedy of Remarriage.* Cambridge, Mass., 1981.

Cervantes Saavedra, Miguel de. *The Adventures of Don Quixote.* Trans. J. M. Cohen. Harmondsworth, Middlesex, 1950.

———. *Don Quijote de la Mancha.* Ed. Justo Garcia Soriano and Justo Garcia Morales. Madrid, 1966.

Chamberlain, Frederick. *The Sayings of Queen Elizabeth.* London, 1923.

Chambers, R. W. *Thomas More.* New York, 1936.

Champneys, John. *The Harvest is at hand.* London, 1548.

Chapman, George. *Chapman's Homer: The Iliad, the Odyssey; and the Lesser Homerica.* Ed. Allardyce Nicoll. Princeton, 1967.

Chapman, Robert, ed. *New Dictionary of American Slang.* [Rev. ed. of H. Wentworth and S. B. Flexner, *Dictionary of American Slang,* 2nd ed., 1975]. New York, 1986.

Chapra, M. Umar. "The Economic System of Islam: A Discussion of its Goals and Nature." *Islamic Quarterly* 14 (1970): 13–18.

Charcot, Jean Martin. *Leçons du mardi à la Salpêtrière: Policlinique, 1887–1888.* Paris, 1892.

———. "Des troubles de la vision chez les hystériques." In *Progrès médical* 3 (1878): 37–39.

Charles de Grassaille. *Regalium Franciae libri duo.* Paris, 1545.

Charlesworth, M. P. "Nero: Some Aspects." *Journal of Roman Studies* 40 (1950): 69–76.

Charron, Pierre. *Of Wisdome.* Trans. Samson Lennard. London, 1606.

Chateaubriand, François Auguste René de. *Atala/René.* Trans. Irving Putter. Berkeley, 1952.

———. *Le Génie du christianisme* and *Les Martyrs.* Vol. 3 of Chateaubriand, *Oeuvres.* 37 vols. Paris, 1837.

———. *Mémoires d'outre-tombe.* Ed. George Moulinier. 2 vols. Paris, 1951.

———. *Oeuvres romanesques et voyages.* Vol. 1. Ed. Maurice Regard. Paris, 1969.

———. *"René" de Chateaubriand: Un Nouveau Roman.* Ed. Pierre Barbéris. Paris, 1973.

Chateaubriand, Lucile de. *Lucile de Chateaubriand: Ses Contes, ses poèmes, ses lettres.* Ed. Anatole France. Paris, 1879.

Châteillon, Sébastien. *Concerning Heretics; whether they are to be persecuted and how they are to be treated; a collection of the opinions of learned men, both ancient and modern.* Ed. and trans. by Roland H. Bainton. New York, 1935.

Chaucer, Geoffrey. *The Riverside Chaucer.* Third Edition. Ed. Larry D. Benson. [Based on *The Works of Geoffrey Chaucer,* ed. F. N. Robinson]. Boston, 1987.

———. "The Parson's Prologue and Tale (exclusive of the Retraction)." Vol. 4, pp. 361–476. In John M. Manly and Edith Rickert. *The Text of the Canterbury Tales.* 8 vols. Chicago, 1940.

Cherry, C. L. *The Most Unvaluedst Purchase: Women in the Plays of Thomas Middleton.* Salzburg, 1973.

Cholakian, Patricia Francis. *Rape and Writing in the "Heptameron" of Marguerite de Navarre.* Carbondale, 1991.

Choppin, René. *De Domanio Franciae.* Paris, 1605.

Christine de Pisan. *The Book of the City of Ladies.* Trans. Earl Jeffrey Richards. Foreword Marina Warner. New York, 1982.

Chronicle of King Henry VIII of England. Ed. M. A. S. Hume. London, 1889.

Church, William Farr. *Constitutional Thought in Sixteenth-Century France.* Cambridge, Mass., 1941.

Cicero, Marcus Tullius. *De divinatione.* Latin Text and Commentary. Ed. Arthur Stanley Pease (Urbana, Ill., 1920). Rpr. New York, 1979.

Clarke, George Frederick. *Expulsion of the Acadians: The True Story (documented).* Fredericton, 1955.

Clarkson, Paul S., and Clyde T. Warren. *The Law of Property in Shakespeare and the Elizabethan Drama.* 2nd ed. New York, 1969.

Clemens, Samuel. *No. 44, The Mysterious Stranger; Being an Ancient Tale Found in a Jug and Freshly Translated from the Jug.* Forew. and notes by John S. Tuckey. Text estab. by William M. Gibson. Berkeley and Los Angeles, 1982.

———. "People and Things." *Buffalo Express.* September 2, 1869.

———. "Personal Habits of the Siamese Twins." In *The Writings of Mark Twain*, pp. 248–53. Vol. 7. Definitive Edition. New York, 1922–25.

———. *Pudd'nhead Wilson* and *Those Extraordinary Twins.* Authorized Edition. New York, 1922.

———. *Pudd'nhead Wilson* and *Those Extraordinary Twins.* Ed. Sidney Berger. New York, 1980.

Clement of Alexandria. *Stromata.* Ed. O. Stahlin. In *Die griechischen christlichen Schriftsteller der ersten drei jahrhunderte* 3,61. Leipzig, 1906–9.

Clement [Pseudo-]. *Epistolae ad virgines. Patrologiae* (Graeca) 1:349–416.

Clermont-Tonnerre, Comte Stanislas de. *Opinion relativement aux persécutions qui menacent les juifs d'Alsace.* Versailles, 1789.

Cloonan, William J. *Racine's Theatre: The Politics of Love.* University, Mississippi, 1977.

Clutton-Brock, Juliet. *Domesticated Animals from Early Times.* Austin, 1981.

Cobbe, Frances Power. *Italics: Brief Notes on Politics, People, and Places in Italy, in 1864.* London, 1864.

Cocteau, Jean. *Beauty and the Beast: Diary of a Film.* Trans. Ronald Duncan. New introd. George Amberg. New York, 1972.

———. *La Belle et la Bête.* Scenario and dialogues by Jean Cocteau. Directed by Jean Cocteau. Produced by André Paulvé. Ed. with notes by Robert M. Hammond. New York, 1970.

Codex Justiniani. In *Corpus iuris civilis.* 5 vols. Venice, 1584.

Cognet, Louis. *Claude Lancelot, Solitaire de Port-Royal.* Paris, 1950.

Cohen, Hermann. *Die Religion der Vernunft aus den Quellen des Judentums.* 2d ed. Frankfurt, 1929.

Cohen, Martin A. *The Martyr: The Story of a Secret Jew and the Mexican Inquisition in the Sixteenth Century.* Philadelphia, 1973.

Cohn, Norman. "The Cult of the Free Spirit: A Medieval Heresy Reconstructed." *Psychoanalysis and the Psychoanalytic Review* 48 (1961): 51–68.

———. *The Pursuit of the Millennium: A History of Popular Religious and Social Movements in Europe from the Eleventh to the Sixteenth Century*. 2nd ed. New York, 1961.

Coke, Edward. *Commentary upon Littleton*. 16th ed. Ed. Francis Hargrave and Charles Butler. London and Dublin, 1809.

Colección de canones de la iglesia españa. Madrid, 1849–55.

Coleman, W. "The Class Basis of Language Policy in Québec, 1949–1975." *Studies in Political Economy* 3 (Spring 1980): 80–112.

Coleridge, Samuel Taylor. *Biographia Literaria*. London, 1845.

———. *Collected Letters*. Ed. E. L. Griggs. 6 vols. London, 1956–71.

———. *Collected Works*. Ed. Kathleen Coburn. 16 vols. Princeton, 1969.

———. *Complete Poetical Works*. Ed. Ernest Hartley Coleridge, 2 vols. Oxford, 1912.

———. *Lectures and Notes on Shakespeare and Other English Poets*. Ed. T. Ashe. London, 1888.

———. *Letters*. Ed. E. H. Coleridge. London, 1895.

———. *Notebooks*. Vol. 1. Ed. Kathleen Coburn. Princeton, 1957.

———. *Shakespeare Criticism*. Ed. T. M. Rayser. 2 Vols. (1936); rpt. Folcroft, Pa., 1969.

Colie, Rosalie L. "John Locke in the Republic of Letters," pp. 111–29. In *Britain and the Netherlands: Papers Presented to the Oxford-Netherlands Historical Conference, 1959*. Ed. J.S. Bromley and E.H. Kossmann. London, 1960.

Collin de Plancy, J. A. S. *Dictionnaire critique des reliques et des images miraculeuses*. 3 vols. Paris, 1821–22.

Collins, Adela Yarbro. "The Function of 'Excommunication' in Paul." *Harvard Theological Review* 73 (1980): 251–63.

Commission de surveillance de la langue française. Rapport d'activité 1982–83. Québec, 1983.

Compagnon, Antoine. "Proust on Racine." *Yale French Studies* 76 (1989): 21–58.

Comte, Auguste. *Oeuvres d'Auguste Comte*. Paris, 1969–70.

Condorcet, Marie Jean Antoine Nicolas de Caritat, Marquis de. "On the Admission of Women to the Rights of Citizenship." In *Condorcet, Selected Writings*, pp. 97–104. Ed. Keith Michael Baker. Indianapolis, 1976.

Connolly, Cyril. *The Unquiet Grave: A Word Cycle by Palinurus*. New York, 1945.

Conrad, J. R. "The Bullfight: The Cultural History of an Institution." Ph.D. diss., Duke University, 1954.

Conzelmann, Hans. *1 Corinthians*. Philadelphia, 1975.

Cooper, J. M. "Incest Prohibitions in Primitive Culture." *Ecclesiastical Review* 33 (1932): 4–7.

Copleston, Frederick. *A History of Philosophy*. Rev. ed. 4 vols. New York, 1962.

Coquille, Guy. *Les Oeuvres*. 2 vols. Paris, 1666.

Corpus Christianorum (Series Latina). Turnhout, Belgium, 1953.

Corpus inscriptorum latinorum. Ed. Koenigliche Akademie der Wissenschaften. Berlin, 1863.

Corpus scriptorium ecclesiasticorum latinorum. Vienna, 1866–.

Corson, S. A., E. O. Corson, and R. Gunsett. "Pet Animals as Socializing Catalysts in Geriatrics: An Experiment in Non-verbal Communication Therapy." In *Society, Stress, and Disease*. New York, 1987, pp. 305–22.

Cossio, J. M. de. *Los Toros*. 4 vols. Madrid, 1943–61.

Costa Mattos, Vincente da. *Breve discurso contra a heretica perifidia do Iudaismo*. Lisbon, 1623.

Cousin, Victor. *Études sur Pascal*. Paris, 1876.

———. *Jacqueline Pascal*. Paris, 1862.

Cousins, Ewert H. *Bonaventure and the Coincidence of Opposites*. Chicago, 1978.

Cranmer, Thomas. *Determinations . . . that it is unlawful for a man to marry his brother's wife*. London, 1531.

Craster, H. H. E. "An Unknown Translation by Queen Elizabeth." *The English Historical Review* 29 (1914): 721–23.

Crawley, Ernest. *The Mystic Rose: A Study of Primitive Marriage and of Primitive Thought in Its Bearing on Marriage.* Ed. T. Bestermann. 2 vols. London, 1927.

Crèvecoeur, Michel-Guillaume-Jean de. (J. Hector St. John.) *Letters from an American Farmer* (1782). New York, 1957.

————. *Lettres d'un cultivateur américain.* Enlarged French edition. With frontispiece designed by C. Bornée and engraved by P. Martini. Paris, 1787.

Crónica del Rey Enrico Otava de Inglaterra. Ed. Marquis de Molins. Madrid, 1874.

Crosse, Robert. *The Lover, or Nuptial Love.* London, 1638.

Crowe, Michael J. *The Extraterrestrial Life Debate, 1750–1900: The Idea of a Plurality of Worlds from Kant to Lowell.* Cambridge, England, 1986.

Curtius, Ernst Robert. *European Literature and the Latin Middle Ages.* Trans. Willard R. Trask. New York, 1953.

Cyprian, *Epistolae.* Vol. 2 of Cyprian, *Opera omnia.* Ed. Guilemus Hartel. Vienna, 1868–71.

Dagens, Jean. "Le *Miroir des simples ames* et Marguerite de Navarre." In *La Mystique Rhénane: Colloque de Strasbourg 16–19 mai 1961,* pp. 281–89. Paris, 1963.

Dahn, F. *Lex Visigothorum. Westgotische Studien.* Würzburg, 1872.

Dalke, Anne French. *"Had I known her to be my sister, my love would have been more regular": Incest in Nineteenth Century American Fiction.* Ph.D. diss., University of Pennsylvania, 1982.

Damasus I. *Epigrammata Damasiana.* Ed. Antonius Ferrua. Vatican, 1942.

Damon, S. Foster. "Pierre the Ambiguous." *The Hound and Horn: A Harvard Miscellany* 2 (1928): 107–18.

d'Anglure, B. Saladin. *L'Organisation social traditionelle des Esquimaux de Kangirsujuaaq (Nouveau Québec).* Centre d'Études Nordiques, Laval University, no. 17. Laval, Quebec, 1967.

Dante Alighieri. *De Monarchia.* Ed. Gustavo Vinay. Florence, 1950.

————. *Divine Comedy.* Trans. John D. Sinclair. 3 vols. New York, 1961.

Darwin, Charles. *The Descent of Man and Selection in Relation to Sex.* (Facsimile of the 1871 edition.) Intro. by John Tyler Bonner and Robert M. May. Princeton, 1981.

Daube, David. "Lex Talionis." In *Studies in Biblical Law.* Cambridge, 1947.

Daus, John (tr.). *A Famous Chronicle of Our Time Called Sleidanes [J.] Commentaries.* London, 1560.

Davis, Cushman Kellogg. *The Law in Shakespeare.* 2nd ed. St. Paul, 1884.

Davis, Natalie Zemon. "The Rites of Violence." In Davis, *Society and Culture in Early Modern France: Eight Essays,* Stanford, 1975, pp. 152–88.

De Quincey, Thomas. *Autobiographical Sketches.* Vols. 1–2 of *The Collected Writings of Thomas De Quincey.* Ed. David Masson. 14 vols. Edinburgh, 1889–90.

————. *The Last Days of Immanuel Kant.* In *Collected Writings.* Vol. 4, pp. 323–79.

————. *Selections Grave and Gay. Writings Published and Unpublished.* 14 vols. Edinburgh, 1853–60.

Decker, Carl L. *The Declaration of Independence: A Study in the History of Political Ideas.* New York, 1922.

Del Mar, Alexander. *Ancient Britain in the Light of Modern Archeological Discoveries.* New York, 1900.

————. *The Worship of Augustus Caesar Derived from a Study of Coins, Monuments, Calendars, Eras, and Astrological Cycles.* New York, 1900.

Delany, Martin Robinson. *The Condition, Elevation, Emigration, and Destiny of the Coloured People of the United States* (1852). New York, 1969.

Delcroix, Maurice. *Le sacré dans les tragédies profanes de Racine*. Paris, 1970.
Des Pres, Terence. *The Survivor*. New York, 1976.
Désessartz. *Traité de l'education corporelle des enfans en bas age*. Paris, 1760.
Desmond, A. *The Ape's Reflection*. New York, 1979.
Detter, Ferd. "Die Hamletsage." *Zeitschrift für Deutsches Altertum und Deutsche Literatur* 26 (1892): 1–25.
Deutsch, Helen. "A Case of Hen Phobia." In *Psychoanalysis of the Neuroses*, pp. 113–26. London, 1932.
Dewey, John. "The Historic Background of Corporate Legal Personality." *Yale Law Journal* 35 (1926): 655–73.
Dewhurst, J. "The Alleged Miscarriages of Catherine of Aragon and Anne Boleyn." *Medical History* 28 (1984): 49–56.
Dewick, E. S., ed. *The Coronation Book of Charles V of France*. Bradshaw Society, XVI. London, 1899.
Diderot, Denis. *La Religieuse*. Paris, [1796].
Diogenes Laertius. *Lives of the Eminent Philosophers*. With trans. by Robert Drew Hicks. 2 vols. London & New York, 1925.
Dion, Léon. "La loi 178: acceptable mais d'application aléatoire." *Le Devoir* [Montreal] (January 6, 1989): 7.
Dionysius Carthusianus [Denis de Leewis]. *The mirroure of golde for the synfull soule*. Trans. Margaret Countess of Richmond. London, 1522.
Disraeli, Benjamin. *Contarini Fleming*. New York, 1832.
Dixon, William Hepworth. *Spiritual Wives*. 2 vols. London, 1868.
Documentary Source Book of American History, 1606–1898. Ed. with notes by William McDonald. New York, 1908.
Doiron, Marilyn. "The Middle English Translation of *Le Mirouer des simples ames*." In *Dr. L. Reypens-Album*. Ed. L. Reypens and Alb. Ampe, p. 131–51. Antwerp, 1964.
Dölger, F. "Brüderlichkeit der Fürsten." In *Reallexikon für Antike und Christentum*, ed. Theodor Klauser. 2 vols. Stuttgart, 1954.
Döllinger, I. I. von. *Beiträge zur Sektengeschichte des Mittelalters*. 2 vols. Munich, 1890.
Dominguez, Virginia P. *White By Definition: Social Classification in Creole Louisiana*. New Brunswick, New Jersey, 1986.
Dostoevsky, Feodor. *The Brothers Karamazov*. Trans. Constance Garnett. Revised by Ralph Matlaw. New York, 1976.
———. "The Dream of a Ridiculous Man: A Fantastic Story." In *The Best Short Stories of Dostoevsky*, pp. 297–322. Trans. David Magarshack. New York, 1955.
Douglas, Mary. *Implicit Meanings: Essays in Anthropology*. Boston, 1975.
Dournes, Jacques "L'Inceste Préférentiel: Étude de mythologie," *L'Homme* 11 (1971): 5–19.
Dreiser, Theodore. "Mark the Double Twain." *The English Journal* 24 (1935): 615–27.
Dryden, Edgar A. "The Entangled Text: Melville's *Pierre* and the Problem of Reading." *Boundary 2* 7 (1979): 145–73.
DuBois, W. E. B. *Darkwater: Voices from Within the Veil*. New York, 1970.
Duff, I. F. Grant. "Die Beziehung Elizabeth-Essex: eine psychoanalytische Betrachtung." *Psychoanalytische Bewegung* 3 (1931): 457–74.
Dumm, Demetrius. *The Theological Basis of Virginity According to Saint Jerome*. Latrobe, Pa., 1961.
Durandus, Gulielmus. *Speculum iuris*. 4 vols. Venice, 1602.
Durbach, Errol. "The Geschwister-Komplex: Romantic Attitudes to Brother-Sister Incest in Ibsen, Byron, and Emily Brontë." *Mosaic* 12 (1979): 61–73.
Durham, John George Lambton [First Earl of Durham]. *Report on the Affairs of British North*

America. Laid before the British Parliament on January 31, 1839. Ed. with an introd. by Charles Lucas. 3 vols. Oxford, 1912. Repr. New York, 1970.
Dusinberre, Juliet. *Shakespeare and the Nature of Women.* London and Basingstoke, 1975.
Easton, Lloyd D. *Hegel's First American Followers: The Ohio Hegelians—J. B. Stallo, Peter Kaufman, Moncure Conway, August Willich.* Athens, Ohio, 1966.
Ebreo, Leone. *Dialoghi d'amore.* Ed. Carl Gebhardt. Heidelberg, 1929.
Edwards. Michael. *La Tragédie racinienne.* Paris, 1972.
Eekhoud, G. *Les Libertins d'Anvers: Légendes et histoire des Loisistes.* Paris, 1912.
Egido, Téofanes. "The Historical Setting of St. Teresa's Life." *Carmelite Studies* (Washington, D.C.) 1980 [1].
Eichmann, Eduard. *Die Kaiserkrönung im Abendland.* 2 vols. Oxford, 1921.
Eisenstadt, Shmuel N. *From Generation to Generation: Age Groups and Social Structure.* With new intro. by Eisenstadt. New York, 1971.
————. "Ritualized Personal Relations." *Man* 56 (1956): 90–95.
Eliot, T. S. *Essays in Elizabethan Drama.* New York, 1956.
————. *"The Idea of a Christian Society." In Christianity and Culture,* New York, 1949.
————. *Selected Essays.* London, 1932.
Elizabeth I. *Englishings of Boethius, De Consolatione Philosophiae, A.D. 1593, Plutarch, De Curiositate, A.D. 1598, and Horace, De Arte Poetica (part), 1598. Ed. Caroline Pemberton.* Early English Text Society. Original Series, No. 113. London, 1899.
————. "The Glass of the Sinful Soul" (1544). Bodleain Library Manuscript, Cherry 36. *See* photographic copy and transcription in Shell, ed., *Elizabeth's Glass;* photographic copy in Ames, ed., *Mirror;* transcription in Salminen, ed., *Miroir.* Published as *A Godly Medytacyon of the Christen Sowle, Concerning a Love Towardes God and His Christe, Compiled in French by Lady Margarete Quene of Nauerre, and Aptely Translated into English by the Ryght Vertuouse Lady Elyzabeth Daughter to Our Late Soverayne King Henry the VIII.* "Epistle Dedicatory" and "Conclusion" by John Bale. Marburg, 1548.
————. *Letters of Queen Elizabeth.* Ed. G. B. Harrison. London, 1935.
————. *Poems of Queen Elizabeth I.* Ed. Leicester Bradner. Providence, Rhode Island, 1964.
————. *Public Speaking of Queen Elizabeth.* Ed. George P. Rice. New York, 1951.
Other manuscripts:
————. Aphoristic texts and mottos embroidered, apparently by Elizabeth herself, on the black cloth covers of a tiny sextodecimo New Testament containing part of Laurence Tomson's *New Testament* (1578). Bodl. MS e. Mus. 242.
————. English translation of John Calvin's *Institution Chrétienne.* The Scottish Record Office. MS.
————. Latin, French, and Italian translations of Catherine Parr's *Prayers, or Meditations.* Entitled *Precationes sev meditationes.* British Library, MS Royal 7 D. X. MS. 1545.
————. Latin translation of Bernardino's Ochino's *De Christo Sermo.* MS Bodley 6.
Ellis, John B. *Free Love and its Votaries, or American Socialism Unmasked.* New York, 1870.
Ellul, Jacques. *Propaganda: The Formations of Men's Attitudes.* New York, 1965.
Elyot, Thomas. *The Image of Governaunce compiled by the actes and sentences notable of the most noble Emperor Alexandre Seuerus.* London, 1544.
Emerson, Ralph Waldo. *Essays and Lectures.* Ed. Joel Porte. New York, 1983.
————. *Journals.* Ed. William Gilman and J. E. Parsons. Cambridge, Mass., 1970.
Encyclopaedia Britannica. 11th ed. 32 vols. 1910–11.
Encyclopedia of Islam. Luzac, 1913.
Encyclopaedia of Islam. 2nd ed. 7 vols. Leiden, 1960–.
Encyclopedia Judaica. 14 vols. Jerusalem, 1971–72.

Encyclopedia of Philosophy. Ed. Paul Edwards. 8 vols. New York, 1967.

Encyclopedia of Religion and Ethics. Ed. James Hastings. New York, 1924–27.

Engdahl, Sylvia L. *Planet-Girded Suns*. New York, 1974.

Engels, Friedrich. *The Origin of the Family, Private Property and the State*. Introd. Michèle Barrett. Harmondsworth, 1985.

Epictetus. *The Discourses, as Reported by Arrian the Manual; and Fragments*. 2 vols. Trans. W. A. Oldfather. London, 1925.

Epstein, Louis M. *Marriage Laws in the Bible and the Talmud*. Cambridge, Mass., 1942.

Erasmus, Desiderius. *Institutio Principis Christiani*. In *Opera Omnia Des. Erasmi Roterodami*, ed. O. Harding, pp. 95–220. vol. 4, pt. 1. Amsterdam, 1974.

Estèbes, Janine. *Tocsin pour un massacre: La saison des Saint-Barthélemy*. Paris, 1968.

Estève, Edmond. "Le théâtre 'Monacal' sous la Révolution, ses précédents et ses suites." In Estève, *Etudes de littérature préromantique*. Paris, 1923.

Etiemble, René. *Parlez-vous franglais?* Paris. 1964.

Eusebius, Pamphili of Caesarea. *The History of the Church from Christ to Constantine*. Trans. G. A. Williamson. New York, 1966.

Evans, Frederick Williams. *Autobiography of a Shaker, and Revelation of the Apocalypse* (1888); Rpr. New York, 1973.

Fairfield, Leslie P. *John Bale: Mythmaker for the English Reformation*. West Lafayette, Ind., 1976.

Farías, Victor. *Heidegger and Nazism*. Ed. with forew. by Joseph Margolis and Tom Rockmore. French trans. Paul Burrell and Dominic Di Bernardi. German trans. Gabriel R. Ricci. Philadelphia, 1989.

Farnell, Lewis Richard. *Greek Hero-Cults and Ideas of Immortality*. Oxford, 1921.

Farr, Edward, ed. *Selected Poetry, Chiefly Devotional, of the Reign of Queen Elizabeth*. 2 vols. Cambridge, 1845.

Faye, Jean-Pierre, Francis Cohen, et al. *Colloque de Cluny* II, 2–4 April 1970. (= *La Nouvelle Critique* special 39 bis.)

Feije, H. *De impedimentis et dispensationibus matrimonialibus*. Louvain, 1885.

Ferenczi, Sandor. *Sex in Psychoanalysis*. Boston, 1916.

Ferran, Vincent Isidore Pierre. *Du vomissement de sang dans l'hystérie*. Paris, 1874.

Feuer, Lewis Samuel. *Lawless Sensations and Categorical Defences: The Unconscious Sources of Kant's Philosophy*. New York, 1970.

Feuerbach, Ludwig. *Die Philosophie der Zukunft*. Ed. with notes by H. Ehrenberg, in *Frommanns Taschenbücher*, Stuttgart, 1922.

———. *The Essence of Christianity*. Trans. George Eliot. Introd. Karl Barth. Forew. H. Richard Niebuhr. New York, 1957.

Fiedler, Leslie. *Freaks: Myths and Images of the Secret Self*. New York, 1978.

———. *Love and Death in the American Novel*. New York, 1966.

Figueroa, Rafael Olivares. *Folklore Venezolano* (1954). Ed. with Prologue by Armando José Sequera. Caracas, 1988.

Fildes, Valerie. *Wet Nursing: A History from Antiquity to the Present*. Oxford, 1988.

Filmer, Robert. *Partriarcha and Other Political Works*. Ed. Peter Laslett. Oxford, 1949.

Finkelstein, Louis. *The Pharisees: The Sociological Background of Their Faith*. Philadelphia, 1983.

Fitzhugh, George. *Cannibals All!, or, Slaves without Masters*. Richmond, Va., 1857.

Fly, Richard. "Accommodating Death: The Ending of *Hamlet*." *Studies in English Literature* 24 (1984): 257–74.

Foissac [pseudonym of Jean-Baptiste Annibal Aubert-Dubayet?]. *Le cri du citoyen contre les juifs*. Lausanne, Metz, 1786.

Foner, Philip S. *The Life and Writings of Frederick Douglass. Volume II: Pre-Civil War Decade—1850–1860.* New York, 1950.
Ford, J. Massingberd. *A Trilogy on Wisdom and Celibacy.* Notre Dame, 1967.
Ford, John. *'Tis Pity She's a Whore.* Ed. Derek Roper. London, 1975.
———. *'Tis Pity She's a Whore.* Ed. S. P. Sherman. Boston, 1915.
Forsythe, R. S. "Tacitus, 'Henry VI, Part III,' and 'Nero.'" *Modern Language Notes* 42 (1927): 25–27.
Fortescue, Sir John. *De laudibus legum Angliae.* Ed. S. B. Chrimes. Oxford, 1885.
Fowler, John Howard. "The Development of Incest Regulations in the Early Middle Ages: Family, Nurturance, and Aggression in the Making of the Medieval West." Ph.D. diss., Rice University, 1981.
Fox, M. W. "Pet Animals and Human Well-being." *See* in Kay et al., eds., *Pet Loss,* pp. 16–21.
———. *Returning to Eden: Animal Rights and Human Responsibility.* New York, 1980.
———. *The Soul of the Wolf.* Boston, 1980.
Fraenger, Wilhelm. *The Millennium of Hieronymus Bosch: Outlines of a New Interpretation.* Trans. Eithne Wilkins and Ernst Kaiser. London, 1952.
Francis of Assisi. *Gli Scritti di San Francesco d'Assisi.* Societe Editrice Vita e Pensiero. Milan, 1954–.
———. *Saint Francis: Omnibus of Sources.* Ed. Marion A. Habig. Chicago, 1973.
Frankl, O. *Der Jude in den deutschen Dichtungen des 15., 16. un 17. Jahrhunderts.* Leipzig, 1905.
Frazer, James G. *The Golden Bough.* 12 vols. London, 1911–15.
Frederichs, J. "Un Luthérien français devenu libertin spirituel: Christophe Herault et les Loisistes d'Anvers (1490–1544)." *Bulletin de la société de l'histoire du protestantisme français* 41 (1892): 250–69.
Frederickson, George M. *White Supremacy: A Comparative Study in American and South African History.* New York, 1981.
Freeman, Kathleen. *The Pre-Socratic Philosophers.* London, 1956.
French, Richard D. *Antivivisection and Medical Science in Victorian Society.* Princeton, 1975.
Freud, Sigmund. "Analysis of a Phobia in a Five Year Old Boy." *Collected Papers.* 3:149–64.
———. "A Case of Successful Treatment by Hypnotism." *Standard Edition* 1:115–30.
———. *Civilization and Its Discontents.* Trans. J. Strachey. New York, 1962.
———. *Collected Papers of Sigmund Freud.* Ed. J. Riviere and James Strachey. 5 vols. New York, 1959.
———. "Footnotes to Charcot's *Tuesday Lectures.*" *Standard Edition* 1:137–46.
———. "From the History of an Infantile Neurosis." *Standard Edition* 17:3–122.
———. *The Interpretation of Dreams. First Part. Standard Edition.* 4.
———. "Obsessive Acts and Religious Practices." *Collected Papers* 2:25–35.
———. "On the Psychical Mechanism of Hysterical Phenomena: A Lecture." *Standard Edition* 3:25–40.
———. *Standard Edition of the Complete Psychological Works of Sigmund Freud.* Edited & trans. by James Strachey, Anna Freud, Alan Tyson, and Alix Strachey. 24 vols. London, 1953–74.
———. "The Taboo on Virginity." *Standard Edition* 11:191–208.
———. *Totem and Taboo.* Trans. J. Strachey. New York, 1950.
Freud, Sigmund, and Joseph Breuer. "On the Psychical Mechanism of Hysterical Phenomena: Preliminary Communication." *Standard Edition* 2:1–18.
Friedman, Albert B. "'When Adam Delved . . .': Contexts of a Historic Proverb." In *The Learned and the Lewed: Studies in Chaucer and Medieval Literature,* ed. Larry D. Benson, pp. 213–30. *Harvard English Studies* 5. Cambridge, 1974.

Friedmann, Erika., "Pet Ownership and Coronary Heart Disease—Patient Survival." *Circulation* 58 (1978): 11–168.

Friedmann, Erika, Aaron A. Katcher, Sue A. Thomas, and James J. Lynch. "Health Consequences of Pet Ownership." In *Pet Loss and Human Bereavement,* ed. William J. Kay, Herbert A. Nieburg, Austin H. Kutscher, Ross M. Grey, and Carole E. Fudin, pp. 22–30. Ames, Iowa, 1984.

Friedmann, Herbert. *The Cowbirds: A Study in the Biology of Social Parasitism.* Springfield, Ill., 1929.

Friedmann, Paul. *Anne Boleyn: A Chapter of English History, 1527–1536.* 2 vols. (London, 1884). Rpr. New York, 1973.

Fry, Annette Riley. "The Children's Migration: The Children's Aid Society's Great Project." *American Heritage* 26 (1974): 4–10, 79–81.

Fuller, Lon L. *Legal Fictions.* Stanford, 1967.

Galton, Francis. *Finger Prints.* New York, 1892.

Garber, Marjorie. *Dream in Shakespeare: From Metaphor to Metamorphosis.* New Haven, 1974.

————. Ed. *Cannibals, Witches, and Divorce: Estranging the Renaissance.* Selected Papers from the English Institute, 1985. New Series, no. 11. Baltimore, 1987.

Garden, M. *Lyon et les Lyonnais au XVIIIe siècle.* Paris, 1970.

Gardiner, Samuel Rawson. *History of the Great Civil War, 1642–1649.* 4 vols. London, 1886–91.

Gardiner, Stephen. *Letters.* New York, 1933.

Garrison, William Lloyd. *Life of William Lloyd Garrison.* 4 vols. New York, 1885–1889.

Gartenberg, P., and N. Thames Whittemore, "A Checklist of English Women in Print, 1475–1640." *Bulletin of Bibliography* 34 (1977): 1–13.

Gauvin, Lise. "From Octave Crémazie to Victor-Lévy Beaulieu: "Language, Literature, and Ideology." Trans. Emma Henderson. *Yale French Studies* 65 (1983): 30–49.

Gazier, Augustin. *Étude sur 'Racine et Port Royal'.* In Gazier, *Mélanges de littérature et d'histoire.* Paris, 1904.

Gazier, Cécile. *Ces Messieurs de Port-Royal.* Paris, 1932.

Geertz, C. "Deep Play: Notes on the Balinese Cockfight." In C. Geertz, ed., *Myth, Symbol, and Culture,* pp. 1–37. New York, 1971.

Gellius, Aulus. *Les Nuits attiques.* Ed. René Marache. Paris, 1967.

Gendron, Jean-Denis. "Les Attitudes des groupes ethniques." Extract from *Rapport de la commission d'enquête sur la situation de la langue françaises et sur les droits linguistiques au Québec. Livre III: le groupes ethniques.* Québec, 1972, pp. 89–106. *See* in Boismenu et al., *Le Québec en textes,* 539–49.

————. "Évolution de la conscience linguistique des Franco-Québécois depuis la Révolution tranquille." Extracted from *Devoir,* 4 may 1985, p. 9. *See* in Boismenu et al., *Le Québec en textes,* pp. 434–44.

Genesis Rabba. *Bereschit Rabba mit kritischem Apparat und Kommentar* by Julius Theodor and Chanoch Albeck. 4 vols., Berlin, 1912–36.

Geoffrey of Monmouth. *Historia Regum Britanniae.* Trans. Robert Ellis Jones. London, 1929.

Gesenius, William. *A Hebrew and English Lexicon of the Old Testament.* Trans. Edward Robinson. Ed. Francis Brown. Oxford, 1972.

Gessain, R. "*Vagina Dentata* dans la clinique et la mythologie." *La Psychanalyse* 3 (1956): 247–95.

Ghazali, Abu Hamid Muhammad. *Réfutation excellente de la divinité de Jésus-Christ d'après les Évangiles.* Trans. and with comment. by R. Chidiac. Paris, 1939.

Gibbon, Edward. *The History of the Decline and Fall of the Roman Empire.* 5 vols. London, 1776–1788.

Gifford, G. *A Discourse of the Subtle Practices of Devils by Witches and Sorcerers.* London, 1587.

Gilberg, Arnold. "The Ecumenical Movement and the Treatment of Nuns." *International Journal of Psycho-Analysis* 49 (1968): 481–83.

Gillman, Susan. *Dark Twins: Imposture and Identity in Mark Twain's America.* Chicago, 1989.

Gilman, Stephen. *The Spain of Fernando de Rojas: The Intellectual and Social Landscape of La Celestina.* Princeton, 1972.

Gilpérez García, L., and M. Fraile Sanz. *Reglamentación Taurina Vigente: Diccionario Comentada.* Seville, 1972.

Gilson, Étienne. *Dante et la philosophie.* Paris, 1939.

———. *Dante the Philosopher.* Trans. David Moore. New York, 1949.

Girard, René. *Things Hidden since the Foundation of the World.* Trans. Stephen Bann and Michael Metteer. Stanford, 1987.

Giraud, Victor. *Soeurs de grands hommes.* Paris, 1926.

Giraudoux, Jean. *Racine.* Paris, 1950.

Gobineau, Joseph Arthur de. *Histoire d'Ottar-Jarl, pirate norvégien, conquérant du pays de Bray en Normandie et de sa descendance.* Paris, 1879.

———. *Selected Political Writings.* Ed. and introd. by Michael D. Biddiss. Trans. Michael D. Biddiss, Adrian Collins, and Brian Nelson. New York, 1970.

Godbout, Jacques. *Le Réformiste.* Montréal, 1975.

Godin, Gérald. "Le joual politique." *Parti pris* 2 (March 1965): 51–59.

Godefroy, Théodore. *Le Cérémonial de France.* Paris, 1619.

Godwin, Francis. *Man in the Moone.* 1638.

Godwin, William. *Enquiry Concerning Political Justice, and Its Influence on Modern Morals.* Ed. Isaac Kramnick. Baltimore, 1976.

Godwyn, Morgan. *The Negro and Indians Advocate, Suing for Their Admission into the Church; or, a Persuasive to the Instructing and Baptizing of the Negro's and Indians in Our plantations.* London, 1680.

Goethe, Johann Wolfgang von. *Elective Affinities.* Trans. R. J. Hollingdale. Baltimore, 1971.

———. *Italian Journey.* Trans. W. H. Auden and Elizabeth Mayer. New York, 1962.

———. "Das Römische Karneval." In *Werke, Hamburger Ausgabe,* 11:484–515.

Goitein, S. D., "The Concept of Mankind in Islam." In *History and the Idea of Mankind,* ed. W. Wager, pp. 72–91. Albuquerque, 1971.

Goldberg, Jonathan. *James I and the Politics of Literature: Jonson, Shakespeare, Donne, and Their Contemporaries.* Baltimore, 1983.

———. *Endlesse Worke: Spencer and the Structures of Discourse.* Baltimore, 1981.

Goldmann, Lucien. *Le Dieu caché: Etude sur la vision tragique dans les "Pensées" de Pascal et dans le théâtre de Racine.* Paris, 1959.

Goldsmith, Oliver. *The Citizen of the World.* Vol. 2 of *Collected Works,* ed. A. Friedman. Oxford, 1966.

Gollancz, Israel, ed. *The Sources of "Hamlet," with an Essay on the Legend.* London, 1926.

Gomes de Solis, Duarte. *Alegación en favor de la Compañia de la Indias Orientales.* 1628.

Gomperz, Lewis. *Moral Inquiries on the Situation of Man and of Brutes. On the Crime of Committing Cruelty on Brutes, and of Sacrificing Them to the Purposes of Man. Etc..* London, 1824.

Gordan, Cyrus H. "Fratriarchy in the Old Testament." *Journal of Biblical Literature* 54 (1935): 223–31.

Gouges, Olympe de. *The Declaration of the Rights of Women.* In *Women in Revolutionary Paris, 1789–1795, Selected Documents.* Ed. and trans. by Darlene Gay Levy, Harriet Branson Applewhite, and Mary Durham Johnson, pp. 87–96. Urbana, 1979.

Gould, Karen, ed. *Focus on Québec 1970 and its Aftermath. Québec Studies* 11 (1990–91).

Gould, Stephen Jay, "Living With Connections." In Gould, S.J., *The Flamingo's Smile: Reflections in Natural History,* pp. 65–75. New York, 1985.

Gower, John. *Confessio Amantis.* Vols. 2 and 3 of *The Complete Works of John Gower.* Ed. G. C. M. Macaulay. 4 vols. London, 1901.

Grandes Chroniques de France. Ed. Jules Édouard M. Viard. Paris, 1920–53.

Gratian. *Decretum. (Concordia Discordantium Canonum).* Ed. A. L. Richter. Rev. A. Friedberg. In *Corpus Iuris Canonici* 1. Leipzig, 1879.

Grayzel, Solomon. *A History of the Jews: From the Babylonian Exile to the Establishment of Israel.* Philadelphia, 1961.

"The Great American Petmania." *Time Magazine* (January 6, 1975): 58.

Grebanier, Bernard. *The Heart of Hamlet.* New York, 1960.

Greenblatt, Stephen. *Shakespearean Negotiations: The Circulation of Social Energy in Renaissance England.* Berkeley and Los and Angeles, 1988.

Greene, Thomas. *The Descent from Heaven: A Study in Epic Continuity.* New Haven, 1963.

Greenfield, Liah. "The Closing of the Russian Mind." *The New Republic* 202 (1990): 30–34.

Grégoire, Henri [l'Abbé]. *Essai sur la regénération physique, morale, et politique des Juifs, ouvrage couronné par la Société Royale des Sciences et des Arts de Metz, le 23 Août 1788, par M. Grégoire, Curé du Diocèse de Metz.* Metz, 1789.

Grégoire, Pierre. *De Republica.* Lyon, 1578.

Gregory of Nazianzus. *Epistolae.* Ed. and trans. P. Gailay. 2 Vols. Paris, 1964–67. Also in *Patrologiae* (Graeca) 37:1–387.

Gregory of Nyssa. *Vita S. Macrinae. Patrologiae* (Graeca) 6:959–99.

Gregory the Great. *Vita S. Benedicti. Patrologiae* (Latina) 66:125–215.

Die Griechischen christlichen Schriftsteller der ersten drei Jahrhunderte. Leipzig, 1897.

Grillparzer, Franz. *Die Ahnfrau.* Vienna, 1817.

Grimm, Jakob, and Wilhelm Grimm. *Kinder- und Haus-Märchen.* 3 vols. 1812–22.

Gröber, Conrad. "Der Altkatholicismus in Messkirch: Die Geschichte seiner Entwicklung und Bekämpfung." *Freiburger Diözesan-Archiv* (Freiburg im Breisgau), New Series 13 (1912): 135–198.

Grossman, Allen. "A Primer of the Commonplaces in Speculative Poetics." *Western Humanities Review* 44 (1990): 5–138.

Guarnieri, Romana. *Il Movimento del libero spirito: Testi e documenti.* Roma, 1965. (Reissue of *Archivio italiano per la storia della pietà* 4 [1964]: 353–708.)

Gudeman, S. "The Compadrazago as a Reflection of the Natural and Spiritual Person." *Proceedings of the Royal Anthropological Institute of Great Britain and Ireland* (1971): 45–71.

Guyon, René. *Réflexions sur la tolérance.* Paris, 1930.

———. *Sex Life and Sex Ethics.* London, 1933.

Haberman, A. M., ed. *Iggeret Baalei Hayyim.* Jerusalem, 1949.

Hadewijch. *De Visioenen.* Ed. Joseph van Mierlo. Louvain, 1924–25.

Hahn, August. *Bibliothek der Symbole und Glaubensregln der alten Kirche.* 2nd ed. Breslau, 1877.

Halevi, Yehuda. *Book of the Kazari.* Trans., introd., and notes by Hartwig Hirschfeld. New York, 1946.

Halévy, Elie. *Instruction religieuse et morale à l'usage de la jeunesse israélite.* Metz, 1820.

Halley, Janet E. "Heresy, Orthodoxy, and the Politics of Religious Discourse: The Case of the English Family of Love." *Representations* 15 (1986): 98–120.

Hallywell, Henry. *An Account of Familism as it is Revised by the Quakers.* 1673.

Hamilton, John. *Catechism* [1552]. London, 1884.

Hamlet. An Attempt to Ascertain Whether the Queen Were an Accessory, Before the Fact, in the Murder of her First Husband. London, 1856. *See* in Furness, ed., *Hamlet,* 2:265–66.

Hansen, W. F. *Saxo Grammaticus and the Life of Hamlet: A Tradition, History, and Commentary.* Lincoln, 1983.

Harding, Anthony John. *Coleridge and the Idea of Love: Aspects of relationship in Coleridge's thought and writing.* Cambridge, Eng., 1974.

Harding, Mary Esther. *Woman's Mysteries, Ancient and Modern.* London, 1935.

Harington, John. *Letters and Epigrams.* Ed. N. E. McClure. Philadelphia, 1930.

Harney, Martin P. *Brother and Sister Saints.* Paterson, New Jersey, 1957.

Harris, Jesse W. *John Bale: A Study in the Minor Literature of the Reformation.* Urbana, Ill., 1940.

Hartmann von Aue. *Gregorius: A Medieval Oedipus Legend.* Trans. Edwin H. Zeydel. Chapel Hill, North Carolina, 1955.

Hastings, James, ed. *Encyclopedia of Religion and Ethics.* 13 vols. New York, 1924–27.

Haugaard, William P. "Elizabeth Tudor's *Book of Devotions:* A Neglected Clue to the Queen's Life and Character." *Sixteenth Century Journal* 12, no. 2 (1981): 79–106.

Haugen, Einar. "Bilingualism, Language Contact, and Immigrant Languages in the United States: A Research Report 1956–1970." In *Advances in the Study of Societal Multi-lingualism,* ed. J. Fishman, pp. 19–111. The Hague, 1978.

Haupt, H. "Ein Beghardenprozess in Eichstadt vom Jahre 1381." *Zeitschrift für Kirchengeschichte* 5 (1882): 487–98.

Hauser, Henri. *Études sur la réforme française.* Paris, 1909.

Havens, Richie. "Freedom." A song adapted from "Motherless Child." In *Woodstock* [record album]. Produced by Eric Blackstead. Cotillion Records, 1970.

Hawthorne, Nathaniel. *The Marble Faun.* Chicago, 1959.

———. *Passages from the French and Italian Note-Books.* 1864. 1871.

Hearne, Vicki. *Adam's Task: Calling Animals by Name.* New York, 1986.

———. "How to Say 'Fetch!'" *Raritan* 3 (Fall 1983): 1–33.

———. "The Moral Transformation of the Dog, and Other Thoughts on the Animals Among Us." *Harper's* 268 (January 1984): 57–67.

Hébert, Anne. *Kamouraska.* Paris, 1970.

Hefele, Carl Joseph von. *Conciliengeschichte.* 9 vols. Freiburg-im-Breisgau, 1855–87.

———. *Histoire des conciles.* Ed. H. LeClercq. Paris, 1907.

Hegel, Georg Wilhelm Friedrich. *Werke.* 20 Vols. Frankfurt, 1969–71.

Heidegger, Martin. "Abraham a Sancta Clara" Zur Enthüllung seines Denkmals in Kreenheinstetten am 15. August 1910." *Allgemeine Rundschau: Wochenschrift für Politik und Kultur* 35 (1910).

———. "Brief über 'Humanismus.'" In *Wegmarken,* Frankfurt am Main, 1967, pp. 145–94.

———. *German Existentialism.* Trans. Dagobert D. Runes. New York, 1965.

———. *Der Selbstbehauptung der deutschen Universität.* 2nd ed. Breslau, 1934.

———. *Über Abraham a Sancta Clara.* Messkirch, 1964.

Heilbrun, Carolyn H. *Toward a Recognition of Androgyny.* New York, 1973.

Heiman, Marcel. "The Relationship Between Man and Dog." *Psychoanalytic Quarterly* 25 (1956).

Hemingway, Ernest. *The Dangerous Summer.* London, 1985.

Henry VIII. *Letters and Papers, Foreign and Domestic, of the Reign of Henry VIII, 1509–1547.* Ed. James Gairdner and R. H. Brodie. 21 vols. London, 1862–1910.

Henzen, Wilhelm, Eugene Bormann, et al., eds. *Inscriptiones urbis Romae Latinae.* In Vol. 2 of *Corpus inscriptorum latinorum.*

Heraclitus. In *Fragmente der Vorsokratiker.* 5th ed. Ed. H. Diels, with additions by Walter Kranz. Berlin, 1934.

Herodotus. *The History.* Trans. David Grene. Chicago, 1987.

Herz, Marcus. "An den D. Dohmeyer, Leibarzt des Prinzen von England, über die Brutalimpfung und deren Vergleichung mit der Humanen." Berlin, 1801.

Heuschner, Julius E. *A Psychiatric Study of Myths and Fairy Tales: Their Origin, Meaning, and Usefulness.* 2nd enlarged and rev. ed. Springfield, Ill., 1974.

Hexter, J.H. *More's Utopia: The Biography of an Idea.* New York, 1965.

Heywood, Thomas. *Apology for Actors.* Introd. with notes by Richard H. Perkinson. New York, 1941.

————. *The Iron Age.* Ed. Arlene W. Weimar. New York, 1979.

Hillebrand, Harold Newcomb. *The Child Actors: A Chapter in Elizabethan Stage History.* New York, 1964.

Hippel, Theodor Gottlieb von. *On Improving the Status of Women.* Trans. Timothy F. Sellner. Detroit, 1979.

Hippolytus. *Philosophumena. Patrologiae* (Graeca) 16: pt. 3, 3017–454.

Hirsch, S. R. Horeb: *A Philosophy of Jewish Laws and Observances.* London, 1962.

Histoire de Anne Boleyne Jadis Royne d'Angleterre. Paris, 1545.

Hitschmann, E. "Über Nerven- und Geisteskrankheiten beim katholischen Geistlichen und Nonnen." *Internationale Zeitschrift für ärztliche Psychoanalyse* 2 (1914): 270–72.

Hodgson, Richard, and Ralph Sarkonak. "Deux hors-la-loi québécois: Jacques Godbout et Jacques Poulin." *Québec Studies* 8 (Spring 1989): 27–36.

Homer, *Iliad.* With trans. A. T. Murray. 2 vols. London & New York, 1928–29

————. *Odyssey.* With trans. A. T. Murray. 2 vols. Cambridge, Mass., 1924–25.

Honigmann, E. A. J. *Shakespeare: The "Lost Years".* Manchester, Eng., 1985.

Hooper, Laurence. "Identification by DNA Said to Be Better." *Wall Street Journal* (November 21, 1991), Sect. B, pp. 1, 5.

Horn, Jack C., and Jeff Meer. "The Pleasure of Their Company: *PT* Survey Report on Pets." *Psychology Today* (1984): 52–59.

Hotman, François. *Francogallia.* Frankfurt, 1586.

Howe, J. "Fox-hunting as Ritual." *American Ethnologist* 8 (1981): 278–300.

Howitt, Mary Botham. *Ballada and Other Poems.* London, 1847.

Hubbell, Jay B. "The Smith-Pocahontas Story in Literature." *Virginia Magazine* 65 (1957): 275–300.

Hubert, Judd. *Essai d'exégèse racinienne: les secrets témoins.* Nizet, 1956.

Hudson, H. R. *The Epigram in the English Renaissance.* Princeton, 1947.

Hunter, W. A. *A Systematic and Historical Exposition of Roman Law.* 4th ed. Edinburgh, 1903.

Hutter, Horst. *Politics as Friendship: The Origins of Classical Notions of Politics in the Theory and Practice of Friendship.* Waterloo, Ont., 1978.

Huxley, Thomas Henry. "Spontaneous Generation" (An Address delivered before the British Association for the Advancement of Science, 1870). In Huxley, *Lay Sermons, Addresses, and Reviews.* pp. 345–78. New York, 1871.

Ignatius of Antioch. "The Epistle to the Ephesians." In *Corpus Ignatium: A Complete Collection of the Ignatian Epistles.* Ed. William Cureton, pp. 15–38. Berlin, 1849. Also in *Patrologiae* (Graeca) 5:729–56.

IJsseling, Samuel. "Martin Heidegger over Abraham a Santa Clara. Een voordracht can Heidegger, ingeleid en vertaald door Samuel IJsseling." *Streven* (Amsterdam) 20 (1966/67): 743–53.

Ionesco, Eugene. *Story Number 1: For Children Under Three Years of Age.* Trans. Calvin K. Towle with illus. by Etienne Delessert. [U.S.A.], 1968.

Irving, Washington. *A History of New York, from the Beginning of the World to the End of the Dutch Dynasty, by Diedrich Knickerbocker.* 4th American edition. New York, 1825.

Irwin, John T. *Doubling and Incest / Repetition and Revenge: A Speculative Reading of Faulkner.* Baltimore, 1975.

Isidore of Seville. "Historia Gothorum, Wandalorum, Sueborum." In *Monumenta Germaniae historica, auctores antiquissimi. Patrologiae* (Latina), 81:471–81.

————. *Etymologiae* [Augsburg], 1472. Also in PL 82:74–728.

————. *Opera*. Ed. F. Orevalo. 7 vols. Rome, 1797–1803. In *Patrologiae* (Latina) 81–84.

Ives, Eric. *Anne Boleyn*. Oxford, 1986.

Ivo Carnotensis. *Decretum. Patrologiae* (Latina) 161:47–1035.

Jacob's Well: An English Treatise on the Cleansing of Man's Conscience. Ed. Arthur Brandeis. London, 1900.

Jacobus, Mary L. "Incorruptible Milk: Breast-feeding and the French Revolution." Paper delivered March 5, 1990, at the Center for Literary and Cultural Studies (Harvard University).

James I [James VI of Scotland]. *The Political Works of James I*. Ed. and introd. by Charles Howard McIlwain. Cambridge, Mass., 1918.

————. *The Trew Law of Free Monarchies; or, The Reciprock and Mutuall Duetie Between a Free King, and his Naturall Subjects*. (1598). In *Political Works* 2:53–70.

James, M. R., ed. *Apocryphal New Testament*. Oxford, 1924.

Jardine, Lisa. *"Still Harping on Daughters": Women and Drama in the Age of Shakespeare*. 2nd ed. New York, 1989.

Jeanmaire, Henri. *Couroi et Courites: Essai sur l'éducation spartiate et sur les rites d'adolescence dans l'antiquité hellénique* (Lille, 1939). Repr. New York, 1975.

Jeanne des Anges (Soeur). *Autobiographie d'une hystérique possédée*. Ed. Gabriel Legué and Gilles de la Tourette. Préf. Jean Martin Charcot (1886). Paris, 1985.

Jellinek, Adolf [Aaron]. "Shlomo." *Beit ha-Midrash. Sammlung kleiner Midraschim und vermischter Abhandlungen aus älteren jüdischen Literatur* 4 (1867): 140–60.

Jenkins, Harold. *Hamlet and Ophelia*. British Academy Shakespeare Lecture. *Proceedings of the British Academy* 49 (1963): 135–51.

Jenner, Edward. *Inquiry into the Causes and Effects of the Variolae Vaccinae, etc.*. London, 1800.

Jennerian Society (Royal). Third Festival. *Minutes*. 1805.

Jerome. *Commentaria in Ezechielem. Patrologiae* (Latina) 25:9–490.

————. *Epistulae*. Ed. Isidorus Hilberg. *Corpus scriptorium ecclesiasticorum latinorum*, 54–56. Also in PL 22:325–1224.

Jeshua ben Judah. *Sefer ha-Yashar: Das Buch von den verbotenen Verwandschaftsgeraden*. (In Hebrew.) St. Petersburg, 1908.

Jewish Encyclopedia. New York, 1905.

John Chrysostom. *De virginitate. Patrologiae* (Graeca) 48:533–96.

John of the Cross. *The Collected Works of St. John of the Cross*. Trans. Kieran Kavanaugh and Otilio Rodriguez. Introd. Kieran Kavanaugh. Washington, D.C., 1973.

Johnson, Lynn Stanley. "Elizabeth, Bride and Queen: A Study of Spenser's April Eclogue and the Metaphors of English Protestantism." *Spenser Studies* 2 (1981): 75–91.

Johnson, Paul. *Elizabeth I: A Study in Power and Intellect*. London, 1974.

Johnson, Samuel *A Dictionary of the English Language*. 2 vols. London, 1775.

Jonas of Orleans. *De institutione laicali. Patrologiae* (Latina) 106:122–178.

Jones, Ernest. *Hamlet and Oedipus*. New York, 1976.

Jordan, Constance. "Feminism and the Humanists: The Case of Sir Thomas Elyot's *Defence of Good Women*." *Renaissance Quarterly* 36 (1983): 181–201.

Jordan, Winthrop D. *White Over Black: American Attitudes Towards the Negro, 1550–1812*. Chapel Hill, N. C., 1968; Baltimore, 1969.

Joyce, James. *Ulysses*. New York, 1961.

Juliana of Norwich. *Revelation of Divine Love*. Ed. Grace Warrack. 3rd ed. London, 1950.

Julius I. *Decreta julii papae decem. Patrologiae* (Latina) 8:967–71.

Justinian. *The Code of Justinian.* Vol. 6 of *The Civil Law.* Trans. S. P. Scott. 17 vols. Cincinnati, 1932 (rpr. New York, 1973).

————. *Codex.* Vol. 2 of *Corpus Iuris Civilis.* ed. Theodor Mommsen and Paul Krüger. 3 vols. Berlin, 1965.

————. *Dialogue with Tryphon. Patrologiae* (Graeca) 6:471–800.

————. *Digest.* Ed. T. Mommsen. 2 vols. Berlin, 1870.

Kahana, David. *R. Avrhaham b. Ezra.* Warsaw, 1894.

Kamen, Henry. *Inquisition and Society in Spain in the Sixteenth and Seventeenth Centuries.* Bloomington, 1985.

————. *The Rise of Toleration.* New York, 1967.

Kant, Immanuel. *Critique of Practical Reason; and Other Writings in Moral Philosophy.* Trans. Lewis White Beck. Chicago, 1949.

————. *Universal Natural History; and, Theory of the Heavens.* Trans. Stanley L. Jaki. Edinburgh, 1981.

————. "What is Enlightenment?" In Kant, *On History,* pp. 3–10. Trans. and ed. with an introd. by Beck. Indianapolis/New York, 1963.

Kantorowicz, Ernst H. *The King's Two Bodies: A Study in Medieval Political Theology.* Princeton, 1957.

Kaplan, Yosef. *From Christianity to Judaism: The Story of Isaac Orobio de Castro.* New York, 1990.

Katcher, Aaron H., Erika Friedmann, Melissa Goodman, and Laura Goodman. "Men, Women, and Dogs." *California Veterinarian* 3 (1983): 14–17.

Katz, David S. *Philo-Semitism and the Readmission of the Jews to England, 1603–1655.* Oxford, 1982.

Katz, Jacob. *Exclusiveness and Tolerance.* London, 1961.

Kaufman, David. *Geschichte der Attributenlehre in der jüdischen Religions philosophie des Mittelalters.* Gotha, 1877.

Kaufmann, R. J. "Ford's Tragic Perspective." *Texas Studies in Languagae and Literature* 1 (1960): 522–37.

Kaula, David. *Shakespeare and the Archpriest Controversy: A Study of Some New Sources.* The Hague, 1975.

Kay, William J., et al., eds. *Pet Loss and Human Bereavement.* Ames, Ia., 1984.

Keddie, K. "Pathological Mourning After the Death of a Domestic Pet." *British Journal of Psychiatry* 131 (1977): 21–25.

Kelly, J. Thomas. *Thorn on the Tudor Rose: Monks, Rogues, Vagabonds, and Sturdy Beggars.* Jackson, Miss., 1977.

Kelso, R. *Doctrine of the Lady of the Renaissance.* Urbana, 1956.

Kennard, Joseph Spencer. *The Friar in Fiction, Sincerity in Art, and Other Essays.* New York, 1923.

Kern, Louis J. *An Ordered Love: Sex Roles and Sexuality in Victorian Utopias—the Shakers, the Mormons, and the Oneida Community.* Chapel Hill, North Carolina, 1981.

Kersey, John. *Dictionarium Anglo-britannicum; or A General English Dictionary* (1708). 3rd ed. London, 1721.

King, John N. *English Reformation Literature: The Tudor Origin of the Protestant Tradition.* Princeton, 1982.

————. "Patronage and Piety: The Influence of Catherine Parr." In Margaret P. Hannay, ed. *Silent But for the Word,* pp. 43–60. Kent, 1985.

————. "Queen Elizabeth I: Representations of the Virgin Queen." *Renaissance Quarterly* 43 (1990): 30–74.

———. *Tudor Royal Inconography: Literature and Art in an Age of Religious Crisis.* Princeton, 1989.

Kinsey, Alfred C., Wardell B. Pomeroy, and Clyde E. Martin. *Sexual Behavior in the Human Male.* Philadelphia, 1948.

Knight, G. Wilson. *The Christian Renaissance.* New York, 1962.

Knight, R. C. *Racine et la Grèce.* Paris, 1950.

Knights, L. C. "How Many Children Had Lady Macbeth?" In Knights, *Explorations,* pp. 15–54. New York, 1947.

Knowles, David, and R. N. Hadcock. *Medieval Religious Houses: England and Wales.* London, 1971.

Knox, John. *First Blast of the Trumpet. . . .* Geneva, 1558.

Kol Bo. New York, 1945–46. (Hebrew.)

Koran. Trans. George Sale, with introd. by E. D. Ross. London, n.d.

Koschaker, Paul. "Fratriarchät, Hausgemeinshaft und Mutterrecht in Keilschriftexten." *Zeitschrift für Assyriologie* 42 (1933): 1–89.

Kott, Jan. "Hamlet and Orestes." Trans. Boleslaw Taborski. *Publication of the Modern Language Association* (PMLA) 82 (1967): 303–13.

Krantz [Crantz], Albert. *Chronica Regnorum Aquilonarium.* Strasbourg, 1545.

Krapp, George P. *The English Language in America.* New York, 1925.

Kretschmer, Paul W. "Die griechische Benennung des Brüders." In Kretschmer, *Einleitung in die Geschichte der griechischen Sprache.* Göttingen, 1896.

Kristeva, Julia. "Stabat Mater." In Suleiman, ed. The *Female Body in Western Culture,* pp. 99–118. Cambridge, Mass., 1986. Translation by Arthur Goldhammer of Kristeva, "Hérétique de l'amour," *Tel Quel* 74 (1977); also appears in Kristeva. *Histoires d'amour,* Paris, 1983.

Kurath, Hans, Sherman M. Kuhn, and John Reidy, eds. *Middle English Dictionary.* Ann Arbor, Mich., 1956–.

Lacoue-Labarthe, Philippe. *Heidegger, Art and Politics.* Trans. Chris Turner. Oxford and Cambridge, Mass., 1990.

Lacretelle, Pierre de. *La Vie privée de Racine.* Paris, 1949.

Lacy, John [Disputed authorship]. *A Middle English Treatise on the Ten Commandments.* Ed. James Finch Royster. *Studies in Philology* 6 (1910): 9–35.

Lalonde, Michèle. "Speak White." *Change* (March 30–31, 1977): 100–104.

Landes, Joan. *Women and the Public Sphere in the Age of the French Revolution.* Ithaca, New York, 1988.

Langland, William. *'Piers the Plowman', with 'Richard the Redeless'.* Ed. W. W. Skeat. Oxford, 1886.

Lanyer, Aemilia (Bassano). *Salve Deus Rex Judaeorum.* London, 1611.

Laporte, P. "Queen Elizabeth? . . . Jamais!" *L'Action nationale* 154 (April 1955): 73–75.

Laslett, Peter, K. Oosterveen, and R. M. Smith, eds. *Bastardy and Its Comparative History: Illegitimacy and Marital Nonconformism in Britain, France, Germany, Sweden, North America, Jamaica, and Japan.* Cambridge, Mass., 1980.

Laurin, Camille, and Guy Rocher, Fernand Dumont, and Gaston Cholette. *La Politique québécoise de la langue française.* White Paper. Gouvernement du Québec, 1977.

Lawrence, E. Atword. *Rodeo: An Anthropologist Looks at the Wild and the Tame.* Knoxville, 1982.

Layard, John. "The Incest Taboo and the Virgin Archetype." *Eranos Jahrbuch* 12 (1945): 254–307.

Lea, H. C. *History of the Inquisition.* 4 vols. New York, 1907.

Leach, Edmund. "The Animal Category and Verbal Abuse." In Eric H. Lenneberg, ed. *New Directions in the Study of Language,* pp. 23–63. Cambridge, 1966.
Leander. *Regula, sive liber de institutione virginum et contemptu mundi, ad Florentinam sororem. Patrologiae* (Latina) 72:874–94.
Le Clerc, Jean. *Bibliothèque universelle et historique.* 25 vols. 1686–93.
———. *Monks and Love in Twelfth-Century France: Psycho-Historical Essays.* Oxford, 1979.
Le Cours, R. "Bourassa: créer des districts bilingues en étendant la loi 101." *Le Devoir* [Montréal] (November 29, 1986): A-2.
Leff, Gordon. *Heresy in the Later Middle Ages: The Relation of Heterodoxy to Dissent, c.1250–1450.* 2 vols. New York, 1967.
Lefranc, Abel. *A la découverte de Shakespeare.* 2 vols. Paris, 1945.
———. *Les Idées religieuses de Marguerite de Navarre d'après son oeuvre poétique.* Paris, 1898.
———. *Sous le masque de Shakespeare.* Paris, 1918.
Léger, Jean-Marc. *La Francophonie: grand dessein, grande ambiguïté.* [LaSalle, Québec], 1987.
Le Goff, Jacques. *Your Money or Your Life: Economy and Religion in the Middle Ages.* Trans. Patricia Ranum. New York, 1988.
Leibniz, G. S. *New Essays on Human Understanding.* Trans. Peter Remnant and Jonathan Bennett. Cambridge, England, 1981.
Leigh, D. "The Psychology of the Pet Owner." *Journal of Small Animal Practice* 7 (1966): 517–31.
Lemaître, Jules. *Racine.* Paris, 1908.
Lemercier, Dom Grégoire. "Freud in the Cloister." *Atlas* 13 (January 1967): 33–37.
Le Rebours, Marie Anel. *Avis aux meres qui veulent nourir leurs enfants.* Paris, 1767.
Lesellier, J. "Deux Enfants naturels de Rabelais legitimés par le pape Paul III." *Humanisme et Renaissance* 5 (1938): 549–70.
Lessing, G. E. *Zerstreute Anmerkungen über das Epigram.* In *Gesammelte Werke.* Berlin, 1968.
Lessing und die Toleranz: Beitrage der vierten internationalen Konferenz der Lessing Society in Hamburg vom 27. bis 29. Juni 1985. Ed. Peter Freimark, Franklin Kopitzsch, and Helga Slessarev. Detroit, 1986.
Lévesque, René. *Attendez que je me rappelle. . . .* Montréal, 1986.
Levi-Bianchini, Mario. "La neurosi antifallica nell'ambito della vita sociale ed in quella religiosa cattolica romana." *Annali di neuropsichiatria e psicoanalisi* 3 (1956): 39–46.
Lévi-Strauss, Claude. *The Elementary Structures of Kinship.* Ed. Rodney Needham. Trans. James Harle Bell and John Richard von Sturmer. Boston, 1969.
———. "Language and the analysis of social laws." In Lévi-Strauss, *Structural Anthropology.* Trans. Claire Jacobson and Brooke Grundfest Schoepf, pp. 55–66. New York, 1967.
———. "Religion, langue, et histoire: A propos d'un texte inédit de Ferdinand de Saussure." In *Mélanges en l'honneur de Fernand Braudel: Méthodologie de l'Histoire et des sciences humaines,* pp. 325–33. Vol. 2. Toulouse, 1973.
Levin, Harry. "An Explication of the Player's Speech." In Levin, *The Question of Hamlet,* pp. 139–64. New York, 1959.
Levine, Marc V. "Language Policy and Québec's *visage français:* New Directions in *la question linguistique.*" *Québec Studies* 8 (1989): 1–16.
———. "The Language Question in Québec: A Selected Annotated Bibliography." *Québec Studies* 8 (1989): 37–41.
Levinson, B. M. *Pet-Oriented Child Psychotherapy.* Springfield, Ill., 1969.
Levitan, William. "Seneca in Racine." *Yale French Studies* 76 (1989): 185–210.
Levy, Michael B., ed. *Political Thought in America: An Anthology.* 2nd ed. Chicago, 1988.

Lewalski, Barbara K. "Of God and Good Women: The Poems of Aemelia Lanyer." In Margaret P. Hannay, ed. *Silent But for the Word*, pp. 203–24. Kent, 1985.
Lewis, Anna Robinson. *"Child of the Sea" and Other Poems*. New York, 1848.
Lewis, Bernard. *The Jews of Islam*. Princeton, 1984.
————. *The Political Language of Islam*. Chicago, 1988.
————. *Race and Slavery in the Middle East: An Historical Inquiry*. New York, 1990.
Limborch, Philip van. *De veritate religionis Christianae amica collatio cum erudito Judaeo*. Gouda, 1687.
————. *Historia Inquisitionis*. Amsterdam, 1692.
Lincoln, Abraham. *Speeches and Writings 1832–58 (Speeches, Letters, and Miscellaneous Writings; The Lincoln-Douglas Debates)*. Notes by Don E. Fehrenbacher. New York, 1984.
Lincoln, Victoria. *Teresa: A Woman. A Biography of Teresa of Avila*. Introd. Antonio T. de Nicholás. Albany, N.Y., 1984.
Lingard, John. *The History of England from the First Invasion by the Romans to the Accession of William and Mary in 1688*. 10 vols. London, 1883.
Link, Mary "Helping Emotionally Disturbed Children Cope with Loss of a Pet." *See* in Kay et al., eds. *Pet Loss*, pp. 82–88.
————. "Pets and Personality Development." *Psychological Reports* 42 (1978): 1031–38.
Linnaeus, Charles. *The Animal Kingdom, or Zoological System of the Celebrated Sir Charles Linnaeus*. Trans. Robert Kerr. London, 1792.
Lippard, George. *The Quaker City, or The Monks of Monk Hall; A Romance of Philadelphia Life, Mystery, and Crime*. Philadelphia, 1845. Ed. with an introd. by Leslie Fiedler. New York, 1970.
Livy, Titus. *The Romane History*. Trans. Philemon Holland. London, 1600.
————. *Livy*. English translation by B. O. Foster, Alfred C. Schlesinger, Evan T. Sage, and Frank Garner Moore. 14 vols. London, 1922–.
Llorente, Juan Antonio. *Memoria Histórica sobre qual ha sido la opinión nacional de España acerca del tribunal de la Inquisición*. Madrid, 1812.
Locher, Uli. "La Minorisation des anglophones au Québec." Originally published in *Conjoncture politique au Québec* 4 (Autumn, 1983): 95–106. *See* in Boismenu et al., *Le Québec en textes*, 522–30.
Locke, John. *A Letter Concerning Toleration*. Trans. with pref. by William Popple (1689). Ed. with an introd. by Patrick Romanell. Indianapolis, 1983.
————. *Epistola de Tolerantia*. Gouda, 1689.
————. *Fundamental Constitutions of Carolina* (1669). Rpr. fr. *Carroll's Historical Collections of South Carolina*. Old South Leaflets. General Series 7, 172. Boston, 1906.
————. *Two Treatises of Civil Government*. Introd. W. S. Carpenter (1924) Rpr. London, 1953.
————. *Two Treatises of Government*. Ed. Peter Laslett, with introd. and notes. New York, 1965.
Logan, James. *The Scotish Gael; or, Celtic Manners, as Preserved Among the Highlanders*. 5th American ed. Hartford, 1846.
Lucas de Penna. *Commentaria in Tres Libros Codicis*. Lyon, 1544.
Lucretius. *De Rerum Natura*. Ed. Cyril Bailey. Oxford, 1922.
————. *On the Nature of the Universe*. Trans. R. E. Latham. Harmondsworth, 1977.
Luis de Granada. *Libro de la Oracion* (1554). Salamanca, 1579.
Lupton, Thomas. *A Persuasion from Papistrie*. London, 1581.
Luther, Martin. "Against the Antinomians." Trans. Martin H. Bertram with an introd. by Franklin Sherman. In *Luther's Works* 47:101–20.

———. "The Estate of Marriage." Trans. Walther I. Brandt. In *Luther's Works* 45:17–49.

———. "Exhortation to All Clergy Assembled at Augsburg." Trans. Lewis W. Spitz. In *Luther's Works* 34:9–62.

———. "An Exhortation to the Knights of the Teutonic Order That They Lay Aside False Chastity and Assume the True Chastity of Wedlock." Trans. Albert T. W. Steinhaeuser, rev. Walther I. Brandt. In *Luther's Works* 45:141–58.

———. *Luther's Works.* Ed. Jaroslav Pelikan and Helmut T. Lehmann. 55 vols. Philadelphia, 1960–62.

———. "The Persons Related by Consanguinity and Affinity Who are Forbidden to Marry According to the Scriptures, Leviticus 18." Trans. Walther I. Brandt. In *Luther's Works* 45:7–9.

———. *Tischreden.* 6 vols. Weimar, 1912–21.

Lydgate, John. *Fall of Princes (1431–38).* Ed. Henry Bergen. 4 vols. Washington, 1923.

Lyttelton, George, (Lord). *Dialogues of the Dead.* London, 1795.

Macarius the Elder [of Egypt]. *Homiliae spirituales. Patrologiae* (Graeca) 34:450–822.

MacCannell, Juliet Flower. *The Regime of the Brother: After the Patriarchy.* London and New York, 1991.

McCusker, H.C. *John Bale, Dramatist and Antiquary.* Freeport, New York, 1971.

McDonnell, Ernest W. *The Beguines and Beghards: With Special Emphasis on the Belgian Scene.* New York, 1969.

Machiavelli, Niccolò. *Discourses on the First Decade of Titus Livius.* In Vol. 1 of Machiavelli, *The Chief Works and Others,* pp. 175–532. Trans. Allan Gilbert. 3 vols. Durham, N.C., 1965.

McKitrick, Eric L., ed. *Slavery Defended: The Views of the Old South.* Englewood Cliffs, New Jersey, 1963.

McOllough, Verlene. "The Orphan Train Comes to Clarion." *Palimpsest* 69 (1988): 144–50.

McWilliams, Wilson C. *The Idea of Fraternity in America.* Berkeley, 1973.

Madariaga, Salvador de. "Cervantes y su tiempo." *Cuadernos* 40 (1960): 39–48.

Maheu, Pierre. "Le Pouvoir clérical." Excerpt from *Les Québécois.* Montreal, 1971, pp. 175–83. *See* in Boismenu et al., *Le Québec en Textes,* pp. 102–10.

Maimonides, *Moreh Nebukhim.* [Hebrew.] Warsaw, 1872.

———. *The Guide of the Perplexed.* Trans. S. Pines. Chicago, 1963.

———. *Code of Maimonides: The Book of Holiness.* Trans. Louis Rabinowitz and Philip Grossman. New Haven, 1965.

Maine, Sir Henry Sumner. *Village-Communities in the East and West . . . to which are added other addresses and essays.* 3rd enlarged ed. New York, 1876.

Malinowski, Bronislaw. "Parenthood, the Basis of Social Structure." In V. F. Calverton and S. D. Schmalhausen, eds., *The New Generation,* pp. 112–68. New York, 1930.

Manifeste du Front de libération du Québec [FLQ]. 1970. Extract from the manifesto as it appears in *Change,* March 30, 1977, pp. 63–68. *See* in Boismenu et al., *Le Québec en textes,* pp. 357–61.

Mankowicz, Wolf, and Reginald G. Haggar. *Encyclopedia of English Pottery and Porcelain.* 2nd ed. London, 1968.

Mannix, Daniel R. *The History of Torture.* London, 1970.

Mansi, J. D., ed. *Sacrorum conciliorum nova et amplissima collectio.* 31 vols. Florence, 1759–98.

Marbeck, John. *A Book of Notes and Common Places with Their Expositions, Collected and Gathered out of the Workes of Divers Singular Writers.* London, 1581.

Marcuse, Herbert. "Repressive Tolerance." In Robert Paul Wolff, Barrington Moore, Jr., and Herbert Marcuse, *A Critique of Pure Tolerance,* Boston, 1965.

Marguerite de Navarre (Marguerite d'Angoulême). *Heptameron*. Trans. with an introd. by G. Saintsbury. 5 vols. London, 1894.
————. *L'Heptameron des nouvelles*. Ed. Le Roux de Lincy and Anatole de Montaiglon. 4 vols. Paris, 1880.
————. *Marguerites de la Marguerite des Princesses tres illustre Royne de Navarre*. Ed. S. Sylvius. Lyon, 1547.
————. *Miroir de l'âme pécheresse [Le]: Discord étant en l'homme par contrariété de l'espirit et de la chair* (1531; 1535). Ed. Joseph L. Allaire. Munich, 1972. *See also* in Salminen, ed., *Miroir*.
————. *The queene of Nauarres tales*. Now newly tr. Pref. A. B. Oxford, 1597.
Marlowe, Christopher. *The Jew of Malta*. Ed. N. W. Bawcutt. Manchester and Baltimore, 1978.
————. *Tamburlaine the Great*. Ed. U. M. Ellis-Fermor. London, 1930.
————. *The Tragedy of Dido*. London, 1594.
Marsden [Marston], John. *Antonio's Revenge*. Ed. W. Reavley Gair. Baltimore, 1978.
————. *The Scourge of Villanie*. London, 1599.
Martin, Edward. *A History of the Iconoclast Controversy*. New York, 1930.
Marti-Rolli, C. *La Liberté de la langue en droit suisse*. Zurich, 1978.
Marvin, Garry. *Bullfight*. Oxford, 1988.
Marx, Karl. *Capital*. Trans. S. Moore and E. Aveling. 3 vols. New York, 1967.
Marx, Karl, and Friedrich Engels. *Werke*. Ed. Institut für Marxismus-Leninismus beim ZK der SED. Berlin, 1956–68.
Matthews, C. M. "The True Cymbeline." *History Today* 7 (1957): 755–59.
Matthiessen, F.O. *Translation: An Elizabethan Art*. New York, 1965.
Maurais, J., and Plamandon, P. *Le Visage français du Québec: Enquête sur l'affichage*. Québec, 1986.
Mauriac, François. *Blaise Pascal et sa soeur Jacqueline*. Paris, 1931.
Mauron, Charles. *L'Inconscient dans l'oeuvre et la vie de Racine*. Paris, 1957.
Mayer, Anton L. "*Mater et filia*." *Jahrbuch für Liturgiewissenschaft* 7 (1927): 60–82.
Meer, Jeff. "Pet Theories: Facts and Figures to Chew On." *Psychology Today* 18 (1984): 60–67.
Melville, Herman. *Billy Budd*. In *Selected Tales and Poems*, ed. with an introd. by Richard Chase. New York, 1950.
————. *Mardi and a Voyage Thither*. 1849. 2 vols. London, 1922.
————. *Moby-Dick; or, The Whale*. Ed. Kuther S. Mansfield and Howard P. Vincent. New York, 1962.
————. *Pierre; or, The Ambiguities*. Foreword Lawrance Thompson. Based on the edition by Henry A. Murray. New York, 1979.
————. *Pierre; or The Ambiguities*. Ed. Harrison Hayford, Herschel Parker, and G. Thomas Tanselle. Historical Notes by Leon Howard and Herschel Parker. Evanston, Ill., 1971.
Mendell, Clarence W. *Tacitus: The Man and His Work*. New Haven, 1957.
Mendelssohn, Moses. *Jerusalem, and other Jewish Writings*. Trans. Alfred Jospe. New York, 1969.
Merry, Joseph. *A Conscious View of Circumstances and Proceedings Respecting Vaccine Inoculation*. London, 1806.
Meyer, Jean. "Illegitimates and foundlings in pre-industrial France." In Laslett et al., eds., *Bastardy*, pp. 249–63.
Mezei, Kathy. "Speaking White: Literary Translation as a Vehicle of Assimilation in Québec." *Canadian Literature* 117 (1988): 11–23.
Michelet, Jules. *History of the French Revolution*. Trans. Charles Cocks. Intro. Gordon Wright. Chicago, 1967.

————. *Histoire de la Révolution française*. Illus. [Daniel Urrabieta] Vierge. 9 vols. Paris, [1883–87].

Middleton, Thomas. *The Family of Love*. Ed. Simon Shepherd. Nottingham, 1979.

Middletown, Russell. "Brother-Sister and Father-Daughter Marriage in Ancient Egypt." *American Sociological Review* 27 (1962): 603–11.

Mill, John Stuart. "The Subjection of Women." In Mill, *On Liberty; Representative Government; the Subjection of Women*, pp. 427–548. London, 1912.

Millin, Aubin Louis. *Antiquités nationales*. 5 vols. Paris, 1790–99.

Milton, John. *Treatise of Civil Power in Ecclesiastical Causes showing that it is not lawful to compel in Matters of Religion*. 1659.

Mintz, Sidney W., and Eric R. Wolf. "An Analysis of Ritual Co-Godparentage." *Southwestern Journal of Anthropology* 6 (1950): 341–68.

Mintz, Thomas. "The Meaning of the Rose in 'Beauty and the Beast.' " *Psychoanalytic Review* 56 (1969–70): 617–20.

Mirabeau (Riquetti, Honoré Gabriel, Count of Mirabeau). *Des Lettres de cachet et des prisons d'état*. Hamburg, 1782.

————. *Erotica biblion*. Rome, 1783.

————. *De la monarchie prussienne sous Frédéric le Grand*. 4 volumes. Paris, 1787.

————. *Sur Moses Mendelssohn, sur la réforme politique des Juifs*. London, 1787.

Miron, Gaston. "Chus Tanné." *See* in *L'Avenir de Français au Québec*, pp. 175–87.

————. "Décoloniser la langue." An "Interview/témoignage" with Gaston Miron. *Maintenant* 125 (April 1973): 12–14.

Missel Quotidien et Vespéral. Ed. Gaspar Lefebvre and the Benedictine Monks of the Abbey of Saint-André. Scriptural trans. E. Osty. Bruges, 1958.

Mitchel, Don. "Advice to the First-Time Butcher. Including Some Thoughts on the Twenty-Third Psalm." [Blair & Ketchum's] *Country Journal* 3 (1976): 42–45.

Mogan, Joseph J., Jr. "*Pierre* and *Manfred:* Melville's Study of the Byronic Hero." *Papers on English Language and Literature* 1 (1965): 230–40.

Molanus, Johannes. *De Fide Haereticis Servanda*. Cologne, 1584.

Monette, Pierre. "Mon Français mais Montréal." *See* in *L'Avenir de Français au Québec*, pp. 109–16.

Montaigne, Michel Eyquem de. *The Complete Essays of Montaigne*. Trans. Donald M. Frame. Stanford, 1965.

————. *Essays*. Trans. Charles Cotten. 3 vols. London, 1738.

————. *Essays*. Trans. John Florio. London, 1603.

————. *Oeuvres complètes*. Ed. Albert Thibaudet and Maurice Rat. Paris, 1962.

Montalembert, Charles Forbes René de Tyron. *The Monks of the West from St. Benedict to St. Bernard*. 7 vols. Edinburgh, 1861–79.

Montesquieu, Charles de Secondat (Baron de la Brède). *Considérations sur les causes de la Grandeur des Romains et de leur décadence*. Paris, 1734.

Montgomerie, William. "English Seneca." *Life and Letters Today* 26 (1943): 25–28.

————. "Folk Play and Ritual in *Hamlet*." *Folk-Lore* (1956): 214–27.

————. "More an Antique Roman than a Dane." *Hibbert Journal* 59 (1960): 67–77.

Montgomery, James. *Poetical Works*. 2 vols. Philadelphia, 1849.

Montrose, Louis Adrian. " 'The Place of a Brother' in *As You Like It*." *Shakespeare Quarterly* 32 (1981): 28–54.

————. " 'Shaping Fantasies': Figurations of Gender and Power in Elizabethan Culture." *Representations* 1 (1983): 61–94.

More, Thomas. *Confutacyon with Tindale*. Vol. 8, pt. 1 of *The Complete Works of St. Thomas More*, ed. Louis A. Schuster, Richard C. Marius, James P. Lusardi, and Richard J. Schoeck. New Haven, 1973.

———. *Utopia*. Ed. Joseph Hirst Lupton. Oxford, 1895.

Morison, Samuel Eliot. *The Oxford History of the American People*. New York, 1965.

Morisset, Paul. "La face cachée de la culture Québécoise." Extracted from *L'Actualité* 10 (11 nov. 1985): 81–87. *See* in Boismenu et al., *Le Québec en Textes*, pp. 531–38.

The Morning Chronicle. (Newspaper.) London, 1780–1832.

Moseley, Benjamin. *Medical Tracts*. London, 1799.

———. *Treatise on Lues Bovilla*. 2nd ed. London, 1805.

Moulton, James Hope. *Early Zoroastrianism*. London, 1913.

Mugford, R. A., and I. G. M'Comiskey. "Some Recent Work on the Psychotherapeutic Value of Cage Birds With Old People." In *Pet Animals and Society: A BSAVA [British Small Animal Veterinary Association] Symposium*, ed. R. S. Anderson. London, 1957.

Mullaney, Stephen. "Brothers and Others, or the Art of Alienation." In Garber, ed., *Cannibals, Witches, and Divorce*, pp. 67–89.

Mumford, Lewis. *Herman Melville*. New York, 1929.

Murray, Gilbert. "Hamlet and Orestes: A Study in Traditional Types." In *Classical Tradition in Poetry*. Cambridge, Mass., 1927.

Murray, Henry A. Review of *Herman Melville* by Lewis Mumford. *New England Quarterly* 2 (July 1929): 523–26.

Mutschmann, Heinrich, and Karl Wentersdorf. *Shakespeare and Catholicism*. New York, 1952.

Nachmanides, Moses. *Hiddushei haTorah*. Lisbon, 1489.

Naden, Tony. "Siamese Twins in Mampruli Phonology." *Africana Marburgensia* 13: 52–58.

Napoleon (I). *Correspondance, publieé par ordre de l'Empereur Napoleon III*. 32 vols. Paris, 1858–70.

———. *Napoleon Self-Revealed in Three Hundred Selected Letters*. Trans. J. M. Thompson. New York, 1934.

Nashe, Thomas. *The Works of Thomas Nashe*. Ed. R. B. McKerrow. 5 vols. London, 1910.

Naz, R., ed. *Dictionnaire de droit canonique*. Paris, 1935–.

Neale, John Ernest. *Elizabeth I and Her Parliaments, 1559–1581*. London, 1953.

Needham, Rodney. *Primordial Characters*. Charlottesville, Va., 1978.

Nemoy, Leon. *Karaite Anthology: Excerpts from the Early Literature, Translated from Arabic, Aramaic, and Hebrew Sources*. New Haven, 1952.

Nelson, Benjamin. *The Idea of Usury: From Tribal Brotherhood to Universal Otherhood*. 2nd enlarged ed. Chicago, 1969.

Netanyahu, Benzion. "Américo Castro and his View on the Origins of the *pureza de sangre*." *Proceedings of the American Academy for Jewish Research* 46–47 (1979–80): 397–457.

———. *Don Isaac Abravanel: Statesman and Philosopher*. Philadelphia, 1953.

Neumann, Erich. *Amor and Psyche: The Psychic Development of the Feminine. A Commentary on the Tale of Apuleius*. Trans. Ralph Manheim. New York and Evanston, 1956.

Neville, Alexander. *The Lamentable Tragedy of Oedipus the Sonne of Laius King of Thebes out of Seneca*. London, 1563.

New Catholic Encyclopedia. 15 vols. New York, 1967.

Newman, Horatio Hackett, et al. *Twins: A Study of Heredity and Environment*. Chicago, 1937.

Newman, John Henry [Cardinal]. *The Arians of the Fourth Century*. London, 1919.

Nicholas I. *Responses to the Questions of the Bulgars*. *See* in Mansi, *Sacrorum* 15: 401–33.

Nicole, Pierre. *'Les Imaginaires' et 'Les Visionnaires'*. [With Nicole and A. Arnauld, *Traité de la foy humaine;* Arnauld, *Jugement équitable, tiré des oeuvres de S. Augustin;* and Nicolas Pauillon, *Lettre à Hardouyn Perefixe*.] Paris, 1683.

Nietzsche, Friedrich. *The Birth of Tragedy*. Trans. Francis Golffing. New York, 1956.

———. *Die Geburt der Tragödie*. In vol. 1 of *Werke in drei Bänden*, ed. Karl Schlechta, pp. 7–134. Munich, 1954.

Nigg, Walter. *The Heretics*. New York, 1962.

Nigrelli, R. F. "Modern Ideas on Spontaneous Generation." *Annals of the New York State Academy of Sciences* 69 (1957): 257–376.

Nohrnberg, James. *The Analogy of the Faerie Queene*. Princeton, 1976.

Norbrook, David. *Poetry and Politics in the English Renaissance*. London, 1984.

Nott, Josiah. *Types of Mankind: Or, Ethnological Researches based upon the Ancient Monuments, Paintings, Sculptures, and Crania of Races and upon their Natural Geographical, Philological, and Biblical History*. Philadelphia, 1854. *See* in McKitrick, ed. *Slavery Defended*, pp. 126–38.

Noyes, John. "An Essay on Scientific Propagation." [U.S.], 1875.

Oates, Stephen B. *Let the Trumpet Sound: The Life of Martin Luther King, Jr.* New York, 1983.

Ogburn, Charlton. *The Mysterious William Shakespeare*. New York, 1984.

Onians, Richard Brocton. *The Origin of European Thought*. Cambridge, 1954.

Onslow, Richard William Alan. *Empress Maud*. London, 1939.

Opie, Peter. *The Classic Fairy Tales*. New York and Boston, 1974.

Optatus, Bishop of Milevi. *Contra Parmenianum Donatistam*. In *The Works of Saint Optatus Against the Donatists*, ed. and trans. O. R. Vassall-Phillips. London, 1917. Also in *Patrologiae* (Latina) 11:883–1102; and *Corpus scriptorium ecclesiasticorum latinorum* 56 (1893).

Orcibal, Jean. *La Genèse d'"Esther" et d'"Athalie."* Paris, 1950.

Origen. *De oratione* libellus. Ed. William Ready. London, 1728.

Oropesa, Alonso de. *Luz para conocimiento de los Gentiles* (1465). Ed. Luis A. Díaz y Díaz. Madrid, 1979.

Orphei Hymni. Zurich, 1973.

Ortiz Cañavate, L. "El toreo en España." In F. Carreras y Candi, ed. *Folklore y Costumbres*, pp. 118–40. Barcelona, 1934.

Oxford Classical Dictionary. 2nd ed. Ed. N. G. L. Hammond and H. H. Scullard. Oxford, 1970.

Oxford Dictionary of Latin. Ed. P. G. W. Glare. Oxford, 1982.

Paine, Albert Bigelow. *Mark Twain: A Biography*. 3 vols. New York, 1912.

Paine, Thomas. *The Writings of Thomas Paine*. Ed. Moncure Daniel Conway. 2 vols. New York, 1902–1906.

Paley, William. *Natural Theology; or Evidences of the Existence and Attributes of the Deity*. 2nd ed. London, 1802.

————. *Principles of Moral and Political Philosophy*. London, 1785.

Palm, Etta. *Appel aux françoises sur la régéneration des moeurs et la nécessité de l'influence des femmes dans un gouvernment libre*. Paris, 1791.

Paneth, F. A. *Chemistry and Beyond*. New York, 1964.

Paola, Tomie de (illustrator). *The Friendly Beasts: An Old English Christmas Carol*. New York, 1981.

Paret, Rudi. "Sure 2,256: la ikraha fi d-dini. Toleranze oder Resignation?" *Der Islam* 45 (1969): 299–300.

Parkman, Francis. *Montcalm and Wolfe*. 2 vols. 1884.

Parr, Catherine. *The Lamentacion of a Synner*. London, 1547.

————. *Prayers, or Meditations*. London, 1545.

Parrington, Vernon L. *The Romantic Revolution in America: 1800–1860*. New York, 1954.

Partridge, Marianne. "Good Queen Bess." Book Review. *New York Times*, March 17, 1991, p. 26.

Patristic Greek Lexicon. Ed. G. W. H. Lampe. Oxford, 1964.

Patrologiae cursus completus (Series Graeca). Ed. J. P. Migne. 161 vols. Paris, 1857–99.

Patrologiae cursus completus (Series Latina). Ed. J. P. Migne. 221 vols. Paris, 1844–64.

Patterson, Francine. *Koko's Kitten.* New York, 1985.
Paul, Robert A. *The Tibetan Symbolic World: Psychoanalytic Explorations.* Chicago, 1982.
Paulson, Ronald. *Hogarth: His Life, Art, and Times.* 2 vols. New Haven and London, 1971.
———. *Representations of Revolution (1789–1820).* New Haven, 1983.
Pavel, Thomas. "Racine and Stendhal." *Yale French Studies* 76 (1989): 265–83.
Payne, R. B. "Clutch Size and Number of Eggs Laid by Brown-Head Cowbird." *Condor* 67(1):44–60.
Pearson, John (Bishop). *Exposition of the Creed* (1659). 2 vols. Oxford, 1816.
Peers, E. Allison. *Studies of the Spanish Mystics.* Vol. 1. 2nd and revised ed. London, 1951.
Pellicer de Tovar, Joseph (José). *Anfiteatro de Felipe el Grande, Rey Catolico de la Españas, . . . contiene los elogios que han celebrado la suerte hizo en el Toro, en la Fiesta Agonal. . . .* Madrid, 1631.
Peñalosa y Mondragón, Fray Benito de. *Libro de las cinco excelencias del Español.* Pamplona, 1629.
Penn, William. *Considerations Moving to Toleration, and Liberty of Conscience: With Arguments Inducing to a Cession of the Penal Statute against All Dissenters Whatever.* London, 1685.
———. *Constitution and Select Laws of the Commonwealth of Pennsylvania.* Harrisburg, Penn., c1842.
———. *The Great Case of Liberty of Conscience once more briefly debated & defended, by the authority of reason, scripture, and antiquity: which may serve the place of a general reply to such late discourses, as have oppos'd a tolleration.* London, 1671.
———. *Select Works of William Penn.* 4th ed. 3 vols. (1825). Rpr. New York, 1971.
Perella, Nicolas J. *The Kiss, Sacred and Profane: An Interpretative History of Kiss Symbolism and Related Religio-Erotic Themes.* Berkeley, 1969.
Pérez de Ayala, R. *Política y los Toros.* Madrid, 1925.
Perrault, Charles. *Histoires ou Contes du temps passé.* Paris, 1697.
Perrens, F. T. *Les Libertins en France au XVIIe siècle* (1896). Repr. New York, 1973.
Picard, Raymond. *La Carrière de Jean Racine.* Paris, 1961.
———. *Corpus Racinianum. Recueil-inventaire des textes et documents du XVIIe siècle concernant Jean Racine.* Paris, 1956.
Pikerung, John. *Horestes.* London, 1567.
Pindar. *Works.* Trans. John Sandys. London, 1930.
Pirque Aboth. Sayings of the Jewish fathers. In Hebrew and English. Notes by Ch. Taylor. 2nd ed. Cambridge, 1897.
Pits, John. *Relationum Historicarum de Rebus Anglicis.* Paris, 1619.
Pitt-Rivers, Julian. "El sacrificio del toro." *Revista de occident* 84 (1984): 27–49.
———. "The Kith and the Kin." In Jack Goody, ed., *The Character of Kinship,* pp. 89–105. Cambridge, 1973.
———. "Pseudo-Kinship." In David L. Sills, ed., *International Encyclopedia of the Social Sciences* 8:408–13. New York, 1966.
———. "Spiritual Kinship in Andalusia." In Pitt-Rivers, *The Fate of Schechem, or The Politics of Sex: Essays in the Anthropology of the Mediterranean,* pp. 48–70. Cambridge, 1977.
Pius XII. "Sponsa Christi." *Acta Apostolicae Sedis* 43 (1951), 5–37.
Plato. *Cratylus, Parmenides, Greater Hippias, Lesser Hippias.* With trans. by H. N. Fowler. Cambridge, Mass., 1970.
———. *Euthyphro, Apology, Crito, Phaedo, Phaedrus.* With trans. by H. N. Fowler. Cambridge, Mass., 1960.
———. *Laws.* With trans by Trevor J. Saunders. Harmondsworth, Middlesex, 1975.
———. *Lysis.* In David Bolotin, *Dialogue on Friendship: An Interpretation of the Lysis, with a New Translation,* pp. 15–52. Ithaca, N.Y., 1979.

———. *Lysis, Symposium, Gorgias.* With trans. by W. R. M. Lamb. Cambridge, Mass., 1975.

———. *Republic.* Trans. Allan Bloom. New York, 1968.

———. *Republic.* Trans. Francis M. Cornford. New York, 1961.

———. *Theaetetus and Sophist.* With trans. by H. N. Fowler. Cambridge, Mass., 1970.

———. *Timaeus, Critias, Clitophon, Menexenus, Epistulae.* With trans. by R. G. Bury. Cambridge, Mass., 1966.

Platt, Colin. *Medieval Southampton: The Port and Trading Community, A.D. 1000–1600.* London and Boston, 1973.

Plourde, Michel. *La Politique linguistique du Québec. 1977–1987.* Québec, 1987.

Plutarch. *Lives.* With trans. by Bernadotte Perrin. 11 vols. London and Cambridge, Mass., 1955–62.

Pococke, Edward *A Commentary on the Prophecy of Hosea.* Oxford, 1685.

Pole, Reginald (Cardinal). *Ad Henricum VIII . . . pro ecclesiasticae unitatis defensione* (1536). *See* in Rocaberti, *Bibliotheca pontifica.*

Polish Brethren (The): Documentation of the History and Thought of Unitarianism in the Polish-Lithuanian Commonwealth and in the Diaspora 1601–1685. Ed. and trans. by George Huntston Williams. Missoula, Mont., c1980.

Pollard, J. *Birds in Greek Life and Myth.* Plymouth, England, 1970.

Pollock, F. *Spinoza: His Life and Philosophy.* London, 1880.

Pomerius, Henry. "De origine monasterii Viridvallis." *Analecta Bollandiana* 4 (1885–86): 263–322.

Ponton, Jeanne. *La Religieuse dans la littérature française.* Laval, Quebec, 1969.

Popkin, Richard H. "Medicine, Racism, Anti-Semitism: A Dimension of Enlightenment Culture." In *The Languages of Psyche: Mind and Body in Enlightenment Thought,* G. S. Rousseau, ed., pp. 405–42. Berkeley and Los Angeles, 1990.

Porete, Marguerite. *Miroir des simples ames.* Ed. Romana Guarnieri. In Guarnieri, *Il Movimento del libero spirito,* pp. 150–285. Roma, 1965.

———. Middle English MSS. Bodleian Library (Oxford) MS 505. British Museum MS 37790. Condé Museum (Chantilly) MS F. XIV.26. St. John's College, Cambridge, MS 71.

Post, Gaines. "Two Notes on Nationalism in the Middle Ages: I. *Pugna pro patria,* II. *Rex imperator.*" *Traditio* 9 (1953).

Poulin, Jacques. *Volkswagon Blues.* Montréal, 1984.

Power, Eileen. *Medieval English Nunneries.* Cambridge, 1922.

Praz, Mario. *The Romantic Agony.* Trans. Angus Davidson. London & New York, 1970.

Preisendanz, Karl Lebrecht, ed. and trans. *Papyri Graecae magicae. Die Griechische Zauberpapyri.* 2 Vols. Leipzig, 1928.

Prescott, Anne Lake. "The Pearl of the Valois and Elizabeth I: Marguerite de Navarre's *Miroir* and Tudor England," in Margaret P. Hannay, ed. *Silent But for the Word,* pp. 61–76. Kent, 1985.

Presley, Priscilla Beaulieu. With Sandra Harmon. *Elvis and Me: The Intimate Story That Could Have Been Written Only by the Woman Who Lived It.* New York, 1985.

Prévost, Arthur. *Dictionnaire FrAnglais Dictionary.* Montreal, 1969.

Pringle, Roger, ed. *A Portrait of Elizabeth I in the Words of the Queen and Her Contemporaries.* Totowa, New Jersey, 1980.

Prosser, Elanor. *Hamlet and Revenge.* Stanford, 1967.

Proulx, J.-P. "Le Bilinguisme peut s'afficher san délai dans certains commerces." *Le Devoir,* [Montréal], (December 27, 1988): 1, 14.

Pulzer, Peter. *Die Entstehung des politischen Antisemitismus in Deutschland und Österreich, 1867-1914.* Gütersloh, 1966.

Québec-Acadie: Modernité/Postmodernité du roman contemporain. Sous la responsabilité de Madeleine Frédéric et Jacques Allard. Actes de Colloques Internationals de Bruxelles (27–29 novembre 1985). (Les Cahiers du Département d'études littéraires de l'Université du Québec à Montréal, 1987.)

Rabanus [Hrabanus] Maurus. *Concilium moguntium. See* in Mansi, ed., *Sacrorum* 14:900–912.

Rabelais, François. *Gargantua.* In *Oeuvres complètes,* ed. Jacques Boulanger, with revisions by Lucian Scheler. Paris, 1955.

———. *Gargantua and Pantagruel.* Trans. J. M. Cohen. Harmondsworth, Middlesex, 1955.

———. *Oeuvres complètes.* Ed. with introd. and notes by Pierre Jourda. 2 vols. Paris, 1962.

Racine, Jean. *Britannicus.* Ed. Jean-Pol Caput. Nouveaux Classiques Larousse. Paris, n.d.

———. *Britannicus.* Ed. with intro. by Philip Butler. Cambridge, Eng., 1967.

———. *Esther.* Ed. Jean Borie. Paris, 1975.

———. *Oeuvres complètes.* Préf. Pierre Clarac. Paris, 1962.

———. *Oeuvres complètes.* Ed. Paul Mesnard. 8 vols. Paris, 1885–88.

———. *Oeuvres complètes de Racine.* Notes by Raymond Picard. 2 vols. Paris, 1960.

———. *Phèdre.* Ed. Jean Salles. [Nancy], 1963.

———. *Three Plays of Racine.* Trans. George Dillon. Chicago, 1970.

The Racovian Cathechisme. Amsterdam, 1652.

Rank, Otto. *Das Inzest-Motiv in Dichtung und Sage* (1912). Darmstadt, 1974.

Rankin, H. D. "Catullus and Incest." *Eranos* 74 (1976): 113–21.

Rappaport, Ernest A. "Zoophily and Zooerasty." *Psychoanalytic Quarterly* 38 (1968): 565–66.

Rathéry, François-Roger. "Contribution à l'étude des hémorragies survenant dans le cours de l'hystérie." *Mémoires de la Société médicale des hôpitaux de Paris.* New Series, 16, 1879.

Rattray, R. S. *The Tribes of the Ashanti Hinterland.* Oxford, 1932.

Renan, Ernest. *Qu'est-ce qu'une Nation?* (Conférence faite en Sorbonne 11 mars 1882)." In Vol. 1 of *Oeuvres complètes de Ernest Renan,* ed. Henriette Psichari, pp. 887–907. 10 vols. Paris, 1947.

Révah, I. S. "La controverse sur les statuts de pureté de sang. Un document inédit." *Bulletin Hispanique* 73 (1971).

La Révolution française et l'émancipations des juifs. Editions d'histoire sociale. 8 vols. Paris, 1968.

Rey, Eusebio. "San Ignacio de Loyola y el problema de los *Cristianos Nuevos.*" *Razón y Fe* 153 (1956).

Reynolds, Quentin, Ephraim Katz, and Zwy Aldouby. *Minister of Death.* New York, 1960.

Rice, George P. *The Public Speaking of Queen Elizabeth.* New York, 1951.

Richard, Willy. *Untersuchungen zur Genesis der reformierten Kirchenterminologie der Westschweiz und Frankreichs.* Bern, 1959.

Richer, P. "Etudes cliniques sur la Grande Hystérie ou hystéro-épilepsie." 2nd ed. Paris, 1885.

Richler, Mordecai. "Oh Canada! Lament for a Divided Country." *The Atlantic* 240 (December 1977): 41–55.

Ring, John. *Treatise on Cow Pox.* London, 1801.

Rittenhouse, David. *An Oration.* Philadelphia, 1775. There is a facsimile printing of this in Brooke Hindle, ed., *The Scientific Writing of David Rittenhouse,* New York, 1980.

Robert, Carl. *Oidipus: Geschichte eines Poetischen Stoffs im Griechischen Altertum.* 2 vols. Berlin, 1915.

Roeg, Nicolas (director). *The Man Who Fell to Earth,* with David Bowie. 1976.

Roca Traver, Francisco A. "Un siglo de vida mudéjar en la Valencia medieval (1238–1338)." *Estudios de Edad Media de la Corona de Aragón* 5 (1952): 115–208.

Rocaberti, Juan T. *Bibliotheca pontifica.* Rome, 1689.

Rogers, John. *The Displaying of an Horrible Secte of [Grosse and Wicked] Heretiques, naming themselves the Familie of Love.* London, 1578.

Röhrich, Lutz. *Märchen und Wirklichkeit.* Wiesbaden, 1974.

Rójas, Fernando de. *Celestina.* Ed. Julio Cejador y Franca. Madrid, 1931.

———. *The Celestina.* Trans. L. B. Simpson. Berkeley and Los Angeles, 1971.

Rolle, Richard. *Form of Living* (c. AD 1425). In *Yorkshire Writers,* ed. Carl Horstmann. Vol. 1. London, 1895.

Rosenberg, Meir, ed. *The Jewish Cat Book: A Different Breed.* Illus. Sara Feldman. Marblehead, Mass., 1983.

Rosenblatt, Jason P. "Aspects of the Incest Problems in *Hamlet.*" *Shakespeare Quarterly* 29 (1978): 349–64.

Rosenthal, Judah. "Interest from the Non-Jew." *Talpiot* 5 (1952): 475–92; 6 (1953): 130.

Ross, Philip E. "Hard Words." *Scientific American* 264 (1991): 138–147.

Roth, Cecil. "The Religion of the Marranos." *Jewish Quarterly Review* (N.S.) 22 (1931–2): 1–35.

Rousseau, Jean-Jacques. *Confessions.* Trans. J. M. Cohen. Harmondsworth, 1953.

———. *Correspondance générale de J. J. Rousseau.* Ed. Théophile Dufour. 20 vols. Paris, 1924–34.

———. *Emile, or, On Education* [1762]. Introd., trans., and notes by Allan Bloom. New York, 1979.

———. *Oeuvres complètes.* Eds. Bernard Gagnebin and Marcel Raymond. 4 vols. Paris, 1959.

Rowley, William. *Cow Pox Inoculation No Security Against Small Pox Infection.* London, 1805.

Roy, Bruno. "Ce dur désir de chanter." Pp. 27–34 in *L'Avenir de Français au Québec.*

Rosenzweig, Franz. *The Star of Redemption.* Trans. from 2nd edition (1930) by William W. Hallo. Boston, 1964.

Ruchames, Louis, ed. *Racial Thought in America. Vol. 1. From the Puritans to Abraham Lincoln: A Documentary History.* New York, 1970.

Rumilly, R. *Histoire de Montréal.* Vol. 5 [1939–1957]. Montreal, 1974.

Rushton, William Lowes. *Shakespeare's Legal Maxims.* Liverpool, 1907.

Ruskin, John. *Munera Pulveris: Six Essays on The Elements of Political Economy.* New York, 1872.

Rymer, Thomas. *Critical Works.* Ed. Curt Zimansky. New Haven, 1956.

Rynearson, E. K. "Humans and Pets and Attachment," *British Journal of Psychiatry* 133 (1978): 550–55.

———. "Pets as Family Members: An Illustrative Case History." *International Journal of Family Psychiatry* 1 (2) (1980): 263–68.

Sade, Donatien Alphonse François, Comte de. *La Philosophie dans le boudoir.* Préf. Matthieu Galey. Paris, 1972.

Saffady, William. "Fears of Sexual License During the English Reformation." *History of Childhood Quarterly* 1 (1973): 89–97.

Saint-Victor, Paul Jacques Raymond Binsse. *Deux Masques.* Paris, 1884.

Sainte-Beuve, Charles Augustin. *Portraits littéraires.* 2 vols. Paris, 1924.

———. *Port-Royal* Ed. Maxime LeRoy. 3 vols. Paris, 1955.

Salminen, Renja, ed. *"Le Miroir de l'âme pécheresse": Edition critique et commentaire, suivis de la traduction faite par la princesse Elizabeth, future reigne d'Angleterre. "The Glasse of the Sinful Soule.*" Dissertations Humanarum Litterarum 22. *Annales Academiae Fennicae.* Helsinki, 1979.

Salutati, Coluccio. *Epistolario di Coluccio Salutati.* Ed. F. Novati. Rome, 1891.

Salvianus [Presbyter of Marseilles]. *De Gubernatione Dei.* Trans. Eva M. Sanford. New York, 1966.

————. *De Gubernatione Dei.* Paris, 1617.

Sánchez, Francisco. *Quod Nihil Scitur* [Why Nothing Can Be Known]. 1581. In *Tratados filisóficos.* 2 vols. Lisbon, 1955. English translation: *That Nothing is Known.* Ed. Douglas F. S. Thomson. Introd. Elaine Limbrick. Cambridge, Eng. 1988. French translation: *Il n'est science de rien.* Ed. Andrée Comparot. Pref. André Mandourze. Paris, 1984.

Sandell, Sandra Dianne. "'A very poetic circumstance': Incest and the English Literary Imagination, 1770–1830." Ph.D. diss., Univ. of Minnesota, 1981.

Santiago, Luciano. *The Children of Oedipus: Brother-Sister Incest in Psychiatry, Literature, History, and Mythology.* New York, 1973.

Saraiva, António José. *Inquisição et Cristaos-novos.* Porto, 1969.

Saxl, Fritz, and Rudolph Wittkower. *British Art and the Mediterranean.* Oxford, 1948.

Saxo Grammaticus. *Historiae Danicae.* Latin text (Editio Princeps, 1514) and English trans. by Oliver Elton (1894). *See* in Gollancz, ed., *Sources of "Hamlet,"* pp. 93–165.

Sayles, John (director). *The Brother from Another Planet.* 1984.

Sayre, F. *The Greek Cynics.* Baltimore, 1948.

Schelly, Judith May. "'A Like Unlike': Brother and Sister in the Works of Wordsworth, Byron, George Eliot, Emily Brontë, and Dickens." Ph.D. diss., University of California, Berkeley, 1980.

Schereschewsky, Ben-Zion [Benno]. "Mamzer." In Singer, ed., *Jewish Encyclopedia.*

Schiller, Friedrich von. *Sämtliche Werke.* Ed. Gerhard Fricke and Herbert G. Göpfert. 4 vols. Munich, 1958.

Schleiner, Winfried. "*Divina virago:* Queen Elizabeth as an Amazon." *Studies in Philology* 75 (1978): 163–80.

Schmitt, Carl. *The Concept of the Political.* Trans., introd., and notes by George Schwab with comments on Schmitt's essay by Leo Strauss. New Brunswick, 1976.

————. *Hamlet oder Hekuba. Der Einbruch der Zeit in das Spiel.* Dusseldorf and Cologne, 1956.

Schneck, J. M. "Zooerasty and Incest Fantasy." *International Journal of Clinical and Experimental Hypnosis* 22 (1974): 299–302.

Schneider, Gerhard. *Der Libertin: Zur Geistes- und Sozialgeschichte des Bürgertums im 16. und 17. Jahrhundert.* Stuttgart, 1970.

Schochet, Elijah Judah. *Animal Life in Jewish Tradition: Attitude and Relationships.* New York, 1984.

Schoeps, Hans Joachim. *Israel und Christenheit.* Munich & Frankfurt, 1961.

Schwab, George. "Enemy oder Foe: Der Konflikt der modernen Politik." Trans. J. Zeumer. *Epirrhosis* 2, 665–82.

Schwarzfuchs, Simon. *Napoleon, the Jews, and the Sanhedrin.* London, 1979.

Schweitzer, B. *Herakles. Aufsatze zur griechischen Religions und sagens Geschichte.* Tübingen, 1922.

Sckommodau, H. "Die religiösen Dichtungen Margarete von Navarra." *Arbeitsgemeinschaft für Forschung des Landes NordrheinWestfalen* 36 (1955).

Screech, M. A. *The Rabelaisian Marriage: Aspects of Rabelais's Religion, Ethics, and Comic Philosophy.* London, 1958.

Scriptorum illustrium virinioris Brytanniae, quam nunc Angliam et Scotiam vocant. Basel, 1557.

Searles, Harold F. *The Non-Human Environment in Normal Development and in Schizophrenia.* New York, 1960.
Sebkova, J. "Anxiety Levels as Affected by the Presence of a Dog." Master's thesis, University of Lancaster, England.
See, Fred G. "Kinship of Metaphor: Incest and Language in Melville's *Pierre.*" *Structuralist Review* 1(1978): 55–81.
Sefer haHinnukh. Ed. Chaval. Jerusalem, 1952.
Séguy, Jean. "Une Sociologie des sociétés imaginés: Monachisme et Utopie." *Annales: Sociétés, Civilisations* 26 (1971): 328–54.
Seneca. *Common Minor Dialogues Together with a Dialogue on Clemency.* Trans. Aubrey Stewart. London, 1889.
———. *De Clementia ad Neronem.* Ed. and trans. François Prechac. Paris, 1921.
———. Seneca? *Apokolokyntosis Divi Claudii* [The Pumpkinification of Claudius]. Ed. R. Roncali. Leipzig, 1990.
———. [Pseudo-Seneca]. *Octavia.* In Seneca, *Four Tragedies, and Octavia.* Trans. E. F. Watling. Middlesex, 1966.
Serís, H. "Nueva genealogía de Santa Teresa." *Nueva revista de filología hispánica* 10 (1956): 364–84.
Serres, Étienne Rénaud Augustin. *Recherches d'anatomie transcendente et pathologique: ıнéorie des formations et déformations organiques, appliquée à l'anatomie de Ritta-Christina, et de la duplicité monstrueuses.* Paris, 1833.
Servius Marius. *In Vergilium Commenatarius.* Ed. Georg Thilo. Leipzig, 1878–87.
Seters, John Van. *Abraham in History and Tradition.* New Haven, 1978.
Seymour, William. *Ordeal by Ambition: An English Family in the Shadow of the Tudors.* London, 1972.
Shafarevich, Igor. *The Socialist Phenomenon.* Trans. William Tjalsma. Foreword Aleksandr I. Solzhenitsyn. New York, c.1980.
———. *Sotsializm kak iavlenie mirovo i istorii.* YMCA Press, 1977.
Shakespeare, William. *All's Well That Ends Well.* Ed. Sylvan Barnett. New York, 1965.
———. *Antony and Cleopatra.* Ed. M. R. Ridley. London, 1965.
———. *Coriolanus.* Ed. Reuben Brower. New York, 1966.
———. *Cymbeline.* Ed. J. C. Maxwell. Cambridge, 1960.
———. *Hamlet.* Editions:
Ed. Edward Dowden. London, 1899.
Ed. Willard Farnham. Baltimore, 1966.
Ed. Horace Howard Furness. *A Variorum Edition of Hamlet.* 2 vols. (1877). Rpr. New York, 1963.
Ed. Cyrus Hoy. *Hamlet: An Authoritative Text, Intellectual Backgrounds, Extracts from the Sources, Essays in Criticism.* New York, 1963.
Ed. Harold Jenkins. London, 1982.
Ed. George Lyman Kittredge. *Complete Works of Shakespeare.* Boston, 1936.
Ed. Alexander Pope and William Warburton. *Works of Shakespear* 8. London, 1747.
Ed. Howard Staunton. *The Plays of Shakespeare,* 1858–60. Vol. 3 (1860.)
Ed. Lewis Theobald. *Works of Shakespeare* 7. London, 1733.
Ed. Benno Tschischwitz. *Shakespeare's Hamlet, Prince of Denmark.* Halle, 1869.
Notes by Samuel Johnson. *The Plays of Shakespeare.* Vol. 8 (1765).
———. *Julius Caesar.* Ed. William and Barbara Rosen. New York, 1963.
———. *King Henry the Eighth.* Ed. R. A. Foakes. London, 1968.
———. *King John.* Ed E.A.J. Honigmann. London, 1954.
———. *Love's Labor's Lost.* Ed. Richard David. London, 1966.
———. *Measure for Measure.* Ed. J. W. Lever. London, 1965.

————. *The Merchant of Venice.* Ed. Brents Stirling. Baltimore, 1973.
————. *Pericles, Prince of Tyre.* Ed. F. D. Hoeniger. London, 1963.
————. *The Plays and Poems of William Shakespeare.* Ed. Edmond Malone. Vol. 9 (1790).
————. *Richard the Third.* In *Riverside Shakespeare,* ed. G. Blakemore Evans. Boston, 1974.
————. *The Tempest.* Ed. Robert Langbaum. New York, 1963.
————. *Timon of Athens.* Ed. Maurice Charney. New York, 1965.
————. *Troilus and Cressida.* Ed. Harold N. Hillebrand. Philadelphia and London, 1953.
————. *The Winter's Tale.* Ed. J.H. Pafford. London, 1968.
Shaw, George Bernard. *Back to Methusaleh: A Metabiological Approach.* New York, 1921.
————. *The Doctor's Dilemma.* New York, 1906.
Shek, Ben-Z. "Diglossia and Ideology: Socio-Cultural Aspects of 'Translation' in Québec." *Études sur le texte et ses transformations: Traductions et Cultures* 1 (1988): 85–91.
Shell, Marc. *Art & Money: A Study in Visual and Economic Representation.* Chicago, forthcoming.
————. *The Economy of Literature.* Baltimore, 1978.
————. *Elizabeth's Glass. With "The Glass of the Sinful Soul" (1544) by Elizabeth I and "Epistle Dedicatory" and "Conclusion" (1548) by John Bale.* Lincoln and London, 1993.
————. *The End of Kinship: "Measure for Measure," Incest, and the Ideal of Universal Siblinghood.* Stanford, 1988.
————. "The Forked Tongue: Bilingual Advertisement in Quebec." *Semiotica,* 4, no. 4 (1978): 259–69.
————. "The Lie of the Fox: Rousseau's Theory of Verbal, Monetary, and Political Representation." *Sub-Stance: A Review of Theory and Literary Criticism* 10 (1974): 111–23.
————. *Money, Language, and Thought: Literary and Philosophical Economies from the Medieval to the Modern Era.* Berkeley and Los Angeles, 1982.
————. "La Publicité bilingue au Québec: une langue fourchue." *Journal canadien de recherche sémiotique,* 5, no. 2 (Winter 1977): 55–76.
Shelley, Percy Bysshe. *Poetical Works.* Ed. W. M. Rossetti. 3 vols. London, 1881.
Sheppard, Claude Armand. *The Law of Languages in Canada.* Ottawa, 1971.
Shevoroshkin, Vitaly V. "The Mother Tongue: How Linguists have Reconstructed the Ancestor of All Living Languages." *The Sciences* (New York Academy of Sciences, Spring, 1990): 20–27.
"The Siamese Twins—Chang and Eng." *Downieville* [Calif.] *Mountain Messenger.* April 28, 1866, p. 1.
Sicroff, A. *Les controverses des statuts de "pureté de sang" en Espagne du XVe siècle.* Paris, 1960.
Sidler, Nikolaus. *Zur Universalität des Inzesttabu: Eine kritische Untersuchung der These und der Einwände.* Stuttgart, 1971.
Sieyès, Emmanuel Joseph [Abbé]. *Qu'est-ce que le Tiers État?* (1789). Ed. Roberto Zapperi. Geneva, 1970.
Silva, Rosa J. S. da. "Literatuurlijst." In H. K. Brugmans and A. Frank, eds., *Geschiedenis der Joden in Nederland.* Pt. I. Amsterdam, 1940.
Silverman, Ruth, ed. *The Dog Observed: Photographs 1844–1983.* New York, 1984.
Simonis, Yvan. *Claude Lévi-Strauss, ou "La Passion de l'inceste": Introduction au structuralisme.* Paris, 1968.
Singer, Peter. *Animal Liberation: A New Ethic for Our Treatment of Animals.* New York, 1975.
Skeat, W. W. *Shakespeare's Plutarch.* London, 1904.
Smart, Christopher. *Poems.* Ed. Robert Brittain. Princeton, 1950.
Smith, Adam. *Lectures on Jurisprudence.* Ed. Ronald L. Meek, D.D. Raphael, and P.G. Stein. Oxford, 1978.

Smith, Charles Edward. *Papal Enforcement of Some Medieval Marriage Laws* (1940). Rpr. Port Washington, New York, 1972.
Smith, David J. *Psychological Profiles of Conjoined Twins: Heredity, Environment, and Identity.* Forew. Robert Bogdan. New York, Praeger, 1988.
Smith, Henry. "The Examination of Usury: The First Sermon." In Vol. 1 of *The Works of Henry Smith,* pp. 88–100. 2 vols. Edinburgh, 1866–67.
Smith, Irwin. *Shakespeare's Blackfriars Playhouse.* New York, 1984.
Smith, Nigel, ed. *A Collection of Ranter Writings from the 17th Century.* Foreword by John Carey. London, 1983.
Smith, T. V., and E. C. Lindeman. *The Democratic Way of Life.* New York, 1951.
Smith, Timothy. "Slavery and Theology: The Emergence of Black Christian Consciousness in Nineteenth Century America." *Church History* 41 (1972): 497–512.
Smith, William. *Dictionary of Greek and Roman Biography and Mythology.* New York, 1967.
———. *Smith's Bible Dictionary.* New York, 1976.
Smith, William, W. Wayte, and G. E. Marindin, eds. *A Dictionary of Greek and Roman Antiquities.* 3d ed. 2 vols. London, 1890–91.
Smyth, Herbert Weir. *Greek Melic Poets.* London, 1900.
Snowden, James Ross. *The Coins of the Bible and Its Money Terms.* Philadelphia, 1864.
Sollors, Werner. *Beyond Ethnicity: Consent and Descent in American Culture.* New York, 1986.
Sophocles. *Oedipus the King.* Trans. with notes by Thomas Gould. Englewood Cliffs, 1970.
———. *'Oedipus the King,' 'Oedipus at Colonus,' and 'Antigone.'* Trans. David Grene, Robert Fitzgerald, and Elizabeth Wyckoff. Chicago, 1954.
———. *The Plays and Fragments.* Ed. and trans. by R. C. Jebb. Cambridge, 1881–96.
Southey, Robert. *The Life and Correspondence.* Ed. C. C. Southey. 6 vols. London, 1849–50.
Speiser, E. A. "The Wife-Sister Motif in the Patriarchal Narratives." In Alexander Altman, ed. *Biblical and Other Studies,* pp. 15–28. Cambridge, Mass., 1963.
———. "The Hurrian Participation in the Civilizations of Mesopotamia, Syria, and Palestine." *Journal of World History* 1 (1953): 311–27. Also in *Oriental and Biblical Studies: Collected Writings of E. A. Speiser,* ed. J. J. Finkelstein and Moshe Greenberg, pp. 244–69. Philadelphia, 1967.
Sperling, Susan. *Animal Liberators: Research and Morality.* Berkeley and Los Angeles, 1988.
Speroni degli Alvarotti, Sperone. *Canace.* Florence, 1546.
Spiegelman, Art. *Maus: A Survivor's Tale.* New York, 1986.
Spinoza, Baruch [Benedict de]. *Descartes' Prinzipien der Philosophie auf geometrische Weise begründet.* Leipzig, 1922.
———. *On the Improvement of the Understanding; The Ethics; Correspondence.* Trans. with introd. by R.H.M. Elwes (1883). Rpr. New York, 1955.
———. *A Theologico-Political Treatise; A Political Treatise.* Trans. with introd. by R. H. M. Elwes (1882). Rpr. New York, 1951.
Spurlock, John C. *Free Love: Marriage and Middle-Class Radicalism in America, 1825–1860.* New York & London, 1988.
Squirrel, R. *Observations addressed to the public in general on the cow pox.* London, 1805.
Stabler, A. P. "Elective Monarchy in the Sources of *Hamlet.*" *Studies in Philology* 62 (1965): 654–61.
Stade, Bernard. *Geschichte des Volkes Israel.* 2 vols. Berlin, 1887–88.
Stanley, Diane, and Peter Vennema. *The Story of Elizabeth I of England.* Illus. Diane Stanley. New York, 1990.
Staves, Susan. *Players' Scepters: Fiction of Authority in the Restoration.* Lincoln, Nebr., 1979.
Steele, R. "English Books Printed Abroad, 1525–48." *Transactions of the Bibliographical Society,* o.s. 11 (1909): 189–236.

Steele, Richard. *The Tatler*. Ed. Donald F. Bond. 3 vols. Oxford, 1987.

Steffen, Gustaf F. "Das Zölibat und seine Ursprunge bei den primitiven Völkern." *Politische-Anthropologische Revue* 11 (1912–13): 235–39.

Steinberg, Maxime. *La Question juive, 1940–42*. Vol. 1 in Steinberg, *L'Etoile et le fusil*. 3 vols. Brussels, 1983–86.

Steiner, George. *After Babel*. New York, 1975.

Stendhal [Marie Henri Beyle]. *The Cenci*. In *The Shorter Novels of Stendhal*. Trans. C. K. Scott-Moncrieff. New York, 1946.

Stenton, Doris May. *The English Women in History*. London, 1957.

Stephan of Tournai. *Summa decreti*. Ed. J. F. von Schulte. Giessen, 1891.

Stephen, James Fitzjames. *Liberty, Equality, Fraternity*. Ed. R. J. White. Cambridge, 1967.

Stern, Bernhard J. *Should We Be Vaccinated? A Survey of the Controversy in its Historical and Scientific Aspects*. New York, 1927.

———. *Social Factors in Medical Progress*. New York, 1927.

Stiles, Henry Reed. *Bundling: Its Origin, Progress, & Decline in America*. Albany, 1872.

Stillman, Norman A. *The Jews of Arab Lands: A History and Source Book*. Philadelphia, 1979.

Stone, Lawrence. *Crisis of the Aristocracy, 1558–1641*. Oxford, 1965.

Storm, Joh. "Mélanges étymologiques." *Romania* 5 (1876).

Strabo (Geographus). Ed. G. Kramer. Berlin, 1844–52.

Strachey, Edward. *Shakespeare's 'Hamlet': An Attempt to Find the Key to a Great Moral Problem by Methodological Analysis of the Play*. London, 1848.

Strauss, Leo. *Die Religionskritik Spinozas als Grundlage seiner Bibelwissenschaft Untersuchungen zu Spinozas Theologisch-Politischem Traktat*. Berlin, 1930.

———. *Persecution and the Art of Writing*. Glencoe, Ill., 1952.

———. *Spinoza's Critique of Religion*. With the 1962 pref. to the English trans. New York, 1982.

Strong, Roy. *The Cult of Elizabeth*. London, 1977.

———. *Portraits of Queen Elizabeth*. Oxford, 1963.

Suetonius Tranquillus, Gaius. *The Historie of Twelve Caesars*. Trans. Philemon Holland. London, 1606.

———. The *Lives of the First Twelve Caesars*. Cambridge, Mass., 1950.

Swaim, M. H. "A New Year's Gift from the Princess Elizabeth." *The Connoisseur* (August 1973): 258–66.

Syberberg, Hans-Jürgen. *Hitler: Ein Film aus Deutschland*. Hamburg, 1978.

Tacitus, Publius Cornelius. *Annales*. Trans. R. Grenewey. London, 1598.

———. *Annals*. Trans. John Jackson. Cambridge, Mass., 1956.

———. *The ende of Nero and the beginning of Galba. Fower Bookes of the Histories*. Trans. and ed. by Sir H. Savile. London, 1591.

Taddeo, Donat, and Raymond Taras. *Le Débat linguistique au Québec*. Montréal, 1987.

Takaki, Ronald T. *Iron Cages: Race and Culture in Nineteenth Century America*. Seattle, 1979.

Tama, Diogene. *Transactions of the Paris Sanhedrin*. London, 1807.

Tatlock, J. S. P. *The Legendary History of Britain: Geoffrey of Monmouth's Historia Regum Britanniae and Its Early Vernacular Versions*. Berkeley, 1950.

Taylor, D. "Grandchildren Versus Other Semi-Domesticated Animals." *International Journal of American Linguistics* 27 (1961): 367–70.

Taylor, John. *Wit and Mirth*. 1630. In Vol. 3 of *Shakespeare Jest Books: Reprints of the Early and Very Rare Jest-Books Supposed to have been used by Shakespeare*. Ed. William Hazlitt. 3 vols. London, 1864.

Taylor, Mark. *Shakespeare's Darker Purpose: A Question of Incest*. New York, 1982.

Telle, Emile V. "L'Île des alliances, ou l'Anti-Thélème." *Bibliothèque d'Humanisme et Renaissance*, 14 (1952): 159–75.

———. *L'Oeuvre de Marguerite d'Angoulême, Reine de Navarre, et La Querelle des femmes.* Toulouse, 1937.

Temerlin, Maurice. *Lucy—Growing Up Human: A Chimpanzee Daughter Growing in a Psychotherapist's Family.* Palo Alto, Calif., 1975.

Tennenhouse, Leonard. "Representing Power: *Measure for Measure* in Its Time." *Genre* 15 (1982): 139–56.

Teppe, Julien. *Chamfort: Sa Vie, son oeuvre, sa pensée.* Paris, 1950.

Teresa of Jesus. *The Complete Works of Saint Teresa of Jesus.* Trans. E. Allison Peers. 3 vols. London, 1946.

Terres, John K. *The Audubon Society Encyclopedia of North American Birds.* Foreword Dean Amadon. New York, 1980.

Tertullian, Quintus Septimus Florens. *Apologeticus.* In *Corpus Christianorum* (Series Latina) 1:85–171.

Thayer, James Bradley. *Preliminary Treatise on Evidence at the Common Law.* Boston, 1898.

Thomas, Brook. "The Writer's Procreative Urge in *Pierre:* Fictional Freedom or Convoluted Incest?" *Studies in the Novel,* 11 (1979): 416–30.

Thomas, Keith. *Man and the Natural World: A History of Modern Sensibility.* New York, 1983.

Thompson, W. Meredith, ed. *Ye wohunge of Ure Lauerd . . . , together with On ureisun of Ure Louerde. . . .* Early English Text Society. London, 1958.

Thomson, George. *Aeschylus and Athens: A Study in the Social Origins of Drama* (1940). Repr. New York, 1967.

Thoreau, Henry David. *Civil Disobedience.* In Vol. 4 of *The Writings of H. D. Thoreau,* pp. 356–87. Boston, 1906.

———. *Walden and Other Writings.* Ed. Joseph Wood Krutch. New York, 1981.

Tilander, Gunnar. *Los Fueros de Aragón según el manuscrito 458 de la Biblioteca Nacional de Madrid.* Lund, 1937.

Tilley, Morris Palmer. *A Dictionary of the Proverbs in England in the Sixteenth and Seventeenth Centuries.* 1950.

"Tocci Twins (The)." *Scientific American* 65 (1891): 374.

Torquemada, Juan de. *Tractatus contra Madiantias et Ismaelitas.* Ed. N. López Martinez. Burgos, 1957.

Torrey, C. C. *The Second Isaiah.* New York, 1928.

Toscano, Mario. "The Jews in Italy from the Risorgimento to the Republic." Trans. Meg Shore. In *Gardens and Ghettos: The Art of Jewish Life in Italy,* ed. Vivian B. Mann, with a pref. by Primo Levi, pp. 25–44. Berkeley and Los Angeles, 1989.

Travitsky, Betty. *The Paradise of Women: Writings by Englishwomen of the Renaissance.* New York, 1989.

Trigano, Shmuel. "The French Revolution and the Jews." *Modern Judaism* 10 (1990): 171–90.

Tristan, Flora. *The London Journal of Flora Tristan, 1842; or, The Aristocracy and the Working Class of England. A Translation of "Promenades dans Londres."* Trans. Jean Hawkes. London, 1982.

———. *La Tour de France.* 2 vols. Ed. Jules L. Peuach. Paris, 1980.

———. *Union ouvrière. Suivi de lettres de Flora Tristan.* Ed. Daniel Armogathe and Jacques Grandjonc. Paris, 1986.

———. *The Workers Union.* Trans. Beverly Livingston. Urbana, 1983.

Trudeau, Pierre Elliott. "Un plébiscite de tous les jours." Delivered in the Canadian House of Commons (April 15, 1980). *See* in Boismenu et al., *Le Québec en textes,* pp. 400–405.

Tuan, Yi-Fu. *The Making of Pets.* New Haven, Conn., 1984.

Turgot, Anne Robert Jacques. "Valeurs et monnaies (project d'article) (1769)." In Turgot, *Ecrits économiques,* pp. 231–50. Préf. Bernard Cazes. Paris, 1970.

Twain, Mark. *See* Clemens, Samuel.
Twine, L. *The Patterne of Painefull Adventures* (c1594). In vol. 4 of *Shakespeare's Library*, ed. John Payne Collier, pp. 229–34. 2nd ed. 6 vols. London, 1875.
Tyrrel, William Blake. *Amazons: A Study in Athenian Mythmaking*. Baltimore, 1984.
"Undertaker for Pets." *New York Times*. March 3, 1981. p. 41.
Unna, Isak. *Tierschutz im Judentum*. Frankfurt, 1928.
Vacant, A., E. Mangenot, and É. Amann, eds. *Dictionnaire de théologie catholique*. 23 vols. Paris, 1923–72.
Valerius Maximus. *Factorum et Dictorum Memorabilium*. Ed. Carolus Kempf. Lepizig, 1887.
Vallières, Pierre. *Nègres blancs d'Amérique*. 1968.
Van Der Beets, Richard. "A Note on Dramatic Necessity and the Incest Motif in 'Manfred'." *Notes and Queries* 11 (1964): 26–28.
Van Rooten, Luis d'Antin. *Mots d'heures: gousses, rames; the d'Antin manuscript*. New York, 1967.
Vaughn, Robert A. *Hours with the Mystics*. London, 1860.
Vaunois, Louis. *L'Enfance et la jeunesse de Racine. Documents sur la vie de Racine (Les Papiers de Jean-Baptiste Racine)*. Paris, 1964.
Vautrollier, Thomas. *Luther's Commentarie ypon the epistle to the Galatians*. London, 1577.
Vaux, Roland de. *The Early History of Israel*. Trans. David Smith. Philadelphia, 1978.
Vellacott, P. H. "The Guilt of Oedipus." *Greece and Rome* 11 (1964): 137–48.
Venesoen, Constant. *Jean Racine et le procès de la culpabilité*. Paris, 1981.
Verdé-Delisle, N. M. H. *De la Dégénérescence physique et moral de l'espèce humanine determinée par le vaccin*. Paris, 1855.
Vermes, Geza. "Essenes, Therapeutae, Qumran. Ancient Jewish Mysticism and the Dead Sea Scrolls." *Durham Journal* 52; n.s. 21, no. 3 (1960):97–115.
Vico, Giovanni Battista. *Principles of New Science of Giovanni Battista Vico concerning the Nature of the Nations*. Rev. trans. of the third ed. (1744). Trans. with notes by Thomas Goddard Bergin and Max Harold Fisch. Ithaca, 1968.
Viller, M., and F. Cavallera, J. de Gulbert; continued A. Rayaz, A. Derville, and A. Solignac. *Dictionnaire de spiritualité*. Paris, 1937–.
Virgil. *The Works of Virgil; containing his Pastorals, Georgics, and Aeneis*. Trans. John Dryden. London, 1697.
Vivre la diversité en français: Le défi de l'école pluriethnique de l'île de Montréal. Notes et Documents 64. Conseil de la langue française. Québec, 1987.
Von Arnim, H. *Stoicorum veterum fragmenta*. Vol. 1. Leipzig, 1903.
Vossler, Karl. *Jean Racine*. Trans. Isabel and Florence McHugh. New York, 1972.
Vovelle, Michel. *La Révolution Française: Images et Récits 1789–1799*. 5 vols. Paris, 1986.
Vuillart, Germain. *Lettres de German Vuillart à Louis de Préfontaine*. Notes by R. Clark. Paris, 1951.
Wagenknecht, Edward. *Cavalcade of the American Novel*. New York, 1952.
Walker, Williston. *John Calvin: The Organizer of Reformed Protestantism*. New York, 1969.
Wallace, Amy. *The Two: A Biography*. New York, 1978.
Wallace, W. *Kant*. Edinburgh, 1882.
Walpole, Horace. *A Catalogue of the Royal and Noble Authors of England*. Ed. Thomas Park. 5 vols. London, 1806.
Warner, Marina. *Alone of All Her Sex: The Myth and Cult of the Virgin Mary*. London, 1976.
A Warning for Fair Women. London, 1599.
Watson, Alan. *Rome of the XII Tables: Persons and Property*. Princeton, 1975.
Webber, Evrett. *Escape to Utopia*. New York, 1959.
Weinzierl, Erika. "Der österreichisch-ungarische Raum: A. Katholizismus in Österreich." In

Karl H. Rengstorf and Siegfried von Kortzfleisch, eds. *Kirche und Synagoge: Handbuch zur Geschichte von Christen und Juden,* pp. 483–531. Vol. 2. Stuttgart, 1970.

Weiss, Peter. *Die Verfolgung und Ermordung Jean Paul Marats dargestellt durch die Schauspielgruppe des Hospizes zu Charenton unter Anleitung des Herrn de Sade.* Frankfurt, 1977.

Wells, Herbert George [H. G.]. *The Outline of History: Being a Plain History of Life and Mankind* [1920]. With maps and plans by J. F. Horrabin. Garden City, New York, 1961.

————. *The War of the Worlds; The Time Machine.* Intro. by Isaac Asimov. New York, 1991.

Wells, Robin Headlam. *Spenser's "Faerie Queene" and the Cult of Elizabeth.* London, 1988.

Wels, L. E. *Theologische Streifzüge durch die ältfranzösiche Literatur.* Vechta, 1937.

Werder, Karl. *Vorlesungen über Shakespeares 'Hamlet'.* Berlin, 1875.

Werstine, Paul. "The Textual Mystery of *Hamlet.*" *Shakespeare Quarterly* 39 (1988): 1–26.

Wessels, D. T., Jr., "Family Psychotherapy Methodology: A Model for Veterinarians and Clinicians." *See* in Kay et al., eds., *Pet Loss.*

West, Rebecca. *The Court and the Castle.* New Haven, 1958.

Westermarck, Edward. *The History of Human Marriage.* 5th ed. 3 vols. New York, 1922.

White, Betty, with Thomas J. Watson. *Betty White's Pet-Love: How Pets Take Care of Us.* New York, 1983.

White, Lynn Townsend, Jr., "The Historical Roots of Our Ecologic Crisis." *Science* 155 (1967).

Wilkinson, W. *A Supplication of the Family of love . . . Examined, and found derogatorie in an hie degree.* London, 1606.

Williams, Howard. *Ethics of Diet: A Catena of Authorities Deprecatory of the Practice of Flesh Eating.* London, 1883.

Williams, Neville. *Elizabeth, Queen of England.* London, 1967.

Williams, Roger. *The Bloody Tenent Yet More Bloody.* Ed. S. L. Caldwell. Providence, 1970.

Williamson, George C. *The Money of the Bible.* Oxford, 1894.

Williamson, Joel. *New People: Miscegenation and Mulattoes in the United States.* New York, 1980.

Willibald of Mainz. *Vita S. Bonifacii.* PL 89: 603–32.

Wilson, J. Dover. *The Manuscript of Shakespeare's 'Hamlet' and the Problems of its Transmission,* 2 vols. Cambridge, 1934.

————. *What Happens in 'Hamlet'.* New York, 1935.

Wilson, James D. "Incest and American Romantic Fiction." *Studies in the Literary Imagination* 7 (1974): 31–50.

Winnicott, D. W. "Transitional Object and Transitional Phenomena: A Study of the First Not-Me Possession." *International Journal of Psycho-Analysis* 34 (1953): 89–99.

Wiznitzer, A. *Jews of Colonial Brazil.* New York, 1960.

Woehrling, José. "Réglementation linguistique de l'affichage et la liberté d'expression." *See* in *L'Avenir de Français au Québec,* pp. 81–97.

Wolf, Arthur P. "Adopt a Daughter-in-law, Marry a Sister: A Chinese Solution to the Problem of the Incest Taboo." *American Anthropologist* 70 (1968): 864–74.

Wolf, L. *Crypto-Jews under the Commonwealth.* London, 1894.

Wolfe, Thomas. *Look Homeward, Angel: A Story of the Buried Life.* New York, 1929.

Wolfson, Harry Austryn. *Philo: Foundations of Religious Philosophy in Judaism, Christianity, and Islam.* 2 vols. Cambridge, Mass., 1947.

Wollstonecraft, Mary. *A Vindication of the Rights of Women. An Authoritative Text* (1792). Ed. Carol H. Poston, with backgrounds and criticism. New York, 1975.

Woodgate, M.V. *Jacqueline Pascal and Her Brother.* Dublin, 1944.

Woodhouse, Barbara. *Talking to Animals.* New York, 1975.

Wright, Robert. "Quest for the Mother Tongue." *Atlantic Monthly* 267 (April 1991): 39–68 passim.

Writings of Ed[ward] VI, William Hugh, Queen Catherine Parr, Anne Askew, Lady Jane Grey, Hamilton and Balnaves. Philadelphia, 1862.

Yablonsky, Lewis. *The Hippie Trip*. Baltimore, 1973.

Yates, F. A. *Astraea: the Imperial Theme in the Sixteenth Century*. London, 1975.

Yovel, Yirmiyahu. *Spinoza and Other Heretics: The Marrano of Reason*. Princeton, 1989.

Young, Benjamin S. *The Testimony of Christ's Second Coming*. Lebanon, Ohio, 1808.

Young, Edward. *Night Thoughts*. Ed. George Gilfillan. Edinburgh, 1853.

Young, Philip. "The Mother of Us All: Pocahontas Reconsidered." *Kenyon Review* 24 (1962): 391–415.

Zeller, Edward. *Die Philosophie der Griechen in ihrer geschichtlich Entwicklung*. 3 vols. Leipzig, 1880.

Zelnick, Stephen. "The Incest Theme in *The Great Gatsby*: An Exploration of the False Poetry of Petty Bourgeois Consciousness." In Norman Rudick, ed. *Weapons of Criticism: Marxism in America and the Literary Tradition*, pp. 327–40. Palo Alto, Calif., 1976.

INDEX

N.B.: Italic page numbers refer to illustrations.